GPU Programming and Code Optimization
High Performance Computing for the Masses

GPU 编程与优化
——大众高性能计算

方民权　张卫民
方建滨　周海芳　著
高　畅

清华大学出版社
北京

内 容 简 介

本书第一篇系统地介绍 GPU 编程的相关知识，帮助读者快速入门，并构建 GPU 知识体系；第二篇和第三篇给出大量实例，对每个实例进行循序渐进的并行和优化实践，为读者提供 GPU 编程和优化参考；第四篇总结影响 GPU 性能的关键要素（执行核心和存储体系），量化测评 GPU 的核心性能参数，给出 CPU/GPU 异构系统上覆盖完全的各种混合并行模式及其实践，帮助读者真正透彻理解 GPU。

本书适合作为计算机及相关专业的教材，也可作为 GPU 程序开发人员和科研人员的参考书。

本书封面贴有清华大学出版社防伪标签，无标签者不得销售。
版权所有，侵权必究。举报：010-62782989，beiqinquan@tup.tsinghua.edu.cn。

图书在版编目（CIP）数据

GPU 编程与优化：大众高性能计算/方民权等著. --北京：清华大学出版社，2016（2025.2 重印）
ISBN 978-7-302-44642-2

Ⅰ. ①G… Ⅱ. ①方… Ⅲ. ①图像处理—程序设计 Ⅳ. ①TP391.41

中国版本图书馆 CIP 数据核字（2016）第 179437 号

责任编辑：白立军
封面设计：杨玉兰
责任校对：李建庄
责任印制：刘海龙

出版发行：清华大学出版社
网　　址：https://www.tup.com.cn，https://www.wqxuetang.com
地　　址：北京清华大学学研大厦 A 座　　　　邮　编：100084
社 总 机：010-83470000　　　　　　　　　　邮　购：010-62786544
投稿与读者服务：010-62776969，c-service@tup.tsinghua.edu.cn
质量反馈：010-62772015，zhiliang@tup.tsinghua.edu.cn
课件下载：https://www.tup.com.cn，010-83470236

印 装 者：三河市龙大印装有限公司
经　　销：全国新华书店
开　　本：185mm×260mm　　印　张：27.25　　字　数：660 千字
版　　次：2016 年 9 月第 1 版　　　　　　　印　次：2025 年 2 月第 6 次印刷
定　　价：59.00 元

产品编号：070097-01

前 言

多核与众核异构平台因其超强的浮点运算能力而成为当前高性能计算领域的新贵。2010年以来,已有3台异构超级计算机夺魁TOP 500,分别是搭载CPU/GPU异构系统的天河1A和泰坦超级计算机、搭载CPU/MIC异构系统的天河2号超级计算机。在这两类主流的多核与众核异构平台中,CPU/GPU异构平台在性价比、能耗比等方面表现尤为突出,例如,在Green500前10中有9台采用了这种架构。对于高性能计算用户而言,CPU/GPU异构系统无疑是一个良好的选择。

另一方面,当前PC已普遍装备GPU(独立显卡),使得这种CPU/GPU异构系统的硬件平台随处可见。尽管这类平台的GPU工作重心是游戏娱乐而非浮点计算,但在一些精度要求不高的领域仍然优势显著。此外,基于GPU编程的工具包是免费的,因此可用较低的成本构建合适的CPU/GPU异构并行平台。由于显卡的普及,CPU/GPU异构并行程序也能在几乎所有的PC中广泛应用。

然而,仅有硬件是没有应用价值的,异构系统上的程序开发是实现异构系统价值的直接且唯一的步骤。但是,异构并行软件开发面临着巨大挑战,主要包括异构数据通信、基于GPU体系结构的编程与优化、多编译器的联合编译等,具体到实践则难度更大。编写本书的目的就是辅助用户解决这些GPU异构并行软件开发的难题。

目前市面上已有很多GPU编程书籍,其中一些已经论述相当全面,为什么还要撰写本书呢?作为一名有多年开发经验的GPU程序员,阅读这些书籍总感觉有些不足。

首先,对于刚接触GPU的开发者,由于所要认识的GPU体系结构与常用的CPU体系结构差异巨大,相关的理论知识较难理解,而已有的一些论著为了增强理论性和学术性还对相关理论知识进行了抽象提升,因而不够通俗易懂,即使是GPU编程老手也未必能完全读懂。而本书将从GPU程序员的角度出发,通俗易懂地阐述GPU编程与优化相关的理论知识。

其次,从GPU理论到编程优化实践的过渡是非常关键的,但目前市面上的书籍重理论而轻实践。仅有理论而缺乏实践和直观实践效果,难以对程序员读者产生直接价值。况且很多没有实践论证的理论知识未必正确。本书将紧密结合理论和实践,并试图从实践中总结理论知识,从而帮助读者更好地理解。

此外,程序性能优化是GPU编程的重中之重。当前GPU书籍中提及了大量关于优化方法的理论知识,但很少针对每种优化方法和策略给出应用实例(即使有也可能只有一个,借鉴范围不够广),更没有针对某一个实例进行系统性循序渐进的优化。事实上,对于新手而言,优化时最大的困惑就是知道优化方法却不知道用到哪儿、怎么用。本书试图针对大量经典实例进行循序渐进式的优化,为读者提供详尽的优化参考。

最后，GPU 编程实践时不仅仅只是编程，还涉及编译器、运行环境等相关配套知识，若是没有这些配套知识，即使看懂了、写出了相关代码又有什么意义呢？这无异于纸上谈兵。本书将涉及代码编写、编译，运行时需要涉及的所有配套知识，包括系统环境、Linux 命令、编译选项、性能分析、并行计算相关常识等。

综上所述，本书的定位是帮助 GPU 程序员从"零知识"入门到精通的书籍，书中内容包含通俗易懂的 GPU 理论知识，配套的知识体系，大量代码实例及其循序渐进的优化过程、详细的性能分析和知识点总结，与性能直接相关的 GPU 核心特征获取、分析和论述。对于 GPU 程序的开发人员，本书具有较为全面的参考价值。

本书的结构和阅读建议

本书共计四篇：第一篇共 5 章，主要是 GPU 的理论知识，包括 GPU 的领域背景（高性能计算概述）、GPU 概述、GPU 硬件架构、GPU 软件体系和 CUDA C 编程；第二篇共 4 章，基于 4 个入门级的 GPU 实例展示了其详细的并行和优化过程，分别是向量加法、向量内积、矩阵乘法和矩阵转置；第三篇共 5 章，分别描述了 5 类不同应用的 GPU 编程和优化过程，包括卷积、曼德博罗特集、前缀求和、排序和简单图像处理；第四篇共 4 章，阐述了影响 GPU 程序性能的核心因素，分别从 GPU 执行核心、GPU 存储体系、影响 GPU 性能的关键因素、CPUs 和 GPUs 的协同运算 4 个角度展开探讨。

首遍阅读时，建议按行文顺序阅读，本书已按知识难易程度做了梳理；接着动手实践入门篇和提高篇；若要进一步深入阅读则需结合核心篇章的 GPU 核心知识与入门篇和提高篇的实践，逐步理解提升；最终在本书基础上进一步优化各类应用，进而开发出自己的优化方法。

对于急于从实例运行入手的读者，可根据 4.3 节内容安装环境，然后跳读入门篇和提高篇的实例章节，在获得一定成功经验后再返回阅读理论篇和核心篇。

本书的特点和优势

本书内容丰富，涵盖了系统全面的 GPU 知识体系、循序渐进的实例优化、从实践导出的真实有效的优化方法、影响 GPU 性能的核心因素、GPU 性能测评和 CPU/GPU 异构协同优化等内容。本书语言朴实，通俗易懂，对不易理解的概念定义，通过笔者的理解重新进行阐述。本书还提供简单易读的实例代码，详尽的编译命令和清晰的运行结果数据。

GPU 发展迅猛，GPU 架构几乎每两年就更新换代（就在本书第 3 轮修订期间，Pascal 架构发布，笔者又重新修订了相应章节），目前市面上的书籍暂时没有提供完整的 GPU 架构知识，本书总结了所有的 GPU 架构，阐述了更加系统完善的 GPU 知识体系。

GPU 优化是关键，市面上同类书籍中阐述了很多优化方法，但经历了循序渐进优化的具体实例相对缺乏，甚至有些优化方法存在问题（本书有相应的实验佐证）。

致谢

本书大纲由方民权和方建滨共同商定；方民权完成本书代码的编写、实验数据测试与分析、本书初稿、首轮修订、重要修订（如错误修正、结构调整和章节内容增加等）以及后续修订内容的权衡和更新；张卫民、方建滨、周海芳等完成后续数轮修订，正是有他们的修订才使本书真正可读；高畅提供了 Linux 图形界面 CUDA 安装的文档和验证，验证了本书所有代码的结果正确性和性能准确性，以及进行一些关键的查漏补缺。正是所有作者共同努力，本书才能真正成稿。

本书支撑课题包括国家自然科学基金项目（41375113 和 61272146）、湖南省研究生创新资助项目（CX2015B030）、国防科学技术大学计算机学院联合博导组项目（面向异构平台的海洋预报软件可移植性技术研究）。

文中涉及许多笔者的个人学术观点和经验总结，由于笔者水平有限，所有成果仅供参考，如有不准确甚至错误之处，望读者谅解并批评指正。另因编写时间仓促，尽管已进行 6 轮修订，但书中纰漏和瑕疵在所难免，若发现笔误或有其他意见建议，请致信 admin@hpc6.com，万分感谢！

编　者
2016 年 5 月

笔者的话

2016年1月20日是一个值得纪念的日子,这本书终于完成初稿,尽管有的章节还有待大幅修改。自2015年3月有了初步构思,我就义无反顾地开始撰写本书,到最终完稿,期间也多次调整论述结构。作为一名在校博士研究生,顶着博士毕业的压力,写一本对博士毕业"无用"的书,确实挺不容易的,不过我认为是值得的。

本科,我学的是机械专业,硕士转学计算机,2012年接触GPU编程,当时MIC已然兴起,却还能感受到GPU的"余热"。说实话,刚接触GPU时整个人是懵的,师兄已经毕业,师门就我一人研究GPU,当时只有两本GPU书——《GPU高性能运算之CUDA》和 *CUDA by example* 供参考,书中的知识还有些"过时"(Tesla架构和Fermi架构的区别)、"不系统"、甚至"错误"(见13.3.2节)(当然不可否认这两本书对我的帮助),网上资料就更少了。也许是我个人水平有限,仅Windows的GPU开发环境安装就耗费了半个多月时间,Linux的GPU环境更是失败了不知道多少次(当时网上资料很少,且很多方法尝试后均不可行);首个GPU程序开发完全不知道如何下手,头大了近20天才在某天灵光一闪想通,对GPU优化更是无从着手。其实上述过程只需要一位"老师"带着成功一遍,就能少走很多弯路。说这么多,其实想表达的意思是,新手时的我非常期盼一本系统的、全面的、循序渐进的GPU书籍,这也是我写本书的动因。

2015年,在学术圈GPU早已"过时",为什么还致力于撰写本书呢?首先,GPU确实稳定好用:2014年也曾研究过MIC,"不可捉摸"的ICC编译器耗光了我的耐心,往往逻辑正确的代码编译器就是报错,明明简单向量化的代码编译器就是认为不可向量化;而CUDA代码只要编写正确,均能正常编译和执行,且性能取决于CUDA代码本身。其次,我认为GPU并行是一个大趋势大市场:GPU有3个层次可扩展性(见3.1节),能满足各种市场需求;GPU(显卡)已在PC普遍装备,性价比高,GPU平台已相当普及;GPU的性能不易受其他程序影响,相比之下CPU程序性能受其他程序影响程度较大;GPU发展迅猛,在未来量子计算机或生物计算机出现前(出现后也不可能立刻取代现有的计算机),CPU/GPU异构系统将会长期存在。

本书的定位是辅助"零知识"GPU程序开发人员从入门到精通的书籍,试图做到"知行合一"(最完美的想法是所有的理论知识均有实验结果佐证,但显然还有差距)、系统性强、知识全面、优化循序渐进。希望本书能够帮助每一位GPU程序开发人员。

方民权

目 录

第一篇 理 论 篇

第 1 章 高性能计算概述 …………………………………………………………… 3
1.1 高性能计算概念辨析 …………………………………………………………… 3
    1.1.1 并行计算、高性能计算和超级计算 …………………………………… 3
    1.1.2 超级计算机与超级计算中心 …………………………………………… 4
1.2 计算科学 ………………………………………………………………………… 5
1.3 高性能计算发展史 ……………………………………………………………… 5
1.4 高性能计算简介 ………………………………………………………………… 6
1.5 向量机与阵列机 ………………………………………………………………… 8
1.6 本章小结 ………………………………………………………………………… 9

第 2 章 GPU 概述 ………………………………………………………………… 10
2.1 GPU 是什么 …………………………………………………………………… 10
2.2 协处理器 ……………………………………………………………………… 10
2.3 GPU 与显卡的关系 …………………………………………………………… 11
2.4 GPU/显卡购买注意事项 ……………………………………………………… 11
2.5 为什么要学 GPU 编程 ………………………………………………………… 12
2.6 GPU 与 CPU 辨析 …………………………………………………………… 13
2.7 GPU 发展简史 ………………………………………………………………… 14
2.8 GPU 编程方法 ………………………………………………………………… 14
2.9 CPU/GPU 异构系统 …………………………………………………………… 16

第 3 章 GPU 硬件架构 …………………………………………………………… 17
3.1 GPU 架构 ……………………………………………………………………… 17
    3.1.1 Tesla 架构 ………………………………………………………………… 18
    3.1.2 Fermi 架构 ……………………………………………………………… 20
    3.1.3 Kepler 架构 ……………………………………………………………… 21
    3.1.4 Maxwell 架构 …………………………………………………………… 23
    3.1.5 Pascal 架构 ……………………………………………………………… 24

3.2 Kernel 的硬件映射 …… 28
3.3 GPU 存储体系 …… 29
3.4 GPU 计算能力 …… 30

第 4 章 GPU 软件体系 …… 33
4.1 GPU 软件生态系统 …… 33
4.2 CUDA Toolkit …… 34
    4.2.1 NVCC 编译器 …… 34
    4.2.2 cuobjdump …… 35
4.3 CUDA 环境安装 …… 36
    4.3.1 Windows 7 安装 CUDA 4.2 …… 36
    4.3.2 Linux 下安装 CUDA …… 38

第 5 章 CUDA C 编程 …… 41
5.1 CUDA 编程模型 …… 41
5.2 CUDA 编程七步曲 …… 42
5.3 驱动 API 与运行时 API …… 42
5.4 CUDA 运行时函数 …… 43
    5.4.1 设备管理函数 …… 43
    5.4.2 存储管理函数 …… 45
    5.4.3 数据传输函数 …… 48
    5.4.4 线程管理函数 …… 51
    5.4.5 流管理函数 …… 52
    5.4.6 事件管理函数 …… 52
    5.4.7 纹理管理函数 …… 53
    5.4.8 执行控制函数 …… 55
    5.4.9 错误处理函数 …… 55
    5.4.10 图形学互操作函数 …… 57
    5.4.11 OpenGL 互操作函数 …… 58
    5.4.12 Direct3D 互操作函数 …… 59
5.5 CUDA C 语言扩展 …… 60
5.6 grid-block-thread 三维模型 …… 61

第二篇 入 门 篇

第 6 章 向量加法 …… 67
6.1 向量加法及其串行代码 …… 67
6.2 单 block 单 thread 向量加 …… 68

6.3 单 block 多 thread 向量加 ·············· 68
6.4 多 block 多 thread 向量加 ·············· 69
6.5 CUBLAS 库向量加法 ················ 70
6.6 实验结果分析与结论 ················ 71
    6.6.1 本书实验平台 ················ 71
    6.6.2 实验结果 ··················· 71
    6.6.3 结论 ······················ 71
6.7 知识点总结 ······················ 72
6.8 扩展练习 ························ 75

第 7 章 归约：向量内积 ················ 76
7.1 向量内积及其串行代码 ··············· 76
7.2 单 block 分散归约向量内积 ············ 77
7.3 单 block 低线程归约向量内积 ··········· 78
7.4 多 block 向量内积（CPU 二次归约） ······ 79
7.5 多 block 向量内积（GPU 二次归约） ······ 81
7.6 基于原子操作的多 block 向量内积 ········ 81
7.7 计数法实现多 block 向量内积 ··········· 84
7.8 CUBLAS 库向量内积 ················ 85
7.9 实验结果与结论 ···················· 86
    7.9.1 实验结果 ··················· 86
    7.9.2 结论 ······················ 86
7.10 归约的深入优化探讨 ················ 87
    7.10.1 block 数量和 thread 数量对归约性能的影响 ·· 87
    7.10.2 算术运算优化 ················ 88
    7.10.3 减少同步开销 ················ 89
    7.10.4 循环展开 ··················· 90
    7.10.5 总结 ····················· 91
7.11 知识点总结 ······················ 91
7.12 扩展练习 ························ 94

第 8 章 矩阵乘法 ······················ 95
8.1 矩阵乘法及其 3 种串行代码 ············ 95
    8.1.1 一般矩阵乘法 ················ 95
    8.1.2 循环交换矩阵乘法 ············· 97
    8.1.3 转置矩阵乘法 ················ 98
    8.1.4 实验结果与最优串行矩阵乘 ········ 99
8.2 grid 线程循环矩阵乘法 ··············· 100

8.3 block 线程循环矩阵乘法 ………………………………………………………… 101
8.4 行共享存储矩阵乘法 …………………………………………………………… 101
8.5 棋盘阵列矩阵乘法 ……………………………………………………………… 103
8.6 判断移除 ………………………………………………………………………… 105
8.7 CUBLAS 矩阵乘法 ……………………………………………………………… 106
8.8 实验结果分析与结论 …………………………………………………………… 108
  8.8.1 矩阵乘精度分析 ………………………………………………………… 108
  8.8.2 实验结果分析 …………………………………………………………… 110
  8.8.3 浮点运算能力分析 ……………………………………………………… 111
8.9 行共享存储矩阵乘法改进 ……………………………………………………… 111
8.10 知识点总结 …………………………………………………………………… 113
8.11 扩展练习 ……………………………………………………………………… 115

第 9 章 矩阵转置 …………………………………………………………………… 116

9.1 矩阵转置及其串行代码 ………………………………………………………… 116
9.2 1D 矩阵转置 …………………………………………………………………… 117
9.3 2D 矩阵转置 …………………………………………………………………… 118
9.4 共享存储 2D 矩阵转置 ………………………………………………………… 119
9.5 共享存储 2D 矩阵转置 diagonal 优化 ………………………………………… 120
9.6 实验结果分析与结论 …………………………………………………………… 121
9.7 共享存储 2D 矩阵转置的深入优化 …………………………………………… 122
9.8 知识点总结 ……………………………………………………………………… 124
9.9 扩展练习 ………………………………………………………………………… 125

第三篇 提 高 篇

第 10 章 卷积 ………………………………………………………………………… 129

10.1 卷积及其串行实现 …………………………………………………………… 129
  10.1.1 一维卷积 ……………………………………………………………… 129
  10.1.2 二维卷积 ……………………………………………………………… 131
10.2 GPU 上 1D 卷积 ……………………………………………………………… 134
10.3 M 常量 1D 卷积 ……………………………………………………………… 135
10.4 M 共享 1D 卷积 ……………………………………………………………… 136
10.5 N 共享 1D 卷积 ……………………………………………………………… 137
10.6 实验结果分析 ………………………………………………………………… 139
  10.6.1 扩展法 1D 卷积实验结果分析 ……………………………………… 139
  10.6.2 判断法与扩展法 1D 卷积对比 ……………………………………… 140
  10.6.3 加速比分析 …………………………………………………………… 141

  10.6.4 线程维度对性能的影响 ·· 141
 10.7 2D 卷积的 GPU 移植与优化 ·· 142
  10.7.1 GPU 上 2D 卷积 ··· 142
  10.7.2 M 常量 2D 卷积 ··· 143
  10.7.3 M 常量 N 共享 2D 卷积 ··· 143
  10.7.4 2D 卷积实验结果分析 ··· 145
 10.8 知识点总结 ·· 145
 10.9 扩展练习 ·· 147

第 11 章 曼德博罗特集 ··· 148
 11.1 曼德博罗特集及其串行实现 ·· 148
 11.2 曼德博罗特集的 GPU 映射 ·· 150
 11.3 一些优化尝试及效果 ·· 152
  11.3.1 访存连续 ··· 152
  11.3.2 uchar4 访存合并 ··· 153
  11.3.3 4 种零拷贝 ··· 153
  11.3.4 总结分析 ··· 155
 11.4 计算通信重叠优化 ·· 156
 11.5 突破 kernel 执行时间限制 ··· 159
 11.6 知识点总结 ·· 160
 11.7 扩展练习 ·· 162

第 12 章 扫描：前缀求和 ··· 163
 12.1 前缀求和及其串行代码 ·· 163
 12.2 Kogge-Stone 并行前缀和 ··· 164
  12.2.1 直接 Kogge-Stone 分段前缀和 ······································· 164
  12.2.2 交错 Kogge-Stone 分段前缀和 ······································· 165
  12.2.3 完整 Kogge-Stone 前缀和 ··· 166
 12.3 Brent-Kung 并行前缀和 ·· 168
  12.3.1 Brent-Kung 分段前缀和 ··· 169
  12.3.2 两倍数据的 Brent-Kung 分段前缀和 ·································· 170
  12.3.3 避免 bank conflict 的两倍数据 Brent-Kung 分段前缀和 ················· 171
  12.3.4 完整 Brent-Kung 前缀和 ·· 173
 12.4 warp 分段的 Kogge-Stone 前缀求和 ·· 174
 12.5 实验结果分析与结论 ·· 177
 12.6 知识点总结 ·· 179
 12.7 扩展练习 ·· 180

第 13 章 排序 ... 181
13.1 串行排序及其性能 ... 181
13.1.1 选择排序 ... 181
13.1.2 冒泡排序 ... 182
13.1.3 快速排序 ... 182
13.1.4 基数排序 ... 183
13.1.5 双调排序网络 ... 185
13.1.6 合并排序 ... 186
13.1.7 串行排序性能对比 ... 187
13.2 基数排序 ... 188
13.2.1 基数排序概述 ... 188
13.2.2 单 block 基数排序 ... 189
13.2.3 基于 thrust 库的基数排序 ... 196
13.3 双调排序网络 ... 197
13.3.1 双调排序网络概述 ... 197
13.3.2 单 block 双调排序网络 ... 199
13.3.3 多 block 双调排序网络 ... 202
13.4 快速排序 ... 206
13.5 合并排序 ... 207
13.6 实验结果分析与结论 ... 208
13.7 知识点总结 ... 209
13.8 扩展练习 ... 210

第 14 章 几种简单图像处理 ... 211
14.1 图像直方图统计 ... 211
14.1.1 串行直方图统计 ... 211
14.1.2 并行直方图统计 ... 211
14.1.3 实验结果与分析 ... 212
14.2 中值滤波 ... 213
14.2.1 串行中值滤波 ... 214
14.2.2 1D 并行中值滤波 ... 215
14.2.3 共享 1D 中值滤波 ... 216
14.2.4 双重共享 1D 中值滤波 ... 218
14.2.5 2D 并行中值滤波 ... 221
14.2.6 共享 2D 中值滤波 ... 222
14.2.7 共享 2D 中值滤波的改进 ... 227
14.2.8 实验结果与分析 ... 229
14.3 均值滤波 ... 231

14.3.1 串行均值滤波 ·· 231
14.3.2 并行均值滤波 ·· 232
14.3.3 实验结果与分析 ·· 233

第四篇 核 心 篇

第15章 GPU执行核心 ·· 237
15.1 概述 ··· 237
15.2 算术运算支持 ·· 238
15.2.1 整数运算 ·· 238
15.2.2 浮点运算 ·· 239
15.3 算术运算性能 ·· 240
15.4 分支处理 ··· 242
15.5 同步与测时 ·· 246
15.5.1 同步 ·· 246
15.5.2 测时 ·· 247
15.6 数学函数 ··· 247
15.7 warp与block原语 ··· 249
15.7.1 warp原语 ··· 249
15.7.2 block原语 ·· 250
15.8 kernel启动、线程切换和循环处理 ·· 251

第16章 GPU存储体系 ·· 254
16.1 概述 ··· 254
16.2 寄存器 ·· 259
16.3 局部存储 ··· 261
16.4 共享存储器 ·· 264
16.4.1 共享存储使用 ·· 264
16.4.2 bank conflict ··· 265
16.4.3 volatile关键字 ··· 266
16.4.4 共享存储原子操作 ··· 267
16.5 常量存储 ··· 268
16.6 全局存储 ··· 269
16.6.1 全局存储的使用 ·· 269
16.6.2 全局存储的合并访问 ··· 271
16.6.3 利用纹理缓存通道访问全局存储 ··· 271
16.7 纹理存储 ··· 273
16.7.1 CUDA数组 ·· 273

16.7.2 纹理存储的操作和限制 274
16.7.3 读取模式、纹理坐标、滤波模式和寻址模式 276
16.7.4 表面存储 278
16.8 主机端内存 281
16.9 零拷贝操作 283

第17章 GPU关键性能测评 284

17.1 GPU性能测评概述 284
17.2 GPU参数获取 286
    17.2.1 GPU选择 286
    17.2.2 详细设备参数获取 287
17.3 精确测时方法汇总 288
    17.3.1 clock测时 289
    17.3.2 gettimeofday测时 289
    17.3.3 CUDA事件测时 289
    17.3.4 cutil库函数测时 290
17.4 GPU预热与启动开销 290
17.5 GPU浮点运算能力 291
17.6 GPU访存带宽 293
17.7 GPU通信带宽 295
17.8 NVIDIA Visual Profiler 296
17.9 程序性能对比约定 298

第18章 CPUs和GPUs协同 299

18.1 协同优化基点 299
    18.1.1 CPU并行矩阵乘基点 299
    18.1.2 GPU并行矩阵乘基点 300
18.2 CPU/GPU协同 300
18.3 多GPU协同 305
    18.3.1 CUDA版本 306
    18.3.2 OpenMP+CUDA 308
    18.3.3 MPI+CUDA 311
18.4 CPUs/GPUs协同 314
    18.4.1 CUDA版本 314
    18.4.2 OpenMP+CUDA 319
    18.4.3 MPI+OpenMP+CUDA 324
18.5 本章小结 329

附 录

附录 A 判断法 1D 卷积代码 ··· 333
附录 A.1 判断法 1D 卷积 basic 版 ································ 333
附录 A.2 判断法 1D 卷积 constant 版 ····························· 334
附录 A.3 判断法 1D 卷积 shared 版 ································ 336
附录 A.4 判断法 1D 卷积 cache 版 ································· 337

附录 B 曼德博罗特集的系列优化代码 ······························· 340
附录 B.1 完整版串行 C 代码 ······································· 340
附录 B.2 cuda_1_0 ··· 343
附录 B.3 cuda_0_2 ··· 345
附录 B.4 cuda_zerocopy ·· 346
附录 B.5 cuda_1_0_zerocopy ·· 348
附录 B.6 cuda_0_0_zerocopy ·· 349
附录 B.7 cuda_0_2_zerocopy ·· 351
附录 B.8 cuda_2 ·· 352
附录 B.9 cuda_1_2 ·· 354

附录 C 几种图像处理完整源码 ·· 357
附录 C.1 BMP 图像读写头文件 ·· 357
附录 C.2 图像直方图串行代码 ·· 373
附录 C.3 串行中值滤波代码 ·· 374
附录 C.4 并行均值滤波相关代码 ······································ 376

附录 D nvprof 帮助菜单 ··· 383

附录 E NVCC 帮助菜单 ·· 388

附录 F 几种排序算法源代码 ·· 399
附录 F.1 bitonic_sort_block 函数 ···································· 399
附录 F.2 GPU 快速排序完整代码 ······································ 400
附录 F.3 GPU 合并排序完整代码 ······································ 408

参考文献 ·· 417

第一篇 理 论 篇

> 第1章 高性能计算概述
> 第2章 GPU 概述
> 第3章 GPU 硬件架构
> 第4章 GPU 软件体系
> 第5章 CUDA C 编程

理论篇主要阐述系统全面的 GPU 知识体系,包括 GPU 并行计算的领域背景(高性能计算概述)、GPU 概述、GPU 硬件架构、GPU 软件体系和 GPU 编程方法(CUDA C 编程)。通过阅读本篇,读者能够较为系统全面地掌握 GPU 相关的基础理论和技术知识,建立必要的知识结构。

本篇内容丰富,笔者推荐几项值得重点关注的内容:①向量机和阵列机的辨别,深入浅出地对比了向量机(CPU)和阵列机(GPU)的区别(1.5节);②GPU 购买注意事项(2.4节);③全面的 GPU 架构知识,包括 Tesla、Fermi、Kepler、Maxwell 和 Pascal 架构(3.1节);④实测可行的 CUDA 环境安装指南,包括 Windows、Linux 命令行和 Linux 图形界面等[①](4.3节);⑤CUDA 编程七步曲,简述了 CUDA 编程的基本步骤(5.2节)。

---

① 同类 CUDA 书籍也有环境安装内容,但笔者亲测大都不可用,也许是笔者水平有限吧。

第 1 章 高性能计算概述

本章主要介绍 GPU 编程与优化所处的领域背景,即高性能计算。从概念辨析、计算科学、高性能计算发展史以及所涉及的相关研究、向量机和阵列机等几个方面综合性地对高性能计算进行了系统阐述。

1.1 高性能计算概念辨析

1.1.1 并行计算、高性能计算和超级计算

并行计算(parallel computing)、高性能计算(high performance computing,HPC)、超级计算(supercomputing)三者的概念在某种程度上可以认为是等价的,都是用并行处理的手段来获得程序的高速运行。在陈国良院士的《并行计算——结构·算法·编程》一书中提到"并行计算就是在并行计算机或分布式计算机(包括网络计算机)等高性能计算系统上所做的超级计算"[1]。

但笔者认为三者不完全相同,相互间存在一种层层递进的关系(见图 1.1)。

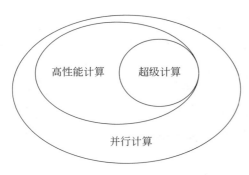

图 1.1 并行计算、高性能计算和超级计算的关系图

(1) 在多个计算核心上同时计算同一任务,即可称为并行计算,从广义上说,任务中存在并行处理即为并行计算。

(2) 高性能计算不能仅仅是多个核心同时运算,还要求高效发挥各计算核心性能,需要程序员针对体系结构进行特殊优化,比如向量化、数

据合并访存、提高 cache 命中率等。显然图形处理单元(graphic processing unit,GPU)程序开发就属于高性能计算范畴。

（3）超级计算的前提是高性能计算（即单节点或少量节点的性能已然发挥到一定水准），在此基础上对计算量、运行规模有一定要求。伴随着处理器和协处理器的高速发展，单节点浮点计算能力持续提高，很难给出一个超级计算的阈值，本书在此简单地给一个区分方法，计算时使用超过 128 个当代计算节点可称为超级计算。

针对某一具体应用的可并行代码部分，通过任务分割和并行程序设计即可实现并行计算，从而获得近线性加速比；进一步根据运行平台的体系结构进行优化，比如向量化、访存优化等，高性能计算可获得超线性加速比；由于问题规模过大，需要将计算任务分配到成百上千的计算节点进行运算，此时的超级计算需要解决可能存在的网络传输堵塞、负载不均衡等问题。

1.1.2　超级计算机与超级计算中心

超级计算机又称为巨型机，是运行超级计算应用的基础硬件平台，也是一种至关重要的国家战略武器。超级计算机是新时代的"核武器"，是美国对中国封锁的重要技术之一。20 世纪 80 年代之前，中国没有自己的巨型机，美国通过技术封锁强迫中国签订带有霸王条款的不平等合约，禁止中国人进入超级计算机机房；邓小平同志要求发展中国自己的巨型机，国防科学技术大学的慈云桂教授（中国巨型机之父）立下军令状，并成功研制中国第一台巨型机——银河Ⅰ号；时至 2010 年，国防科学技术大学研制了天河 1A 超级计算机夺得 TOP 500 桂冠，后被日本、美国超越；2013 年至今，天河 2 号超级计算机（见图 1.2）已连续蝉联 6 届 TOP 500 榜首。国际上超级计算机的竞争异常激烈，高性能计算和超级计算的研究至关重要。

图 1.2　天河 2 号超级计算机

中国的超级计算机主要有 3 个系列，分别是银河/天河、神威、曙光，其中天河和曙光面向民用，目前已在各大超级计算中心部署。超级计算中心是为管理超级计算机而设的，目前主要有天津、广州、长沙和济南四大国家级超级计算中心，部分省市建有自己的超级计算中心，比如上海、甘肃、成都和深圳等，各高校也建有自己的超级计算中心，比如中科院超级计算中心、清华大学计算机系高性能计算研究所、中国科技大学超级运算中心等。

1.2 计算科学

　　计算机出现以前,理论科学和实验科学是学者进行科学研究的两种主要途径。随着计算机的出现和发展,计算科学也随之出现,并成为科学研究的第三种重要研究方法。计算科学是通过计算模拟的方法,利用计算机强大的计算能力,解决复杂的科学问题。理论科学以数学学科为基础,实验科学以物理学科为基础,计算科学的基础包括计算机学科、数学学科和物理学科等。利用计算科学,能够有效解决理论科学中理论模型过于复杂、实验科学中实验代价过于巨大的问题,目前已在物理、化学、生物、地质、气象、材料等众多领域内广泛使用。

　　高性能计算是计算科学的一部分,其所属一级学科是计算机科学与技术,二级学科为大规模科学与工程计算。高性能计算的研究可以促进计算能力提升和计算模拟时间缩减,从而在有限时间内增加实验次数,最终促进各领域的发展。

1.3 高性能计算发展史

　　美国作为计算机出现和持续发展的土壤,在一定程度上代表了世界高性能计算的发展,因此本节主要介绍美国的高性能计算发展史。

　　1942年8月13日,美国开始曼哈顿计划(爱因斯坦提出的制造原子弹计划,涉及理论物理、爆轰物理、中子物理、金属物理、弹体弹道等大量的数值计算),历时3年制造3颗原子弹。

　　1950年,开始研制氢弹,1952年成功,原子弹和氢弹研制过程中许多物理规律必须通过计算机模拟,因而计算物理就自然地产生了。

　　1959年,美国总统颁布命令,可以揭开曼哈顿计划内幕,部分内容解密以"计算物理方法"丛书名义出版。

　　1963年,美国"计算物理方法"丛书首次出现"计算物理"一词;计算物理以计算机为基础,以美国核武器研制为动力迅猛发展。

　　20世纪60年代中期开始推出小型机,20世纪70年代末推出个人计算机,20世纪80年代中期推出高性能超级微机,大型机也得到发展。

　　1981年以哈佛大学普雷斯为首的11位著名科学家联名上书,向美国国家科学基金会(NSF)呈送"发展计算物理的建议书",提出建立国家范围内的网络计算系统,包括管理通信的网络、大型超级计算机、可供用户使用的阵列处理机和图像显示设备等。这给计算物理带来所需要的存储容量和计算能力。

　　1983年,"科学计算"一词首次出现在美国,以数学家拉克斯为首的不同学科的专家委员会向美国政府提出的报告之中,强调"科学计算是关系到国家安全、经济发展和科技进步的关键性环节,是事关国家命脉的大事"。

　　1983年,由于发展星球大战"战略防御倡议"(SDI)计划的需要,制订了"战略计算机计划"(SCP)。到 20 世纪 80 年代中后期这一计划导致了峰值速度达每秒十亿次的 Cray 超级向量计算机和多 CPU 并行计算机等高性能计算机的问世。

1984 年,美国政府大幅度地增加对科学计算经费的支持,NSF 建立了"先进科学计算办公室"(OASC),制定全面高级科学计算发展规划;连续 5 年累计拨款 2.5 亿美元。新建成 5 个国家级超级计算中心(分别在普林斯顿大学、圣地亚哥、伊里诺大学、康奈尔大学、匹兹堡),配备当时最高性能的计算机,建立 NSFnet 新网络。

1987 年,NSF 把"科学与工程计算""生物工程""全局性地科学"作为三大优先支持的领域。

1991 年,以美国总统倡议的名义提出了"高性能计算与通信(HPCC)计划"。这是为了保持和提高美国在计算和网络等所有先进领域中的领导地位而制订的。

1995 年,美国宣布全面禁止核武器条约,为了确保核库存的性能、安全性、可靠性和更新需要而实施的"加速战略计算创新(ASCI)计划",首次出现"战略计算"一词。开始用模拟核试验替代真实核试验。

2000 年,"加速战略计算创新计划"改称为先进模拟和计算(ASC)计划。

2002 年,美国 DARPA 提出 HPCS(High Productivity Computing System)计划,旨在填补 HPC 计算与未来的量子计算之间的空隙。

2008 年,美国 DARPA 信息处理技术办公室(IPTO)公布了 E 级计算研究报告,E 级计划将为大气科学、高能物理、核物理、计算化学及生物学等众多领域提供支撑。

2015 年,美国发布了新的国家战略计算规划(NSCI),旨在攻克传统超算难题的同时融合大数据的处理。

伴随上述政策的发布,其成果是一台又一台跨量级超级计算机的产生。1964 年公认的第一台超级计算机 CDC6600 诞生;1985 年 Cray 公司发布了 8 个处理器的 1.9Gflop 的 Cray-2,是世界上首台 Giga 级别的超级计算机;1996 年底 Intel 公司成功研制 1.453Tflop 的 ASCI-RED 超级计算机;2008 年 IBM 公司成功研制"走鹃(RoadRunner)"宣告超级计算机进入 PetaScale 时代;目前蝉联 6 届 TOP 500 榜首的天河 2 号超级计算机理论峰值性能高达 54.9Pflop。短短 50 年,超级计算机的计算能力已经提升了 10 个数量级。

1.4 高性能计算简介

高性能计算尽管只是计算科学的一个很小的分支,但其涉及的研究范围却非常广泛,包括高性能计算机系统、高性能计算环境、体系结构、并行计算模型、高性能计算网络、性能测评、并行算法设计技术、并行编程方法等。本节主要内容总结自文献[1]。

Flynn 分类法将计算机分为单指令流单数据流(single instruction single data,SISD)、单指令流多数据流(single instruction multiple data,SIMD)、多指令流单数据流(multiple instruction single data,MISD)和多指令流多数据流(multiple instruction multiple data,MIMD)4 类计算机。其中 MISD 并不存在,常见的高性能计算机属于 SIMD 和 MIMD。

常见的高性能计算机系统包括共享存储对称多处理机(symmetric multiprocessor,SMP)、分布存储大规模并行处理机(massively parallel processor,MPP)、分布共享存储计算机(distributed shared memory,DSM)、集群(Cluster)等。

单独的高性能计算节点主要分为同构节点和异构节点两类。同构节点仅采用中央处理器(central processing unit,CPU)作为计算设备,目前主流的有 Intel Xeon CPU 和 AMD Opteron CPU。异构节点分为主机端和设备端,分别注重逻辑处理和浮点计算,主流异构节点类型包括 CPU+GPU 异构系统和 CPU+MIC(many integrated core,集成众核)异构系统。

高性能计算环境也是高性能计算中不可缺少的环节。单节点上高性能计算环境主要是并行开发环境,包括编译器(GCC(GNU C/C++ compiler)、ICC(Intel C/C++ compiler)、PGI、NVCC(NVIDIA CUDA compiler))、驱动程序(GPU、MIC 驱动)等。集群环境包括集群管理系统、集群作业分发系统、并行文件系统等子系统和 MPI 编译包。

并行计算机体系结构包括并行计算机结构模型、访存模型。结构模型包括 SMP、MPP、DSM、Cluster 等,访存模型有均匀存储访问(uniform memory access,UMA)、非均匀存储访问(nonuniform memory access,NUMA)、全高速缓存存储访问(cache only memory access,COMA)、高速缓存一致性非均匀存储访问(coherent-cache nonuniform memory access,CC-NUMA)和非远程存储访问(no remote memory access,NORMA)等。另外,异构系统中协处理器的体系结构也是并行计算机体系结构的研究重点之一,比如 GPU 和 MIC 体系结构。

并行计算模型是硬件与软件间的桥梁,可对并行系统进行性能建模。主流的并行计算模型包括并行随机存取机器(parallel random access machine,PRAM)模型、异步 PRAM 模型、BSP(bulk synchronous parallel)模型和 LogP(latency overhead gap processors)模型等。另外,有很多学者基于这些模型进行了扩展研究,比如 APRAM(分相(Phase)PRAM 模型)、H-BSP、LogGP 和 LogGPS 等。

高性能计算集群一般采用专用高速网络,比如 IB(InfiniBand)网络,也可使用普通以太网。IB 网络是统一的互连结构,可以处理存储 I/O、网络 I/O 和进程间通信(IPC)。高性能网络的相关研究包括网络互连结构、选路方法、通信技术等。

并行算法设计时常采用划分、分治、平衡树、倍增、流水线等设计技术。划分又可分为均匀划分、方根划分、对数划分和功能划分。在并行算法设计过程中,一般可分为 4 步,任务划分(Partitioning)、通信(Communication)、任务组合(Agglomeration)、处理器映射(Mapping),即所谓的 PCAM 设计过程。基本要点是先尽量开拓算法的并发性和扩展性,接着考虑通信成本和局部性,利用局部性相互组合减少通信成本,最后将组合后的任务分配到处理器。

传统串行代码经过编译得到的执行程序仅能使用一个核心运算。要发挥所有处理器核的性能,必须对程序进行并行编程。现有并行编程模型可分为以下几类。

(1) 分布存储编程模型,主要包括 MPI(massage passing interface)等。

(2) 数据并行(数组划分)编程模型,主要有 HPF(high performance fortran)等。

(3) 共享存储编程模型,主流有 OpenMP(open multi-processing)、Pthread(POSIX threads)、TBB(thread building blocks)。

(4) 专用异构编程模型,比如专用于 NVIDIA GPU 开发的 CUDA(compute unified

device architecture）和开发 MIC 程序使用的 LEO（language extensions for offload），Microsoft 公司的 Direct Compute 等。

（5）通用异构编程模型，包括 OpenCL（open computing language）、OpenACC 和 OpenMP 4.0（4.0 以上版本开始支持异构系统）。

针对高性能计算机进行性能测评时，主要的性能指标包括理论峰值性能、实测峰值性能、访存带宽、通信延迟、通信开销等。所使用的加速比性能定律包括 Amdahl 定律、Gustafson 定律、Sun 和 Ni 定律。针对可扩展性的测评标准包括等效率、等速度、平均延迟。测试时采用的基准测试程序也各种各样，主流的有 Linpack、HPL（high performance Linpack）、HPCG（high performance conjugate gradients）等。

1.5 向量机与阵列机

笔者认为高性能计算机从本质上分仅有两种：向量机与阵列机。向量机通过向量处理获得高性能，而阵列机通过大量线程同时执行达到高吞吐量。向量机和阵列机也是目前获得高加速比的主要途径之一。

向量机拥有独特的向量处理单元，可以进行向量运算。向量机利用流水线技术，将计算机的计算单元和控制单元按序组装，利用部件分离和时间重叠思想，形成流水。只有在向量处理单元工作时，向量机才是 SIMD 计算机；而标量运算和分支处理时，向量机是 SISD 计算机。目前主流的向量机包括 Intel Xeon CPU、Intel Xeon Phi（Intel 公司的集成众核 MIC 产品）等。其中，最新的 Xeon CPU 已支持 256 位宽的向量运算，而 Xeon Phi 可支持 512 位向量处理。

阵列机由一个控制器同时控制多个运算器，控制器进行指令解析和指令分发，而处理器只负责计算，处于相同控制器控制下的运算器执行相同指令操作（见图 1.3）。控制器执行两类指令：一类是控制指令，其本身解释执行；另一类是运算指令，控制器将该指令分发给所有处理器，所有处理器同时执行该条指令。显然，阵列机是一种典型的 SIMD 计算机。

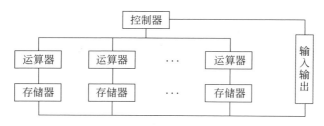

图 1.3　阵列机原理图

虽然某些文献将 GPU 判断为向量机，但笔者认为 GPU 是典型的阵列机。GPU 执行采用单指令多线程（single instruction multiple thread，SIMT）模式，即每个 warp（线程束）执行同一条指令，线程 ID 为 0~31 的 thread 同时执行同一条指令。

为了形象地理解向量机和阵列机，本书做一个比喻：将向量机（CPU、MIC）看成武林大侠，阵列机（GPU）看成军团战阵。武林大侠（CPU、MIC）拥有强大的内力（ICC 编译器),并且掌握着一门威力巨大的刀芒绝招(向量运算),此绝招一出可以形成刀芒同时砍翻 8 或 16 个对手(同时处理 8 或 16 个浮点运算),当刀芒范围内的敌人排列整齐时,刀芒的杀伤力才能发挥(数据对齐),但大侠不会主动使用绝招,需要玩家操纵(程序员编程),由于玩家自身没有内力,不懂得如何掌控内力(ICC 编译器),也不能保证成功激发刀芒绝招(程序员只能给编译器一些编译指导,而无法保证编译成可向量运算的可执行程序)。军团战阵(GPU)拥有大量士兵(计算核心),并且提供给玩家(程序员)一本战阵秘籍(CUDA 编程),战阵秘籍记载了如何排兵布阵,玩家需要熟练掌握战阵秘籍才能玩转战阵,但对于新手而言这本战阵秘籍的难度显得比较大。需要注意的是,战阵中的一排士兵(warp)是同时行动的,比如同时挥刀砍杀(warp 内所有线程执行相同指令),当其对面敌人(数据)排列整齐时,能保证所有士兵都砍中敌人,杀伤力自然巨大(性能好)。

那么战阵(阵列机)和大侠(向量机)到底哪个厉害呢?首先,如果完全不懂战阵的新兵去操纵战阵,自然一团乱,别说发挥战阵能力,可能连里面一个士兵都无法用好,自然毫无战斗力可言(GPU 入门较难,性能优化不易);而一个菜鸟操纵大侠,由于大侠有内力,尽管菜鸟啥也不会,也有一定几率发出刀芒绝招(CPU 开发入门容易,ICC 编译器功能强大,提供了较好的代码向量化优化能力,向量化几率还是比较大的)。当新兵进行了战阵秘籍培训,进阶为先锋官,懂得基本操纵战阵,能令战阵发挥绝大多数功能,杀伤力强大(基本学会 GPU 编程后,GPU 性能发挥主要看程序员编写的代码,程序性能主要由代码决定,编译器基本不会影响其性能,因此写好代码就能发挥 GPU 性能);而内力太过玄幻(ICC 编译器太复杂,很难真正掌握),即使不断修炼也很难操纵自如,刀芒的发挥仍然无法得心应手(学会 CPU 编程后想继续进阶就变得相对困难了,由于 ICC 编译器太过复杂,无法真正学会掌握,而真正对 CPU 或 MIC 程序性能起决定性作用的并不是程序员写的代码本身,而是编译器如何编译;如果编译时编译器就是认为代码无法向量优化,即使程序员很轻易地找到了向量化方案,真正的可执行程序依然只能标量执行)。经过积年累月的修炼,菜鸟终成大将军和大宗师(如 NVIDIA 和 Intel 的天才数学库开发人员),就能完全发挥战阵和武林大侠的能力,此时应该说不相上下。

本书对战阵秘籍(CUDA 编程)进行简化和实例注解,期望通过该秘籍更快更好地训练好新兵,使其迅速成长为先锋官乃至大将军,充分发挥战阵杀伤力,更快更好地战胜敌军(本书简化 CUDA 编程,利用大量实例阐述 GPU 编程和性能优化,能更好更快地帮助 GPU 程序员迅速入门和提升,达到一个较高的水平)。

1.6 本章小结

本章主要概述 GPU 编程所处的领域背景——高性能计算,包括概念辨析、计算科学、发展历史、相关研究、向量机和阵列机等几个部分。首先辨析并行计算、高性能计算和超级计算间的差异;接着阐述国内超级计算机的起源与发展、超级计算中心的配置;再阐明高性能计算所属学科,以及高性能计算发展史;然后简单系统地介绍高性能计算领域涉及的主要研究内容;最后通过一个形象生动的比喻分析向量机和阵列机的区别,并界定了 GPU 的定位。

第 2 章 GPU 概述

本章将阐述 GPU 是什么、协处理器、GPU 与显卡的关系、GPU/显卡购买注意事项、为什么要学 GPU 编程、GPU 与 CPU 辨析、GPU 发展史、GPU 编程方法和 CPU/GPU 异构系统等内容。

2.1 GPU 是什么

GPU(graphic processing unit)又名图形处理单元,是一种扩展的计算设备,称为协处理器,外接在 PCI-E 接口上。GPU 的主要厂商是 NVIDIA 和 ATI(已被 AMD 公司收购),这两家公司生产的 GPU 分别称为 N 卡和 A 卡。本书采用 CUDA C 编写的代码只能在 N 卡上运行。若要在 A 卡上运行,需要将代码改为 OpenCL 代码。值得一提的是,在最新的 SC2015 会议(international conference for high performance computing, networking, storage, and analysis)上,AMD 公司宣布 A 卡 GPU 将支持 CUDA,这意味着现在用 CUDA 书写的代码能在未来不做修改直接运行在 A 卡上。

2.2 协处理器

协处理器是一种辅助处理器,实现某种特定功能,比如数值计算,目前主流的协处理器有 GPU 和 Intel Xeon Phi。

最早的协处理器是 Intel 公司推出的 8087 协处理器,其目的是弥补 8086 处理器浮点计算能力差的弱点。其后由于工艺水平的提升,从 486DX 开始,为这种目的设计的协处理器被集成到了 CPU 中,协处理器的概念也随之消失在了历史长河中。后来由于功耗的限制以及对计算需求的不断增长,协处理器重新登上历史舞台,例如,GPU 从专业图形处理到科学计算的转型。2012 年,Intel 公司推出集成众核协处理器产品 Xeon Phi。GPU 和 MIC 协处理器已成为超级计算领域的新宠,2013 年 6 月,基于 CPU/MIC 异构系统的天河二号超级计算机在 TOP 500 超级计算机榜单排名第一,并已连续蝉联 6 届。

2.3 GPU 与显卡的关系

相信大多数人都听说过显卡,而 GPU 较少被关注;说到 GPU,人们往往认为 GPU 就是通常所说的显卡;事实上,两者是有区别的。显卡是 GPU,而 GPU 并不完全是显卡,两者是统属关系,显然 GPU 范围更大,包含显卡。比如 NVIDIA 公司有 3 个主流的 GPU 产品系列,分别是 Tesla、GeForce 和 Quadro(还有嵌入式(移动端,Tegra)、虚拟化 (Grid)和多显示屏(NVS)产品,具体可参考 NVIDIA 官网)。其中,Tesla 系列是专为计算设计的,Tesla K20c GPU 就没有提供显示接口(见图 2.1),故不能称为显卡;另外两个系列产品均提供了显示输出功能,俗称显卡,但其侧重点各不相同,Geforce 系列专注游戏、娱乐,而 Quadro 系列专为图形图像处理而生。

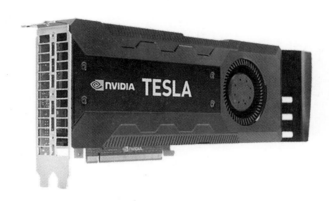

图 2.1 NVIDIA Tesla GPU

2.4 GPU/显卡购买注意事项

GPU 选购时有些要点需要特别注意。

首先 Tesla 系列专为计算设计,由 NVIDIA 官方生产,其计算性能无须质疑,一般情况下产品越新性能越好,但需要一定的购买力保证,因为该系列的 GPU 价格昂贵,普遍价格超过万元。

GeForce 系列是无力购买 Tesla 系列 GPU 情况下的最佳替代产品。尽管它是为游戏设计的,但其单精度计算能力还是能满足计算需求的(但双精度计算能力极弱),同代高端 GeForce 显卡的单精度浮点运算性能可以媲美 Tesla 系列 GPU,而价格仅为其 1/5 左右。计算机市场上的显卡基本上都是 GeForce 系列,型号五花八门,厂商多种多样,该系列产品只有计算核心是 NVIDIA 生产的,其他部件的生产包括最终组装都是由代理厂商完成。代理显卡厂商为了获得经济利益,偷工减料对显卡进行性能阉割也是可以理解的,而商家宣传销售时一般都会宣扬其显存大小,而很少提及浮点计算性能、访存带宽等真正影响 GPU 性能的关键指标,因此在选购时需要一双火眼金睛(说个亲身经历,笔者在最开始接触 CUDA 时,完全没有经济能力,但又确实需要支持 CUDA 的显卡,查看官方 CUDA 支持列表后发现本人计算机上的 9500GT 是支持 CUDA 计算的,但 GPU-Z 测试结果显示不支持,经过一番调查发现其核心根本不是 9500GT,已忘记是什么了,后来去找商家理论半天,才给换了个 9600GT)。

选择显卡(性能)时,如果资金充裕,一般情况下价格越贵性能越好。如果资金有限,则重点关注以下两个方面。首先是 GPU 核心型号,现在市面上显卡的型号一般类似 GTX 980,前面的 GTX(也可能是 GT、GTS,其中 GTX 一般更注重性能)表示为 GeForce 系列产品,数值 980 中第一个数 9 可指代产品年份,一般每过 1 年加 1,某种程度上可视为产品先进程度;80 可以理解为产品等级,其值从 05 到 90 不等,数值越大计算性能越好,一般 40 以下为低端卡,40 到 60 为中端卡,70 及其以上为高端卡(高端卡在购置前需考虑电源负载、有无电源线,其中 50 及其以下无须电源线,适用面更广);型号可反映性能,但不绝对,一个有效方法是通过 GPU-Z 和 CUDA-Z 等测试工具测试显卡参数,确认其核心、工艺、浮点计算性能、架构、计算能力、通信速率、显存、访存带宽等详细参数,其中,浮点计算性能和访存带宽是关键。

2.5 为什么要学 GPU 编程

从第 1 章高性能计算概述中可知,GPU 编程仅仅是高性能计算领域内一个小的分支,那么为什么要学习 GPU 开发呢? 主要原因是 GPU 有以下优点。

(1) 浮点计算能力强。GPU 的设计目标是其超高的浮点计算能力,因而 GPU 相对多核 CPU 在体系结构设计上有天然的区别。GPU 作为众核协处理器的典型,GPU 上一个 warp 的 32 个线程由一个控制器控制,同时处理同一条指令。这种设计为 GPU 提供了超强的浮点计算能力,从而使得这种 CPU/GPU 异构系统在超级计算机领域表现突出:2010 年基于 CPU/GPU 异构系统的天河 1A 超级计算机夺得 TOP 500 桂冠,2012 年相同结构的泰坦超级计算机在 TOP 500 夺魁。

(2) 超高性价比。在 CPU/GPU 异构系统中,多核 CPU 在提供 256 位宽向量处理的基础上,主要负责复杂的逻辑处理,而众核 GPU 协处理器的设计目的是超高的浮点计算能力,特别是双精度浮点计算能力。由于多核 CPU 和 GPU 的设计理念不同,导致其内部结构、元器件成本上的重要区别。GPU 在设计时避免或减弱了类似分支处理、逻辑控制等与浮点计算无关的复杂功能,专注于浮点计算,因此在制造成本上有着巨大的优势。众核 GPU 协处理器仅需 1/10 的成本即可达到与多核 CPU 同等的浮点运算能力。

(3) 绿色功耗比。GPU 集成了大量的轻量级微处理器单元,这些处理单元功能简单(仅做浮点运算)、时钟频率有限,使得运算产生的功耗极小。比如 NVIDIA Tesla K20c GPU 在休眠状态仅需 15W 左右功耗,其满载运转时功耗约为 150W,能提供超过 2Tflops 单精度浮点运算能力;而 Intel Xeon Phi 31S1P 协处理器在低功耗状态的能耗为 100W,在满载运行提供 2Tflops 单精度浮点运算能力时需要 250W 供能。因此 GPU 比 MIC 在能耗上更有优势。在 2015 年 11 月发布的 Green500 榜单前 10 名中有 9 台超级计算机使用了 CPU/GPU 异构系统,其中一台使用了 AMD FirePro 协处理器,其余均使用

NVIDIA Tesla 协处理器。

（4）普及度广。部分 GPU 产品（显卡）已在绝大多数 PC 上装备，购置新 GPU 也非常方便。GPU 拥有完善的产品体系，其产品价格从数百到上万不等。无论经济实力如何，任何人都能买得起合适的 GPU 来搭建 GPU 编程平台。这一点非常重要，有了 GPU 编程实践平台才能深入学习该项技术，而不局限于看书想象。

另外，从并行计算领域来说，该领域存在以下几个重大问题：存储墙、编程墙、功耗墙和不平衡的计算机科学生态系统。本书介绍 GPU 编程开发技术，试图帮助程序员突破编程墙；另外，GPU 提供程序员可控制的层次式存储，在一定程度上可以帮助程序员突破存储墙；GPU 的低功耗优势在一定程度上朝着功耗墙发起了冲击；开发 GPU 并行应用，特别是大规模并行应用，让更多的程序员学会编写 GPU 并行代码，充分发挥 CPU/GPU 异构系统和 GPU 集群性能，从而真正扭转计算机科学生态系统"重硬轻软"的不平衡。

2.6　GPU 与 CPU 辨析

CPU 是中央处理单元，GPU 是图形处理单元，从名字上已经可以了解部分差别，CPU 就像一个中央大脑，控制整个计算机（包括 GPU），而 GPU 专注于图形图像处理（通用计算）。

从设计目的上说，CPU 的设计目标是使执行单元能够以很低的延迟获得数据和指令，因此采用了复杂的控制逻辑和分支预测，以及大量的缓存来提高执行效率；而 GPU 必须在有限的面积上实现超强的运算能力和极高的存储器带宽，因此需要大量执行单元来运行更多相对简单的线程，在当前线程等待数据时切换到另一个处于就绪状态等待计算的线程。简而言之，CPU 对延迟更敏感，而 GPU 则侧重于提高整体的数据吞吐量。CPU 与 GPU 设计目标的不同决定了两者在架构和性能上的巨大差距，图 2.2 对 CPU 与 GPU 中部件的数量及用途进行了比较。

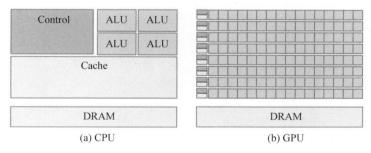

图 2.2　CPU 与 GPU 对比

从本质上说，CPU 和 GPU 的区别是：CPU 是向量机，GPU 是阵列机。关于向量机和阵列机，详见 1.5 节。向量机通过向量处理获得性能，因此优化 CPU 程序时，向量化是至关重要的指标（编译时查看向量化报告开关-vec-report6，等级 0～6 自选）；阵列机是由同一个控制器控制下的多个处理单元组成的，各处理单元没有指令控制部件，在控制器控制下，各处理单元各自对分配的数据并发完成同一指令运算，阵列机的优化原则有数据对齐访问、避免分支分离等。

2.7 GPU 发展简史

NVIDIA 公司作为 GPU 界龙头老大,其 GPU 产品占领了 80% 左右的市场,本节将从 NVIDIA 公司的 GPU 硬件发展来阐述 GPU 发展简史。

1993 年,NVIDIA 成立,相继开发显示芯片 NV-1 和 NV-2。

1997 年,NVIDIA 发布 NV-3(RIVA-128)、RIVA-128ZX、TNT。

1999 年 2 月,NVIDIA 发布 TNT-2。

1999 年 8 月,NVIDIA 发布第一颗 GPU——GeForce-256,支持 DirectX 7 的有 Geforce 2 系列。

2001 年,NVIDIA 推出第一款拥有可编程顶点着色器的 GPU——GeForce 3,支持 DirectX 8 的有 GeForce 3 和 GeForce 4 系列。

2002 年,NVIDIA 推出第一款用 32 位浮点流水线作为可编程的顶点处理器 GPU——GeForce Fx,支持 DirectX 9.0 的有 GeForce 6 系列和 GeForce 7 系列。

2006 年,GeForce 8 系列采用统一渲染架构,用通用渲染单元替代原来分离的顶点着色单元和图像着色单元,支持 DirectX 10.0,支持 CUDA。

2007 年,CUDA 正式发布。

2008 年,NVIDIA 发布支持 CUDA 计算能力 1.1(关于 GPU 的计算能力,详见 3.4 节)的 GeForce 9 系列 GPU(这之前的 GPU 型号类似 9800GT),以及计算能力 1.3 的 GT200 GPU(此后 GPU 型号类似 GTX480)。

2010 年,NVIDIA 推出 Fermi 架构产品 M2050、GTX480 等(计算能力为 2.X),搭载 Tesla M2050 GPU 的天河 1A 超级计算机成为 TOP 500 榜首。

2012 年,NVIDIA 发布 Kepler 架构产品 Tesla K20(计算能力为 3.5),搭载该产品的泰坦超级计算机在 TOP 500 夺冠;其后相继发布 K40 和 K80。

2015 年,基于 Maxwell 架构的 Tesla M40 等相关产品已经上市,其针对当下研究火热的深度学习进行了优化,设计了 16 位半精度数据类型。

2016 年 4 月,NVIDIA 在 GTC 大会上发布了基于 Pascal 架构的产品 Tesla P100,该系列产品采用 NvLink、CoWoS HBM2 堆叠内存两项关键技术,极大地提升了访存带宽和通信带宽。

2.8 GPU 编程方法

GPU 从最初的不可编程,到使用着色器语言编程,再到 CUDA、Brook 等的提出,OpenCL、OpenACC 等通用编程模型的推广,可编程性获得了逐步提高。本节将分别对这些 GPU 编程方法展开简单介绍。

1. 图形学 API 编程

最早 GPU 只能执行固定的几类操作,没有可编程 GPU 的概念。利用 DirectX 和

OpenGL 等图形 API 进行程序映射，将需要计算的科学问题转换为图形处理问题，然后调用相应的图形处理接口完成计算，即所谓的可编程着色器（programmable shader）。其后出现了相对高级的着色器语言（shader language），比如基于 DirectX 的 HLSL 和基于 OpenGL 后端的 GLSL，以及同时支持 DirectX 和 OpenGL 的 Cg。

2. Brook 源到源编译器

由斯坦福大学的 Ian Buck 等人在 2003 年开发。Brook 是对 ANSI C 的扩展，是一个基于 Cg 的源到源编译器，可以将类 C 的 brook C 语言通过 brcc 编译器编译为 Cg 代码，很好地掩藏了图形学 API 的实现细节，大大简化了 GPU 程序开发过程。由于早期 Brook 只能使用像素着色器运算，且缺乏有效的数据通信机制，导致效率低下。

3. Brook＋

AMD/ATI 在 Brook 基础上结合 GPU 计算抽象层（compute abstraction layer，CAL），推出 Brook＋。Brook＋利用流与内核的概念，在编程指定流和内核后，由编译器完成流数据和 GPU 的通信，运行时自动加载内核到 GPU 执行。内核程序再编译为 AMD 流处理器设备代码 IL，运行时由 CAL 执行。Brook＋相对 Brook 有了巨大改进，但仍存在数据传输和流程序优化困难的缺陷。AMD/ATI 公司的 Stream SDK 中采用了 Brook＋作为高级开发语言，用于 AMD 的 Firestream 系列 GPU 编程开发。但目前 Stream SDK 和 Brook＋都已弃用，AMD 产品主要支持 OpenCL 为主。

4. CUDA

由 NVIDIA 在 2007 年发布，无须图形学 API，采用类 C 语言，开发简单，到目前已发布 CUDA 7.5。CUDA 支持 C/C++、FORTRAN 语言的扩展；提供了丰富的高性能数学函数库，比如 CUBLAS、CUFFT、CUDNN 等。CUDA 定义了 GPU 上执行的数据和核函数，通过运行时 API 或设备 API 进行设备和数据管理。CUDA 结合 GPU 底层体系结构特性，为用户提供更底层的控制，程序优化具有巨大优势。目前 CUDA 程序仅能在 NVIDIA 的 GPU 上运行。但在 SC2015 会议上，AMD 已宣布 A 卡 GPU 将支持 CUDA 程序。

5. OpenCL

OpenCL 是第一个面向异构系统的通用并行编程标准，也是一个统一的编程环境。OpenCL 最初由苹果公司提出，后由 Khronos Group 发布并制定 OpenCL 行业规范，NVIDIA、Intel、AMD 等 IT 巨头均已支持 OpenCL。OpenCL 并行架构包含宿主机和若干 OpenCL 设备，宿主机与 OpenCL 设备互连并整合为一个统一的并行平台，同时为程序提供 API 和运行库。主流的 OpenCL 设备包括多核 CPU、GPU、DSP（数字信号处理器）、FPGA 和 Intel Xeon Phi 等。

6. OpenACC

OpenACC 最早由 PGI 公司提出并实现，后被 NVIDIA 公司收购。类似于 OpenMP，OpenACC 提供了一系列编译指导指令，通过在程序并行区域外指定编译指导语句，然后由编译器对并行区域内代码进行分析，编译为目标平台上的源代码，这是一种源到源的转换。OpenACC 隐藏了异构系统主机端和设备端间数据传输和执行调度等细节,从而大大简化了异构编程。OpenACC 的执行模型包括 gang、worker 和 vector 三级并行结构: gang 级是粗粒度的,一个加速器可以运行多个 gang;worker 是细粒度的,一个 gang 中有一个或多个 worker;vector 是 worker 内的向量操作或 SIMD。OpenACC 主要支持 CPU + GPU 异构并行计算,目前 NVIDIA 已提供高校和科研用户免费注册使用 OpenACC Toolkit。

7. OpenMP 4.5

OpenMP 4.5 是对标准共享存储编程模型 OpenMP 的扩展,扩展内容主要是支持异构计算。通过在指导命令层指定数据传输和描述加速任务区,实现异构计算。目前 Intel 的 ICC 编译器已经有了 OpenMP 4.0 的实现,其加速器主要支持 Intel Xeon Phi。NVIDIA 预计在 OpenMP 4.5 支持 GPU 加速,GCC 已有初步的支持。

另外,若有多个 GPU 或在 GPU 集群中,可以联用 OpenMP 或 MPI 与 GPU 编程方法(比如 CUDA),进行混合编程。MPI + OpenMP + CUDA 是 GPU 集群上主流的混合编程模型。在本书第 18 章将会具体阐述混合编程的实例和用法。

2.9 CPU/GPU 异构系统

CPU/GPU 异构系统中,CPU 与 GPU 经北桥(Bridge)通过 AGP 或 PCI-E 总线连接,各自有独立的外部存储器,分别是主存(Host Memory)和显存(Device Memory)。其中,CPU 负责逻辑性较强的事务处理,GPU 负责高密集度的浮点计算,图 2.3 展示了典型的 CPU + GPU 异构系统。在使用 GPU 计算前,CPU 必须先通过北桥将数据传到 GPU 显存中;在 GPU 计算完成后,GPU 再将结果数据返回给主机内存。CPU 与 GPU 间的通信是必不可少的,由于接口 PCI-E 的通信带宽限制,优化数据通信开销是必须要考虑的问题。

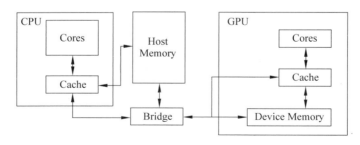

图 2.3 典型的 CPU + GPU 异构系统

此外,有一个比喻可简单认识 CPU、GPU 及其异构系统。GPU 好比 F1 赛车,对跑道(环境)要求较高,但一旦跑起来速度非常快;而 CPU 相当于一般家用轿车,一般公路即可满足条件。如果 CPU 和 GPU 跑在一起,速度往往取决于慢的一方,很可能发生的情况是 GPU 有能力却发挥不出来,跑车夹在轿车间在公路上缓慢行驶。

第 3 章 GPU 硬件架构

本章将介绍 GPU 硬件架构,主要包括 GPU 架构、kernel 函数的 GPU 映射、GPU 存储体系和 GPU 计算能力,帮助读者对 GPU 硬件有一个整体认知。

3.1 GPU 架构

到目前为止,NVIDIA 主要有 5 类不同架构的 GPU 产品。

(1) Tesla 架构。其硬件核心有 G80、G92、GT200,产品有 8800GTX、9800GTX、GTX280 等。

(2) Fermi 架构。其硬件核心有 GF100、GF104,产品有 Tesla M2050、GTX480 等。

(3) Kepler 架构。其硬件核心有 GK104、GK110,产品有 Tesla K10、K20、GTX680 等,其中 GK104 核心只是一个过渡,不支持所有的 Kepler 架构功能。目前市面上能买到的基本上都是 Kepler 架构产品。

(4) Maxwell 架构。其硬件核心有 GM107、GM200、GM204,产品有 GTX 750Ti、Tesla M40、GTX TITAN X、GTX980 等,Maxwell 产品 Tesla M40 针对深度学习进行了专门优化,比如提供了 16 位数据类型。

(5) Pascal 架构。目前发布的核心有 GP100,产品有 Tesla P100 GPU。在 2016 年 4 月的 GPU 技术大会(GPU technology conference,GTC)上,NVIDIA 发布了 Pascal 架构产品 Tesla P100,该产品预计于同年 6 月份全面上市(截至本书定稿还未上市)。根据 NVIDIA 公布的数据,该产品拥有 153 亿个 16 纳米制造工艺的晶体管,计算性能达到双精度 5.3Teraflop 和单精度 10.6Teraflop,若使用 NVIDIA GPU BOOST 技术可达到单精度 21.2Teraflop。此外 P100 还采用 NVLink 技术,令双向通信带宽高达 160GB/s;配置了 CoWoS HBM2 堆叠内存,显存带宽可达 720GB/s。

不同 GPU 架构的设计理念、工艺水平等均不相同,相应的内部体系结构和性能也不一致。接下来本书将展开介绍这几类 GPU 架构。

为了更好地认知 GPU 体系结构,首先,介绍几个 GPU 体系结构相关术语。

流处理器(streaming processor,SP)：也称为 core，是 GPU 运算的最基本计算单元。早期(tesla 架构)只能进行单精度浮点运算和整数运算，符合 IEEE 754-1985 标准；后来(Fermi 架构开始)采用 IEEE 754-2008 标准，增加了对双精度浮点运算的支持，并支持完整的 32 位整数运算。

渲染核(shader core)：SP 的另一个名称，又称为 CUDA core，始于 Fermi 架构。

双精度浮点运算单元(DP)：专用于双精度浮点运算的处理单元。

特殊功能单元(special function unit,SFU)：用来执行超越函数指令，比如正弦、余弦、倒数和平方根等函数。每个 SFU 一次执行一个线程的一条指令需要一个时钟周期，因此 Tesla 架构(2 个 SFU，半 warp 执行)中的一个 warp 需要消耗 8 个时钟周期。SFU 并不始终占用 SM 中的 dispatch，即当 SFU 处于执行状态的时候，指令分发单元可以向其他的执行单元分发相应的指令。

流处理器(streaming multiprocessors,SM)：GPU 架构中的基本计算单元，也是 GPU 性能的源泉，由 SP、DP、SFU 等运算单元组成。这是一个典型的阵列机，其执行方式为 SIMT(单指令多线程)，区别于传统的 SIMD，能够保证多线程的同时执行(向量化)。

SMX：Kepler 架构中的 SM。

SMM：Maxwell 架构中的 SM。

线程处理器簇(thread processing cluster,TPC)：由 SM 和 L1 cache 组成，存在于 Tesla 架构中，后又出现在 Pascal 架构。G80 架构包含 2 个 SM 和 16KB L1 cache；GT200 架构包含 3 个 SM 和 24KB L1 cache。

图形处理器簇(graph processing cluster,GPC)：类似于 TPC，是介于整个 GPU 和 SM 间的硬件单元，Fermi 架构开始有这个称谓。TPC 和 GPC 还是有所区别的，TPC 由 SM 控制器、SM 和纹理缓存组成，而 GPC 由一个光栅单元、4 个 SM 和 SM 控制器组成。GPC 中 SM 数量是可扩展的。

注意：开始笔者认为 Fermi 架构开始用 GPC 取代了 TPC，甚至认为两者是同一个概念的不同叫法。但在最新的 Pascal 架构中，同时出现了 GPC 和 TPC，且 GPC 包含 TPC，故两者并不等价。

流处理器阵列(scalable streaming processor array,SPA)：所有处理核心和高速缓存的总和，包含所有的 SM、TPC 和 GPC。与存储器系统共同组成 GPU 架构。

存储控制器(memory controller,MMC)：顾名思义，控制存储访问的单元，合并访存。每个存储控制器可以支持一定位宽(比如 64 位)的数据合并访存。

光栅操作单元(raster operation processors,POP)。

存取单元(load/store unites,LD/ST)。

3.1.1 Tesla 架构

Tesla 架构的计算能力为 1.X，不同计算能力版本间有些功能上的区别，比如 1.0 不支持原子操作，也不支持双精度运算；1.1 添加了全局存储器的 32 位字原子操作函数；1.2 开始支持共享存储原子操作功能，并添加了对全局存储 64 位字原子操作的支持；1.3

开始支持双精度运算。由于 Tesla 架构中 SP 不支持双精度运算,双精度运算是通过增加 DP 实现的。

Tesla 架构的 SM 如图 3.1 所示,由 8 个 SP 和 2 个 SFU 组成,在 GT200 a/b 中增加了 DP 支持双精度运算,同时,SM 中还包含寄存器、共享存储、常量存储等存储单元。其中,G80 核心拥有 8192 个 32 位寄存器单元,GT200 扩充了一倍,共 16 384 个。

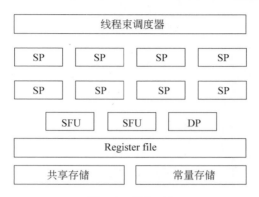

图 3.1 SM 1.X

2~3 个 SM 配合 L1 Cache 构成 TPC,Tesla 架构主要核心型号有 G80 和 GT200,G80 核心中 2 个 SM 和 16KB L1 Cache 组成 1 个 TPC,GT200 包含 3 个 SM 和 24KB L1 Cache,另外,每个 TPC 均由 1 个 SM 控制器进行统一控制(见图 3.2)。

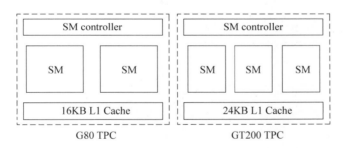

图 3.2 TPC

Tesla 架构从整体上主要由两个部分组成,即 SPA 和存储系统。其中 SPA 由所有的 TPC 构成,存储系统包含一系列可编程存储单元,将在 3.3 节详细介绍。这两部分通过片上高速交叉互联网络连接,具有良好的扩展性。当然,Tesla 架构中 G80 和 GT200 两种核心也存在差异,如图 3.3 和图 3.4 所示。

GPU 可扩展性主要通过增减 TPC 和内存控制器实现,NVIDIA 可以生产出满足不同消费者需求的产品,比如在基于 G80 核心的 GeForce 8800 GTX 中,包含 8 个 TPC,每个 TPC 有 2 个 SM;基于 GT200 的产品中,根据市场定位不同,NVIDIA 推出了诸多不同档次的产品,比如包含 10 个 TPC(30 个 SM)和 8 个存储控制器的 GTX285,包含 10 个 TPC 和 7 个存储控制器的 GTX275,和由 9 个 TPC 和 7 个存储控制器组成的 GTX260+。

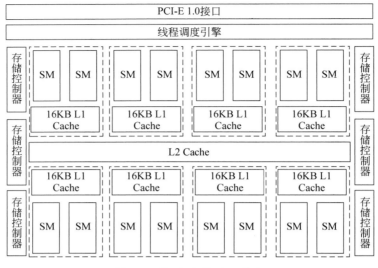

图 3.3 Tesla 架构 G80 核心

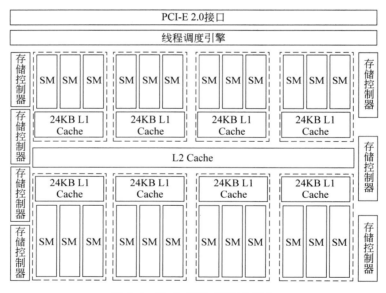

图 3.4 Tesla 架构 GT200 核心

3.1.2 Fermi 架构

Fermi 架构的计算能力是 2.X,主要包括 2.0 和 2.1。在 Fermi 架构开始,正式出现核心(Core)的概念,取代了原来的 SP,主要区别是 core 采用 IEEE 754-2008 标准,增加了对双精度运算和完整 32 位整数运算的支持。此时即使没有单独的 DP,GPU 也能进行双精度运算。另外,Fermi 架构对原子操作进行了巨大改进,使得该功能真正有了应用的可行性(此前的 Tesla 架构原子操作性能极差,不适合实际应用)。Fermi 还增加了 ECC 内存校验、多 kernel 函数并发执行、共享存储可配置(16KB 或 48KB)、双 warp 调度机制等功能。

图 3.5 和图 3.6 分别展示了 SM 2.0 和 SM 2.1 的组成结构，SM 2.0 中包含 32 个 Core、16 个 LD/ST、4 个 SFU、2 个线程束调度器。SM 2.1 对 SM 2.0 再次进行了改进，将 Core 数量增加到 64 个，LD/ST 增加到 32 个，SFU 增加到 8 个。

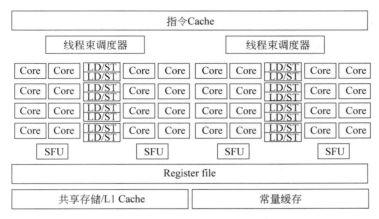

图 3.5　SM 2.0

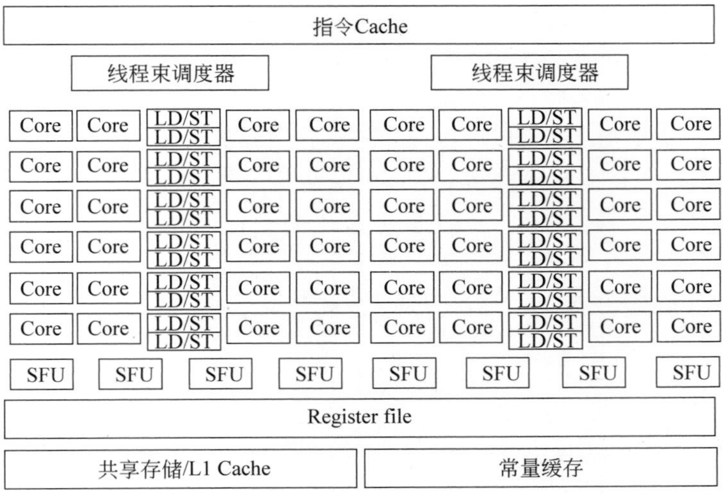

图 3.6　SM 2.1

Fermi 架构（见图 3.7）中，4 个 SM 和 1 个光栅单元组成 1 个 GPC，所有 GPC 构成 SPA。其他部分与 Tesla 类似，包括 L2 Cache、PCI-E 接口、内存控制器（MMC）、线程调度引擎等。每个内存控制器控制 1 个 64b（位）显存分区，即每个内存控制器支持每次 64b 数据合并访存，因此图 3.7 中 6 个内存控制器供支持 384b 显存数据合并访存。

3.1.3　Kepler 架构

本书所描述的 Kepler 架构是 GK110 及以后的 Kepler 架构，计算能力为 3.5 和 3.7。

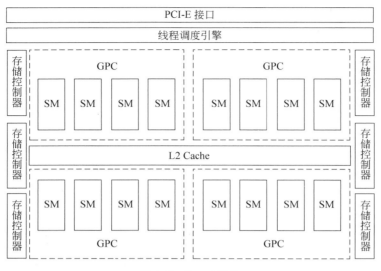

图 3.7　Fermi 架构

不包括 GK104(计算能力为 3.0)。GK110 开始有了重大创新，比如引入 SMX 替代了原来的 SM，增加了动态并行功能，引入 Hyper-Q 技术允许多个 CPU 线程同时在同一个 GPU 上启动 kernel 函数运算，引入 GPUDirect 技术利用远程直接存储访问(remote directly memory accessing，RDMA)支持 GPU 间的直接数据通信等。

SMX(见图 3.8)包含 192 个单精度 CUDA 核心、64 个双精度单元(DP)、32 个特殊功能单元(SFU)和 32 个加载存储单元(LD/ST)，其中，单精度 CUDA 核采用 IEEE 754-2008 标准支持单精度核双精度运算，Kepler 架构额外增加 DP 来增强双精度运算能力(GK110 专为高性能计算设计)。SMX 以单个 warp(32 线程)为单位进行调度。SMX 还

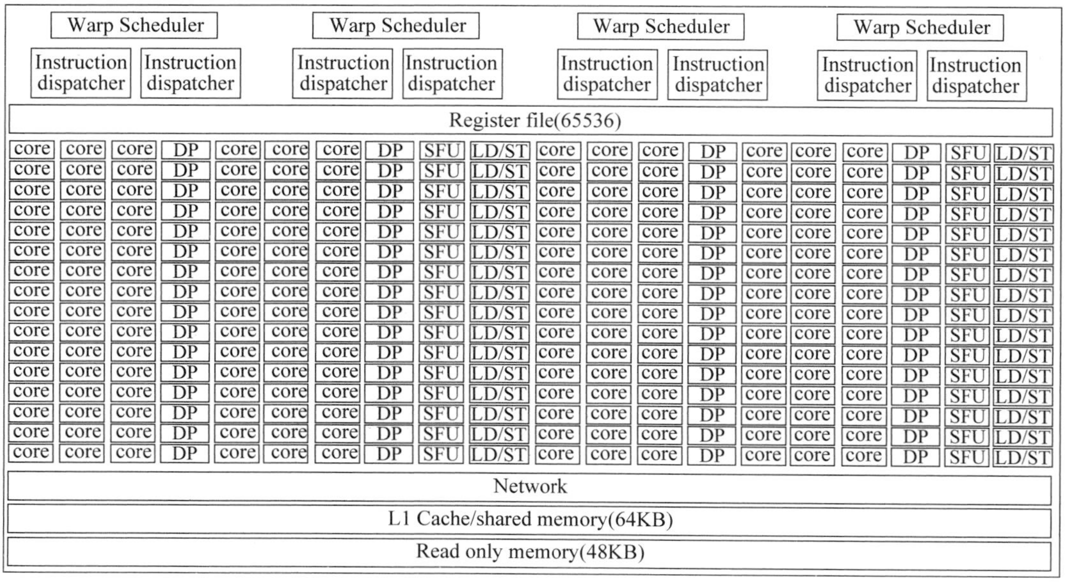

图 3.8　SMX

包含4个指令分发单元、8个指令调度器、可配置共享存储和L1 Cache共64KB(首次支持16/32/48KB三种配置),48KB只读存储。

Kepler GK110架构(见图3.9)有15个SMX和6个64位内存控制器,部分产品有所不同(比如Tesla K20c GPU只有13个SMX和5个MMC)。官方的Kepler架构白皮书中没有提及GPC的概念(笔者推测Kepler架构中不存在GPC,因为13或15个SMX不能平均分配到GPC中)。其他组成部分类似,只是性能比之前架构有了大幅提升,比如支持PCI-E 3.0接口等。

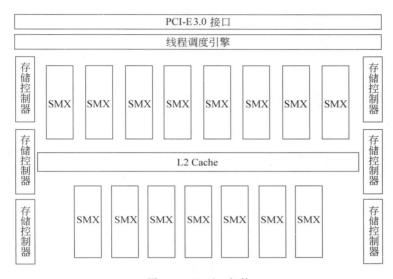

图3.9 Kepler架构

3.1.4 Maxwell架构

Maxwell架构对SM再次做了升级,称谓变为SMM。SMM(见图3.10)中包含128个CUDA核、32个SFU和32个LD/ST、4个指令分发单元和8个调度器(每个调度器每个时钟能启动1条指令)。SMM将L1 Cache从Shared Memory中抽离了出来,提供了专门的Shared Memory,将L1 Cache与Texture相结合。在GM204中提供了96KB共享存储、PolyMorph Engine 3.0,而在GM107中仍提供64KB共享存储和PolyMorph Engine 2.0。

图3.11展示了Maxwell架构最后一代产品GTX980上的GM204核心架构,由内存控制器、GPC、线程调度引擎和PCI-E接口共同组成了整个架构。其中GPC由1个光栅单元和4个SMM组成。

从GTX750Ti架构白皮书中了解到,其GM107核心仅有1个GPC,包含5个SMM。由此可知,GPU架构具有非常优秀的多层次可扩展性,包括TPC(GPC)层可扩展性、SM(SMX、SMM)层可扩展性和SP(Core)层可扩展性。

图 3.10 SMM

3.1.5 Pascal 架构

Pascal 架构其实是 NVIDIA 的第 6 代 GPU 架构,GP100 核心的计算能力为 6.0,但是真正发布的产品仅有 5 代。

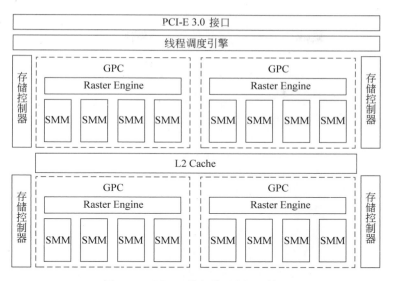

图 3.11　Maxwell 架构 GM204 核心

Pascal 架构（GP100）中的 SM 可分为两个处理块，每个处理块拥有 32 个单精度 CUDA 核、16 个双精度处理单元、8 个 LD/ST 单元、8 个 SFU 单元、1 个指令 buffer、1 个 warp 调度器、2 个指令分发单元、32 768 个 32 位寄存器（见图 3.12）。整个 SM 还拥有 4 个纹理单元和 64KB 共享存储。

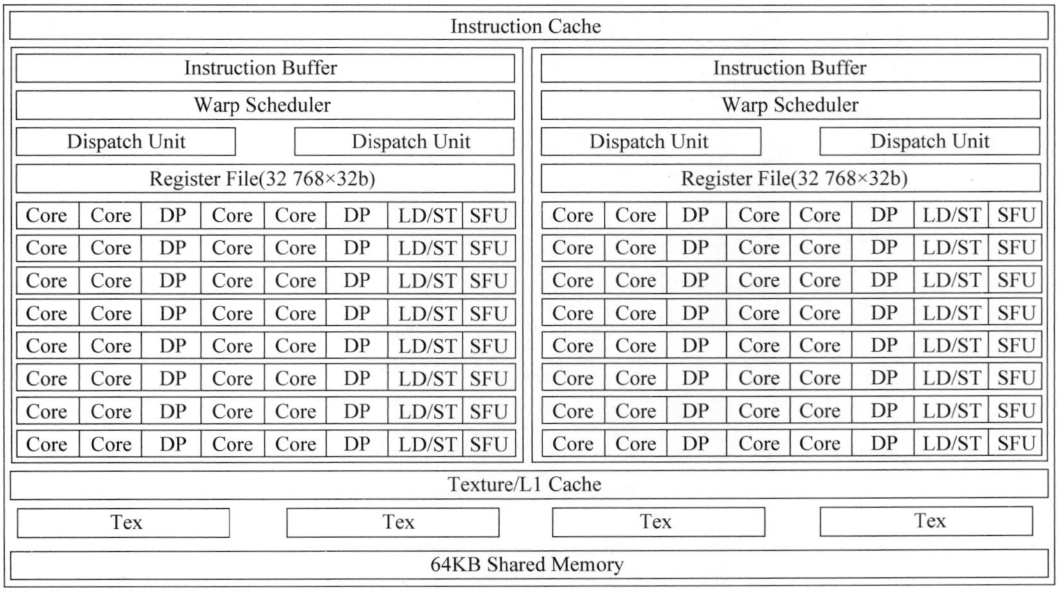

图 3.12　Pascal(GP100) SM

在 Pascal 架构中，2 个 SM 组成 1 个 TPC，5 个 TPC 组成 1 个 GPC（该架构中 TPC 和 GPC 同时出现，且是包含与被包含的关系）。1 个完整的 GP100 核心（见图 3.13）拥有

6个GPC、30个TPC、60个Pascal架构SM、8个512位宽的存储控制器(共4096位)、3840个单精度CUDA核、240个纹理单元和4个NVLink单元。每个存储控制器管理512KB的L2 Cache(共4096KB)。两个存储控制器控制1个HBM2(high bandwidth memory 2) DRAM stack。每个NVLink单元可以提供40GB/s的双向通信带宽。

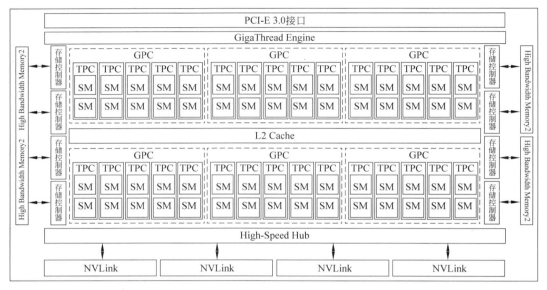

图3.13 完整的GP100 GPU

当然,根据市场定位的不同,不同的GPU产品的SM数量和TPC数量等均有所差异,比如Tesla P100 GPU采用了GP100核心,但并不完整,仅包含28个TPC、56个SM、3584个单精度CUDA核心。

相对于Kepler架构和Maxwell架构,Pascal架构引入了不少新技术,提供了更加强大的性能。Pascal架构的优势如下。

(1) 超高的浮点运算性能。Tesla P100 GPU提供了5.3Tflops双精度运算性能、10.6Tflops单精度运算性能和21.2Tflops半精度运算性能(半精度运算主要是针对当前极火的深度学习开发的),利用GPU Boost技术可达到单精度21.2Tflops运算性能。

(2) NVLink技术。NVLink技术能支持多GPU间或CPU与GPU间的通信,能提供相对PCI-E带宽5倍的通信速率,双向通信带宽可达160GB/s。

NVLink可以将多个GPU交叉连接,构建成网格以获得更好的通信效益。比如图3.14展示了一个由8个P100 GPU组成的立体网格。图中每个虚线表示的NVLink连接可以提供40GB/s的双向通信带宽。若是相互连接的GPU数量下降,那么多出来的NVLink单元可以与现有其他GPU连接,若两个GPU间连接了两个NVLink连接,那么其双向通信带宽为80GB/s。若系统中只有两个GPU通过NVLink连接(4条虚线),则双向通信带宽为160GB/s。

当NVLink被用来进行CPU和GPU通信时,前提条件是CPU和GPU都支持NVLink技术。目前只有Pascal架构的NVIDIA Tesla P100 GPU和IBM公司的

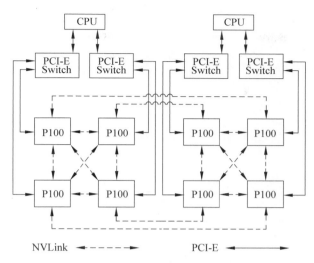

图 3.14　8 个 P100 组成的立体网格

Power8 CPU 支持 NVLink 技术。当 CPU 和 1 个 GPU 利用 NVLink 技术连接时，如图 3.15(a) 所示，有 4 条 NVLink 连接，可提供 160GB/s 的双向通信带宽。而当 CPU 连接两个 GPU 时，如图 3.15(b) 所示，每个 GPU 只能提供 2 条 NVLink 连接与 CPU 相连，GPU 间有两条 NVLink 连接，故双向通信带宽仅为 80GB/s。

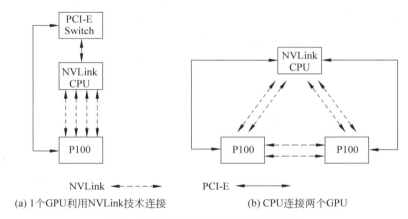

(a) 1 个 GPU 利用 NVLink 技术连接　　　(b) CPU 连接两个 GPU

图 3.15　NVLink 连接 CPU 和 GPU

注意：也别把上述带宽数据想象得太过美好，双向通信带宽在 GPU 程序开发过程中应该极少能够用到，我们能够用到的可能仅仅是其一半。比如在本书实验平台上测得的页锁定存储的 PCI-E 带宽大概是 6GB/s，其双向通信带宽至少是 12GB/s，但极少有应用存在同时双向通信来使用这 12GB/s 的双向通信带宽，多数情况只能用到 6GB/s 带宽，甚至有时由于通信数据量过小导致真实的通信速率远低于 6GB/s。

(3) HBM2 堆叠内存。P100 首次在 GPU 中引入 HBM2 高速 GPU 存储架构，访存带宽同比增长 3 倍，最高可达 720GB/s。在 Pascal 架构中，一个存储控制器的访存位宽为 4096 位，而 Kepler 架构和 Maxwell 架构的 GDDR5 位宽普遍为 384 位。

（4）统一存储空间。P100 首次在 GPU 中引入统一存储空间（在软件上，CUDA 6 开始就引入了统一虚拟存储，不过 Pascal 架构引入的统一存储空间做了重要的扩展和改进，详细内容请查阅 Pascal 架构白皮书），统一存储提供了 CPU 和 GPU 存储的统一地址空间（49 位寻址空间，可容纳 512TB 虚拟存储空间）。利用统一存储，程序员无须关心数据存储位置和设备间通信。

（5）计算抢占。Pascal 架构允许计算任务在指令级被中断，将程序上下文交换到 DRAM，其他程序被调入和执行（此前的 Maxwell 和 Kepler 架构只能在 block 级进行任务切换）。计算抢占技术能避免长时间运行的应用独占系统或发生运行超时。这样，在与类似于交互式图像任务或交互式调试应用协同运行时，该长时运行的应用可以长时间运行来处理大规模数据或等待其他条件。另外，计算抢占可以打破 kernel 函数的运行时间限制，GPU 程序开发时将不必专门考虑 kernel 函数运行是否会超时。

（6）原子操作扩展。Pascal 架构对原子操作进行了扩展，增加了 64 位数据的支持，包括 64 位长整型数据和双精度浮点型数据。

3.2 Kernel 的硬件映射

Kernel 函数是指在 GPU 上执行的函数。一般用 __device__ 和 __global__ 关键字声明，在 CPU 端调用时利用<<<>>>配置相应的线程（块）维度、共享存储容量和 CUDA 流等信息。

Kernel 函数映射到 GPU 上执行，对比普通函数，增加了以下几个概念，分别是 grid（线程格）、block（线程块）、thread（线程）。三者是包含关系，大量的 thread 组成一个 block，大量的 block 组成一个 grid（参见图 5.1）。在执行 kernel 函数时，一个 kernel 函数对应一个 grid（从 Fermi 架构开始，一个 GPU 支持同时多个 kernel 函数执行），grid 内 block 数量和 block 内 thread 数量在 kernel 函数启动时在<<<>>>内指定。声明方法如下：

```
dim3 blocks(bx,by,bz),threads(tx,ty,tz);
gpu_kernel<<<blocks,threads>>>(…);
```

Tesla K20c GPU 中 threads(tx,ty,tz)尺寸维度为(1024,1024,64)，每个 block 最多 1024 个 thread；blocks(bx,by,bz)尺寸维度为(2147483647,65535,65535)。线程设计原则：每个 block 中的 thread 数量合适（一般为 256、512、1024），使得占用率最大化（占用率计算工具在 CUDA 安装目录的 tools 文件夹下，CUDA_Occupancy_Calculator.xls）；block 数量越大越好（GPU 中 block 的切换开销极小，几乎可以忽略不计）。

在执行时，kernel 函数映射到 GPU，对应的 block 映射到 SM(SMX,SMM)，thread 映射到 SP(CUDA core)。很明显，kernel 中设置的 block 数量和 thread 数量都远远超过 GPU 硬件中的 SMX 和 core 数量，GPU 通过频繁的线程切换实现硬件资源的分时使用，切换开销极小，几乎可以忽略。在实际的开发过程中，与 GPU 映射和执行相关的开销主要有两个,分别是 kernel 函数启动开销、GPU 线程(块)切换开销。

从较细粒度的角度看 GPU 执行,GPU 执行采用单指令多线程(single instruction multiple thread,SIMT)模式,即每个 warp 执行同一条指令,比如 ID 为 0~31 的 thread 同时执行同一条指令。借用 1.5 节的比喻,该过程可以看成是战场上 32 个士兵(thread)为战阵的一排(warp),同时挥刀砍向敌军,此时,如果敌方(数据)也是整齐排列(访存对齐),那么效果最佳,否则自然会导致效果不好,条件是战阵中一排士兵只能做相同的动作(指令)。

需要说明的是,在 Tesla 架构中,SIMT 线程调度单位是半 warp,即 16 线程;从 Fermi 架构开始,SIMT 线程调度的单位改成了一个 warp,即 32 线程。

3.3 GPU 存储体系

GPU 存储体系是影响 GPU 程序性能的最重要的因素之一。GPU 设计了鲜明的层次式存储,使用好层次式存储是进行性能优化的关键。GPU 存储结构如图 3.16 所示。

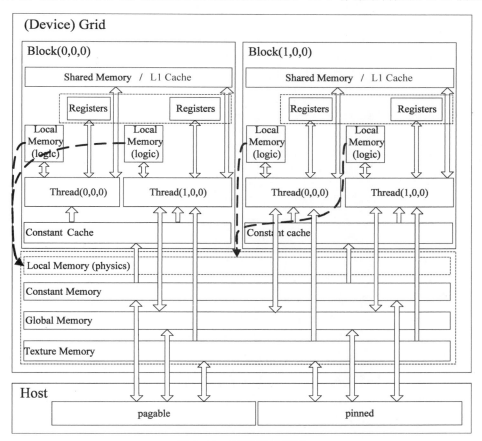

图 3.16 GPU 存储模型

（1）寄存器和局部存储对线程私有，寄存器资源位于SMX上，数量有限，线程运行时动态获得；局部存储逻辑上等同寄存器，但其物理局部存储空间位于显存中，故其访存延迟与全局存储相当。

（2）共享存储与L1 Cache使用相同的硬件资源（在Maxwell架构进行了分离，提供专门的共享存储，将L1 Cache与Texture结合），一般总大小为64KB，其中共享存储可通过编程控制，且对线程块内所有线程共享。

（3）每个SMX上都有常量缓存，主要支持常量存储的缓存和广播功能，常量存储本身位于显存中，通过SMX上的常量缓存实现常量存储快速访问。

（4）全局存储和纹理存储都位于显存中，区别是纹理存储提供了专门的纹理缓存通道。

（5）主机端提供了页锁定内存和可分页内存两种形式的存储类型，可与设备端的常量存储、全局存储和纹理存储进行数据交换。各级存储根据离核距离不同访存延迟也不相同，基本使用原则是尽量使用离核近的、访存延迟小的存储单元，尽可能地避免使用全局存储等访存延迟高的存储单元。（具体的存储单元阐述和使用请阅读本书第16章。）

3.4 GPU 计算能力

在GPU中，计算能力和运算性能是两个不同的概念，在直观上容易令人混淆。

运算性能包括整数运算性能、单精度浮点运算性能和双精度浮点运算性能等，表示GPU处理算术运算的能力，也是衡量GPU好坏的关键标准之一。

GPU计算能力是指GPU架构或GPU支持的功能，而与GPU的浮点运算性能无关（一些旧版本的高端卡性能往往超过新版本的中端卡或低端卡）。随着GPU架构的发展，GPU支持的功能也越来越多。从最初的计算能力1.0到目前最新的计算能力6.0，GPU经历了Tesla架构、Fermi架构、Kepler架构、Maxwell架构和Pascal架构，增加了大量实用功能的支持。

计算能力1.0：主要产品是8800Ultras和8000系列卡，GPU核心为G80，经典产品如8800GTX。笔者姑且将计算能力1.0称为基准版本。尽管此时还不支持比如原子操作、动态并行、多kernel同步执行等功能，但此时基本的GPU框架已经成型，其后所有的GPU架构发展都没有脱离此时的框架。一般情况下，使用NVCC进行程序编译时，无须特殊功能的不需要指定计算能力，此时默认采用的就是计算能力1.0的编译选项，经笔者测试，多数情况往往比采用更高计算能力编译得到的程序性能还好一点。

计算能力1.1：主要产品是9000系列卡，GPU核心为G92，经典产品有9800GTX。计算能力1.1在1.0的基础上增加了计算通信重叠功能，即数据传输和kernel函数可以重叠执行，设备函数中cudaGetDeviceProperties()函数的deviceOverlap属性反映了该项功能。此外，计算能力1.1还添加了全局存储器的32位字原子操作函数。

计算能力1.2：主要产品是GT200系列卡，GPU核心为GT200，经典产品GTX260和GTX280。计算能力1.2大幅度提升了GPU的浮点运算性能，主要表现在：①SM和SP数量增加，片上集成的计算核心数量增加；②SM中并发执行的线程束（warp）从原来的 24 增加到了 32；③大大减少了在计算能力 1.0 和 1.1 中常见的对全局存储器的访问限制和共享存储器的存储体冲突(bank conflict)。此外，计算能力 1.2 开始支持共享存储原子操作功能，并添加了对全局存储 64 位字原子操作的支持。

计算能力 1.3：主要产品是 GT200 系列高端卡，GPU 核心为 GT200 a/b 修订版。相对于计算能力 1.2，计算能力 1.3 的主要改进是开始支持双精度运算，该功能的引入为 GPU 后来在高性能计算领域的卓越成绩做出了不可磨灭的贡献。

计算能力 2.0：计算能力 2.X 开始采用 Fermi 架构的 GPU 核心，计算能力 2.0 采用的是 GF100(SM 2.0)(详见 3.1.2 节)，主要产品是 GT400 和 GT500 系列，经典产品有 GTX480(入选 NVIDIA 提供的 CUDA 入门套包)、GTX580、M2050(天河 1A 超级计算机上装备)。计算能力 2.0 做了大量改进：①L1 Cache 引入 SM 中(此前 Tesla 架构中位于 SM 外 TPC 中，详见 3.1.1 节和 3.1.2 节)，并可选择 16KB 和 48KB 配置；②增加了 ECC(error correcting code)校验；③支持双复制(dual-copy)引擎；④扩充了共享存储容量；⑤共享存储片(体，bank)数从 16 增加到 32；⑥执行时开始以一个线程束(warp)为单位执行(此前是半 warp 执行)；⑦大幅改进了原子操作的性能，使其真正有了实用的可能性。

计算能力 2.1：采用 Fermi 架构的 GF104(SM 2.1)核心，主要产品有 GT400 和 GT500 系列中端卡、GT600 系列低端卡，比如 GTX460、GTX550 Ti、GT610 等。计算能力 2.1 产品的市场定位是中端游戏卡，而非高端计算卡，因此牺牲了其双精度运算来增加 CUDA 核心数量。计算能力 2.1 的主要特色包括：①SM 中 CUDA 核数量从 32 增加到 48；②SM 中 SFU 数量增加 1 倍，从 4 个增加到 8 个；③采用了双束调度器替代单束调度器(单束调度器需要 2 个时钟取出 2 条指令，而双束调度器 2 个时钟可取出 4 条指令执行)。

计算能力 3.0：计算能力 3.X 采用了 Kepler 架构，计算能力 3.0 的核心是 GK104，主要产品是 GT700 系列中高端卡、GT600 系列高端卡，比如 GTX770、GTX690、Tesla K10 等。

计算能力 3.5：NVIDIA 官方给出的 Kepler 架构白皮书所描述的 Kepler 架构就是指代计算能力 3.5 的 GPU 架构。计算能力 3.5 的核心是 GK110，主要产品有 GT700 系列高端卡、TITAN 系列，比如 GTX780、GTX TITAN、GTX TITAN Z、Tesla K20、Tesla K40 等。计算能力 3.5 主要有以下改进：①引入 SMX 替代了原来的 SM；②SMX 中 CUDA 核数量增加到 192；③增加了动态并行功能；④引入 Hyper-Q 技术允许多个 CPU 同时在同一个 GPU 上启动 kernel 函数运算；⑤引入 GPUDirect 技术利用 RDMA 支持 GPU 间的直接数据通信。

计算能力 3.7：核心为 GK210，主要产品 Tesla K80。主要特征有：①GPU Boost(超频)技术；②双倍的共享存储和寄存器；③Zero-power Idle(闲置时零功耗)。

计算能力 5.0：计算能力 5.X 采用了 Maxwell 架构，计算能力 5.0 采用了 GM107 核心，主要产品有 GTX750 Ti、GTX 960M 等。计算能力 5.0 的改进如下：①引入了 SMM 替代 SM；②SMM 中 CUDA 核数量降到 128；③提升了每 Watt 性能；④提供了 2048KB 的 L2 Cache(GK107 中仅有 256KB)；⑤L1 Cache 从共享存储抽离，与 Texture memory 结合。

计算能力 5.2：采用 GM204 核心，主要产品有 GT900 系列高端卡，如 GTX TITAN X、GTX980 等。计算能力 5.2 做了如下改进：①SMM 中 PolyMorph Engine 升级为 3.0；②SMM 中共享存储提升到了 96KB。

计算能力 6.0：采用 GP100 核心，当前产品有 Tesla P100 GPU。计算能力 6.0 属于 Pascal 架构，引入了大量改进：①SM 结构进行调整，单个 SM 中的 CUDA 核调整为 64 个，加入 32 个 DP；②GPU 的组成做了调整，GPC 和 TPC 同时出现，且 GPC 包含 TPC，TPC 包含 2 个 SM；③计算性能极大增强；④采用 NVLink 技术，支持 GPU 间和 CPU 与 GPU 间通信，双向带宽可达 160GB/s；⑤采用 CoWoS HBM2 堆叠内存，访存带宽高达 720GB/s；⑥统一存储空间，简化编程时的显式数据传输；⑦计算抢占，避免应用独占系统或运行超时，应用可以长期运行来处理大规模数据或等待其他条件；⑧扩展了原子操作，增加了对 64 位数据的支持，包括 64 位长整型数据和双精度浮点型数据。

此外，关于 GPU 计算能力，值得一提的是，在 NVCC 编译器中的--gpu-architecture(-arch) 和--gpu-code(-code)选项都需要指定计算能力(关于这两个选项的区别请参见附录 E)，此时可设置的计算能力有两大类：compute_35 和 sm_35。其中，compute_35 指定虚拟编译计算能力，NVCC 编译器利用该指定的计算能力编译 CUDA C 代码为 PTX (parallel thread execution)中间代码；而 sm_35 指定 GPU 架构，NVCC 编译器根据该项数据将 PTX 中间代码编译为二进制执行文件。

第 4 章 GPU 软件体系

本章主要介绍 GPU 软件生态系统、CUDA Toolkit、CUDA 环境安装配置等内容。

4.1 GPU 软件生态系统

时至今日，GPU 并行程序开发已经得到全面发展，形成了有机的软件生态系统。构成 GPU 软件生态系统的成分包括编译器、编程模型、数学函数库、性能分析工具、程序调试工具、代码实例（SDK）、管理软件、应用软件和完整的文档等。

编译器主要有 NVIDIA CUDA Compiler（NVCC）、PGI CUDA Fortran Compiler、ptxas（PTX 汇编工具）、cuobjdump（CUDA 目标文件转储工具）。此外，还有些自动化并行编译器（源到源的转换工具），比如，可将 Fortran 和 C 转换成 CUDA C 的 PGI 编译器和 CAPS HMPP，可将 C 转换为 CUDA C 代码的 Goose 编译器，将 Fortran 转换为 CUDA C 的 NOAA F2C 编译器。

GPU 编程模型主要有 CUDA、OpenCL、OpenACC，其中 CUDA 已支持诸多编程语言，最常用的是 CUDA C（NVCC），其他的还包括 CUDA Fortran（PGI、FLAGON 支持、ArrayFire GPU library）、CUDA Python（PyCUDA、ArrayFire GPU library）、CUDA Java（jCUDA、JaCUDA）、CUDA .NET、CUDA C++（Thrust、CuPP、Libra、ArrayFire GPU library）、CUDA F#。

面向 GPU 的数学函数库种类相当丰富，比如线性代数库 CUBLAS、稀疏矩阵运算 CUSPARSE、快速傅里叶变换 CUFFT、深度学习 CUDNN、AMGX、NPP、FFmpeg、CHOLMOD、CULA Tools、MAGMA、IMSL Fortran 数值库、CUSOLVER、ArrayFire、CURAND、Thrust、NVBIO、NVIDIA VIDEO CODEC SDK、HiPLAR、OpenCV、GPP（Geometry Performance Primitives）、Paralution、Triton Ocean SDK 等。

性能分析工具主要有 NVIDIA Visual Profiler。

程序调试工具有命令行的 CUDA-GBD、界面版开发工具的 Nsight。

代码实例有 NVIDIA 提供的 CUDA SDK Code samples。

管理软件包括 NVIDIA 系统管理工具 nvidia-smi。

另外，针对常用软件如 MATLAB、Mathematica、R、LabView 等提供了二次 GPU 开发支持，比如支持 MATLAB 二次开发的 MathWorks 和 GPUlib 等。

4.2 CUDA Toolkit

在 CUDA 5.0 之前的 CUDA Toolkit 包中不包含 GPU 驱动和 SDK，而 CUDA 5.0 开始三者集成到统一的 CUDA Toolkit 安装包中。

CUDA Toolkit 提供了 CUDA C 程序开发的基本软件环境，包括 NVCC 编译器、CUDA 驱动 API、CUDA 运行时 API、CUDA 库函数等（见图 4.1）。程序员可根据需求选择驱动 API、运行时 API 或 CUDA 库函数进行 CUDA C 编程，然后利用 NVCC 编译器编译即可得到可执行程序。其中，驱动 API 是 CUDA 编程时使用的最底层的 API，使用较为复杂，其函数命名为 cu*()；运行时 API 使用相对简单，其函数命名方式为 cuda*()。

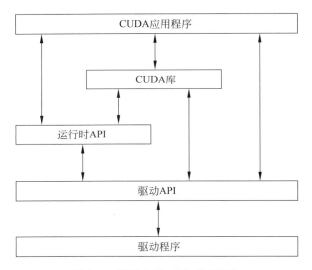

图 4.1 CUDA Toolkit 软件层次

此外，CUDA Toolkit 还提供了其他编译工具，如 ptxas 汇编工具、cuobjdump 等，还有 GPU 管理工具 nvidia-smi、性能分析工具 visual profiler、occupancy 计算工具（excel 文档）、丰富的文档资料等。

4.2.1 NVCC 编译器

NVCC(NVIDIA CUDA Compiler)是编译 CUDA 代码的编译器，其功能是将 CUDA C 源代码编译为一个可执行的 CUDA 应用程序。NVCC 包含编译、链接等多种功能。最简单的使用方法如下：

```
$nvcc abc.cu -o abc
```

除此之外,NVCC 还提供了多种编译开关,详细信息请阅读附录 E。这里仅介绍两个最简单、最实用的编译开关,-O3 是优化等级,等同 ICC、GCC 等编译器,只在编译 CPU 端代码时起作用;-arch 'compute_35' -code 'compute_35' 指定代码编译的计算能力,最好与 GPU 计算能力一致,35 即计算能力为 3.5,与本文代码性能测试的 Tesla K20c GPU 一致。完整指令如下,详细功能说明参考附录 B。如果需要编译驱动 API 程序,需要加入 -lcuda 编译选项,否则链接时会出错。

```
$nvcc abc.cu -o abc -O3 -arch 'compute_35' -code 'compute_35'
$nvcc abc.cu -o abc -O3 -arch 'compute_35' -code 'compute_35' -lcuda
```

注意:附录 E 提供了 NVCC 编译器的帮助菜单供读者参考,CUDA Toolkit 包中提供了 nvcc.pdf 文档。

4.2.2 cuobjdump

PTX(并行线程执行)代码是编译后的 GPU 代码的一种中间形式,可以再编译为原生的 GPU 微码。PTX 编译成 GPU 微码有离线和在线两种方式。离线编译可生成未来可在计算机上执行的目标程序。在线编译指程序在运行时将 PTX 代码在线编译成 GPU 微码。PTX 离线编译采用 ptxas 编译器,生成微码存储在 CUDA 二进制形式的 cubin 文件中。cubin 文件可以反汇编成 SASS 代码,查看 SASS 代码是 GPU 优化的高级手段之一。下面演示了编译和反汇编的相关指令,这里以 6.4 节的代码为例子进行演示。

```
[fangmq@cn18%yhstar SASS_test]$nvcc vectoradd_gpu_3.cu -cubin
[fangmq@cn18%yhstar SASS_test]$cuobjdump --dump-sass vectoradd_gpu_3.cubin
 code for sm_10
    Function : _Z16vector_add_gpu_3PfS_S_i
  .headerflags @ "EF_CUDA_SM10 EF_CUDA_PTX_SM(EF_CUDA_SM10)"
    /*0000*/     MOV.U16 R0H, g [0x1].U16;            /*0x0023c78010004205*/
    /*0008*/     I2I.U32.U16 R1, R0L;                 /*0x04000780a0000005*/
    /*0010*/     IMAD.U16 R0, g [0x6].U16, R0H, R1;   /*0x0020478060014c01*/
    /*0018*/     ISET.S32.C0 o[0x7f], g [0xa], R0, LE; /*0x6c20c7c83000d5fd*/
    /*0020*/     RET C0.NE;                           /*0x0000028030000003*/
    /*0028*/     MOV.U16 R1L, g [0x1].U16;            /*0x0023c78010004209*/
    /*0030*/     SHL R3, R0, 0x2;                     /*0xc41007803002000d*/
    /*0038*/     IMUL32.U16.U16 R5, g [0x4].U16, R1L; /*0x41022814*/
    /*003c*/     IADD32 R4, g [0x4], R3;              /*0x2103e810*/
    /*0040*/     IADD32 R2, g [0x6], R3;              /*0x2103ec08*/
    /*0044*/     IADD32 R1, g [0x8], R3;              /*0x2103f004*/
    /*0048*/     SHL R3, R5, 0x2;                     /*0xc410780300020a0d*/
    /*0050*/     GLD.U32 R7, global14[R4];            /*0x80c00780d00e081d*/
```

```
/*0058*/    GLD.U32 R6, global14[R2];              /*0x80c00780d00e0419*/
/*0060*/    FADD32 R6, R7, R6;                     /*0xb0060e18*/
/*0064*/    IADD32 R0, R0, R5;                     /*0x20058000*/
/*0068*/    GST.U32 global14[R1], R6;              /*0xa0c00780d00e0219*/
/*0070*/    ISET.S32.C0 o[0x7f], g [0xa], R0, GT;  /*0x6c2107c83000d5fd*/
/*0078*/    IADD32 R2, R3, R2;                     /*0x20028608*/
/*007c*/    IADD32 R4, R4, R3;                     /*0x20038810*/
/*0080*/    IADD R1, R3, R1;                       /*0x0400478020000605*/
/*0088*/    BRA C0.NE, 0x50;                       /*0x000002801000a003*/
/*0090*/    NOP;                                   /*0xe0000001f0000001*/
...
```

详情可参考 CUDA Toolkit 提供的 cuobjdump.pdf 文档。

4.3 CUDA 环境安装

4.3.1 Windows 7 安装 CUDA 4.2

本文为什么要在 Windows 环境下选择介绍 CUDA 4.2 的安装流程而非最新版本？首先 CUDA 5 开始已经做了较好的集成，可以直接安装，不需要进行手动配置；而 CUDA 5 之前的版本均需要手动配置。其次是 CUDA 4.2 版本稳定好用，笔者曾安装使用过 CUDA 5 和 CUDA 5.5，使用时曾多次出现 CUDA 环境不明原因的损坏、使用不方便（如编译时出一大堆无用信息）等现象，另外，高版本 CUDA 需要的 VS 版本也越来越高，而笔者认为 CUDA 4.2 已经能够提供几乎所有的 CUDA 编程功能，因此在此推荐使用。

1. 安装前需要准备的软件

（1）VS2008ProEdition90DayTrialCHSX1435983。

（2）VisualAssistX_10.7.1912_Soft711。

（3）310.90-notebook-win8-win7-winvista-32bit-international-whql（选择计算机对应的驱动）。

（4）cudatoolkit_4.2.9_win_32。

（5）gpucomputingsdk_4.2.9_win_32（cudatoolkit 和 skd 版本必须一致）。

（6）CUDA VS Wizard（32 位和 64 位有区别的），32 位：CUDA_VS_Wizard_W32.2.2.exe。

2. 软件安装（按照上面罗列的顺序逐步安装软件）

CUDA 环境配置：64 位请选择相应目录。

1）添加文件包含

依次打开"工具"→"选项"→"项目和解决方案"→"VC++ 目录"，添加包含文件、库文件和源文件。

包含文件：

```
C:\Program Files\NVIDIA GPU Computing Toolkit\CUDA\v4.2\include
C:\ProgramData\NVIDIA Corporation\NVIDIA GPU Computing SDK 4.2\C\common\inc
C:\ProgramData\NVIDIA Corporation\NVIDIA GPU Computing SDK 4.2\shared\inc
```

库文件：

```
C:\Program Files\NVIDIA GPU Computing Toolkit\CUDA\v4.2\lib\Win32
C:\ProgramData\NVIDIA Corporation\NVIDIA GPU Computing SDK 4.2\C\common\lib\Win32
C:\ProgramData\NVIDIA Corporation\NVIDIA GPU Computing SDK 4.2\C\common\lib
C:\ProgramData\NVIDIA Corporation\NVIDIA GPU Computing SDK 4.2\shared\lib\Win32
```

源文件：

```
C:\ProgramData\NVIDIA Corporation\NVIDIA GPU Computing SDK 4.2\C\common\src
C:\ProgramData\NVIDIA Corporation\NVIDIA GPU Computing SDK 4.2\shared\src
```

2）添加扩展名支持

依次打开"工具"→"选项"→"文本编辑器"→"文件扩展名"，在扩展名中添加 cu 和 cuh，在编辑器中选择 Microsoft Visual C++。

依次打开"工具"→"选项"→"项目和解决方案"→"VC++ 项目设置"里面的"C/C++ 文件扩展名"，添加 *.cu 和 *.cuh。

3）配置系统 Path 环境变量

在 VS 中建立一个 CUDA 工程，按照默认为 Console application，不要选择 EmptyProject，建好后直接编译。编译通过但运行时提示无法找到 cutil32D.dll。

解决方法：将 SDK 安装目录下的 C/bin/win32 目录中 Debug 和 Release 两个文件夹加入系统 Path 环境变量中。SDK 的默认安装路径为 C:\ProgramData\NVIDIA Corporation\NVIDIA GPU Computing SDK 4.2。

需要在 path 变量中加入下面两个值，添加环境变量后需要注销一遍系统才能生效。

```
C:\ProgramData\NVIDIA Corporation\NVIDIA GPU Computing SDK 4.2\C\bin\win32\Debug;
C:\ProgramData\NVIDIA Corporation\NVIDIA GPU Computing SDK 4.2\C\bin\win32\Release;
```

4）编译 cutil 库（包括 debug 模式和 release 模式）

若出现错误：无法打开输入文件 cutil32D.lib，则需要编译 cutil 库。

方法：在目录 C:\ProgramData\NVIDIA Corporation\NVIDIA GPU Computing SDK 4.2\C\common\lib\Win32 中查找 cutil32D.lib 文件是否存在，若不存在则编译目录 C:\ProgramData\NVIDIA Corporation\NVIDIA GPU Computing SDK 4.2\C\common 下的 cutil_vs2008.vcproj（根据需要编译 debug 模式还是 release 模式）。然后再添加到库里，至此 CUDA 环境配置成功。

3. 添加 Visual Assist X 支持（可选，辅助 CUDA C 编程）

此前已安装了 Visual Assist X 软件，在此基础上进行以下步骤即可令 Visual Assist X 支持 CUDA 代码高亮显示。

（1）从目录 C:\ProgramData\NVIDIA Corporation\NVIDIA GPU Computing SDK 4.2\C\doc\syntax_highlighting\visual_studio_8 复制 usertype.dat 到目录 C:\Program Files\Microsoft Visual Studio 9.0\Common7\IDE 中。可用记事本打开 usertype.dat 查看里面要求高亮显示的字符。

（2）关闭 VS 2008（注意：一定要关闭 VS，否则注册表中部分项目无法找到）。

（3）进入注册表，在 HKEY_LOCAL_MACHINE/SOFTWARE/Microsoft/VisualStudio/9.0/Languages/File Extensions 下面添加子键.cu，然后复制.cpp 的键值到.cu。

（4）在 HKEY_CURRENT_USER/Software/Whole Tomato/Visual Assist X/VANet 9 下的 ExtSource 键添加键值.cu 和.cuh。

（5）配置 Visual Assist 属性。打开 VS 2008，打开 Visual Assist 属性，在 projects 的 C/C++ Directories custom 下面添加 CUDA 的头文件目录。文件目录如下所示（而在一般情况下，Assist 会自动获取 VS 的工程属性配置，故此步骤通常可省略）。

```
C:\Program Files\NVIDIA GPU Computing Toolkit\CUDA\v4.2\include
C:\ProgramData\NVIDIA Corporation\NVIDIA GPU Computing SDK 4.2\C\common\inc
C:\ProgramData\NVIDIA Corporation\NVIDIA GPU Computing SDK 4.2\shared\inc
```

（6）打开 VS2008，查看 __global__ 等 CUDA 内置关键字是否已被着色，再在代码中敲 cuda，是否有提示以 cuda 开头的函数列表。若有则说明已成功添加了 Visual Assist X 支持。

4.3.2 Linux 下安装 CUDA

Linux 下安装 CUDA 环境首先需要准备 CUDA 软件包，NVIDIA 提供了针对不同系统的软件包，需要知悉系统型号版本等信息。当然了解当前环境是否存在 GPU 也至关重要。下面两个命令分别查看 GPU 和 Linux 系统信息。

```
$lspci | grep -i nvidia
$uname -m && cat /etc/*release
```

Linux 下 CUDA 环境的安装分为命令行（无图像界面）和图像界面两种情况。

在命令行模式运行的 Linux 系统中安装 CUDA 非常简单，直接运行安装（根据自己的情况填写相应配置），然后在环境变量添加相应的目录即可完成，注意安装需要使用 root 权限（Linux 命令行模式安装的是 CUDA 5.5 版本，下面给出了本书实验平台的 CUDA 环境安装流程）。

```
$su
#sh cuda_5.5.22_linux_64.run
$vi ~/.bashrc
$source ~/.bashrc
export PATH=/vol/home/fangmq/gpu_cuda/cuda/cuda_5_5/bin:.:$PATH
export LD_LIBRARY_PATH=/vol/home/fangmq/gpu_cuda/cuda/cuda_5_5/lib:$LD_LIBRARY_PATH
export LD_LIBRARY_PATH=/vol/home/fangmq/gpu_cuda/cuda/cuda_5_5/lib64:$LD_LIBRARY_PATH
```

而在有图形界面的 Linux 下安装 CUDA 环境略微复杂,具体的操作步骤如下。

(1) root 权限下进入命令行模式。

```
#systemctl set-default multi-user.target
#reboot
```

(2) 禁用图形界面。

① 把 Nouveau 加入黑名单:修改/etc/modprobe.d/blacklist.conf 文件,加入 blacklist nouveau。

② 重新建立 initramfs image 文件。

```
$mv /boot/initramfs-$(uname -r).img /boot/initramfs-$(uname -r).img.bak
$dracut -v /boot/initramfs-$(uname -r).img $(uname -r)
```

检查 nouveau 驱动,确保没有被加载 lsmod | grep nouveau。

注意:命令行模式不识别中文,因此,所有需要的软件包必须存放在英文路径下。

(3) 安装 CUDA 环境:与命令行模式下的 CUDA 安装和配置方法相同,笔者在另一个 GPU 平台(centos7)上实践安装时采用了 CUDA 7.0 版本(注意,笔者安装时禁止了 OpenGL 项)。

(4) 重新启动图形界面。

① 还原禁掉的 nouveau:用 vi 命令修改/etc/modprobe.d/blacklist.conf 文件,删除 blacklist nouveau。

② 还原原来的 initramfs image 文件。

```
$sudo cp /boot/initramfs-$(uname -r).img.bat /boot/initramfs-$(uname -r).img
$sudo dracut -v /boot/initramfs-$(uname -r).img $(uname -r) --force
systemctl get-default
设置当前的模式为 systemctl set-default multi-user/graphical.target
```

(5) root 权限下进入图形界面模式。

```
#systemctl get-default                    %查看当前的模式
#systemctl set-default graphical.target
#reboot
```

至此,Linux 图形界面下 CUDA 环境安装完成。

验证 CUDA 是否安装成功:①用 which nvcc 查看 nvcc 命令是否存在,若不存在则修改环境变量;②进入 NVIDIA_Sample 文件夹,随便选择进入一个工程,make 成功,./XX 运行成功就说明 CUDA 环境已经安装成功了。

MAC 下安装 CUDA 环境:笔者本人没用过苹果电脑,但曾见过他人在 MAC 上安装 CUDA 环境,只需要直接安装即可,类似 Linux 无图像界面命令行 CUDA 环境安装,不必像 Linux 图形界面那么麻烦。

第 5 章 CUDA C 编程

本章主要阐述 CUDA C 编程模型、CUDA 编程七步曲、运行时 API 与驱动 API、运行时函数、CUDA C 语言扩展和 grid-block-thread 三维模型等内容。

5.1 CUDA 编程模型

在 CUDA 编程模型中引入主机端和设备端的概念，其中 CPU 作为主机端，GPU 作为设备端，主机端仅有一个，而设备端可以有多个。CPU 负责逻辑处理和运算量少的计算，而 GPU 负责运算量大的并行计算。

图 5.1 展示了 CUDA 编程模型，完整的 CUDA 程序包括主机端和设备端两部分代码，主机端代码在 CPU 上执行，设备端代码又称为 kernel 函数，运行在 GPU 上。其中一个 kernel 函数对应一个 grid，每个 grid 根据需要配置不同的 block 数量和 thread 数量。具体的 kernel 函数声明和使用方法将在第二篇中介绍。

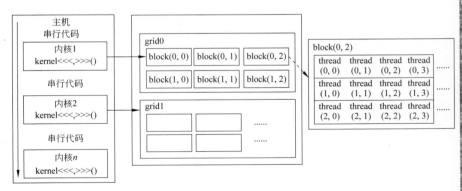

图 5.1 CUDA 编程模型

从 CUDA 编程模型可以看出，CUDA 包含两个并行逻辑层，分别是 block 层和 thread 层。在执行时，block 映射到 SM，thread 映射到 SP (Core)。如何在实际应用程序中高效地开发这两个层次的并行是 CUDA 编程与优化的关键之一。

5.2 CUDA 编程七步曲

一个完整的 CUDA 代码需要包含 7 个关键步骤(笔者称为 CUDA 编程七步曲),特别是多 GPU 情况下;若是单 GPU,可省略为五步曲。下面简单描述 CUDA 代码 7 个关键步骤。

```
cudaSetDevice(0);         //获取设备;只有一个 GPU 时或默认使用 0 号 GPU 时可以省略
cudaMalloc((void**) &d_a, sizeof(float) * n);   //分配显存
cudaMemcpy(d_a,a,sizeof(float) * n,cudaMemcpyHostToDevice);
                                                //数据传输(host to device,H2D)
gpu_kernel<<<blocks,threads>>>(***);            //kernel 函数
cudaMemcpy(a,d_a,sizeof(float) * n,cudaMemcpyDeviceToHost);
                                                //数据传输(device to host,D2H)
cudaFree(d_a);            //释放显存空间
cudaDeviceReset();        //重置设备;可以省略
```

上述 7 个步骤对几乎所有的 CUDA 程序开发都有效,在后续章节中,第 6 章第一个 CUDA 代码就体现了五步曲,第 18 章的多 GPU 方案中体现了七步曲;当然也有例外,比如在第 11 章曼德博罗特集中,使用了 zero-copy 技术,省略了数据传输过程。

5.3 驱动 API 与运行时 API

CUDA 驱动 API(driver API)和运行时 API(runtime API)提供了 CUDA 编程支持,实现了包括设备管理、上下文管理、存储管理、代码块管理、执行控制、纹理索引管理、与 OpenGL 和 Direct3D 的互操作接口等功能。

CUDA 驱动 API 是基于句柄的底层接口,可加载二进制或汇编的 kernel 函数模块。由于二进制和汇编代码的使用,驱动 API 可以被各种语言调用。驱动 API 的所有函数声明以 cu 为前缀。由于驱动 API 涉及大量硬件实现细节,其编程相对复杂,但可以获得更高的性能或用于实现特殊功能。驱动 API 程序编译时需要添加选项-lcuda,否则出错。

CUDA 运行时 API 在驱动 API 基础上进行了封装,隐藏了部分细节,使得编程更加简单方便。运行时 API 初始化在第一次调用运行时函数时完成,而不必专门进行初始化。运行时 API 存放在 CUDArt 包中,其函数前缀为 cuda,编译时(特别是利用其他编译器链接时)需要添加选项-lcudart。

本书旨在讲解 GPU 编程和优化,为了方便读者阅读和理解,所有代码均采用 CUDA 运行时 API 书写。5.4 节也仅介绍 CUDA 运行时函数。

5.4 CUDA 运行时函数

CUDA 运行时提供了丰富的函数,功能涉及设备管理、存储管理、数据传输、线程管理、流管理、事件管理、纹理管理、执行控制、与 OpenGL 和 Direct3D 互操作等。

5.4.1 设备管理函数

cudaGetDeviceCount():返回计算能力大于等于 1.0 的 GPU 数量,赋值给 count。该函数的声明和调用方式如下。

```
extern __host__ cudaError_t CUDARTAPI cudaGetDeviceCount(int *count);
int count;
cudaGetDeviceCount(&count);
```

cudaSetDevice():根据 GPU 索引号 device 设置需要调用的 GPU 设备。CUDA 编程时如果不调用此函数指定运行设备,则默认选择索引为 0 的 GPU。

```
extern __host__ cudaError_t CUDARTAPI cudaSetDevice(int device);
int gpuid=0;//选择 0 号 GPU
cudaSetDevice(gpuid);
```

cudaGetDevice():获得当前线程所使用的 GPU 索引号,赋值给 device。

```
extern __host__ cudaError_t CUDARTAPI cudaGetDevice(int *device);
int gpuid;
cudaGetDevice(&gpuid);
```

cudaGetDeviceProperties():获取 GPU 索引号为 device 的参数信息到结构体数据 prop 中。其中,结构体 cudaDeviceProp 的具体参数信息(CUDA 4.2)如下。

```
extern __host__ cudaError_t CUDARTAPI cudaGetDeviceProperties(struct
cudaDeviceProp *prop, int device);
struct __device_builtin__ cudaDeviceProp
{
    char      name[256];            /**<GPU 型号 */
    size_t    totalGlobalMem;       /**<Global memory(B) */
    size_t    sharedMemPerBlock;    /**<每块共享存储容量(B) */
    int       regsPerBlock;         /**<每块注册器数量 */
    int       warpSize;             /**<Warp size */
    size_t    memPitch;             /**<最大内存复制步长 */
```

```c
int         maxThreadsPerBlock;              /**< 每块最大线程数量 */
int         maxThreadsDim[3];                /**< 线程块三维 */
int         maxGridSize[3];                  /**< 线程格三维 */
int         clockRate;                       /**< 计算核心时钟频率(kHz) */
size_t      totalConstMem;                   /**< 常量存储容量(B) */
int         major;                           /**< 主计算能力(小数点前的值) */
int         minor;                           /**< 次计算能力(小数点后的值) */
size_t      textureAlignment;                /**< 纹理对齐要求 */
size_t      texturePitchAlignment;           /**< 绑定到等步长内存的纹理满足的要求 */
int         deviceOverlap;                   /**< GPU是否支持并发内存复制和kernel执行 */
int         multiProcessorCount;             /**< SMX数量 */
int         kernelExecTimeoutEnabled;        /**< 是否有运行时限制 */
int         integrated;                      /**< 设备是否集成(否则独立) */
int         canMapHostMemory;                /**< 可否对主机内存进行映射 */
int         computeMode;                     /**< 计算模式 */
int         maxTexture1D;                    /**< 最大1D纹理尺寸 */
int         maxTexture1DLinear;              /**< 线性内存相关的最大1D纹理尺寸 */
int         maxTexture2D[2];                 /**< 最大2D纹理维度 */
int         maxTexture2DLinear[3];           /**< 最大2D纹理维度(width, height, pitch) */
int         maxTexture2DGather[2];           /**< 纹理聚集时的最大纹理维度 */
int         maxTexture3D[3];                 /**< 最大3D纹理维度 */
int         maxTextureCubemap;               /**< 最大立方图纹理维度 */
int         maxTexture1DLayered[2];          /**< 最大1D分层纹理维度 */
int         maxTexture2DLayered[3];          /**< 最大2D分层纹理维度 */
int         maxTextureCubemapLayered[2];     /**< 最大立方图分层纹理维度 */
int         maxSurface1D;                    /**< 最大1D表面尺寸 */
int         maxSurface2D[2];                 /**< 最大2D表面维度 */
int         maxSurface3D[3];                 /**< 最大3D表面维度 */
int         maxSurface1DLayered[2];          /**< 最大1D分层表面维度 */
int         maxSurface2DLayered[3];          /**< 最大2D分层表面维度 */
int         maxSurfaceCubemap;               /**< 最大立方图表面维度 */
int         maxSurfaceCubemapLayered[2];     /**< 最大立方图分层表面维度 */
size_t      surfaceAlignment;                /**< 表面对齐要求 */
int         concurrentKernels;               /**< 设备能并发的kernel数量 */
int         ECCEnabled;                      /**< 是否打开ECC校验 */
int         pciBusID;                        /**< PCI总线ID */
int         pciDeviceID;                     /**< PCI设备ID */
int         pciDomainID;                     /**< PCI域ID */
int         tccDriver;                       /**< 是否支持TCC(Tesla集群) */
int         asyncEngineCount;                /**< 异步引擎数量 */
int         unifiedAddressing;               /**< 主机和设备共享同一地址空间 */
int         memoryClockRate;                 /**< 存储时钟频率 */
```

```
    int      memoryBusWidth;                /**<Global memory 总线带宽 */
    int      l2CacheSize;                   /**<L2 Cache 尺寸(B) */
    int      maxThreadsPerMultiProcessor;   /**<每个 SMX 驻留的最大线程数量 */
};
```

cudaChooseDevice()：根据 prop 的参数信息，选择设备参数最匹配的 GPU，返回其索引号到 device。

```
extern __host__ cudaError_t CUDARTAPI cudaChooseDevice(int *device, const
struct cudaDeviceProp *prop);
```

cudaSetValidDevices()：设置 GPU 设备列表，len 为设备数量，device_arr 是具体的 GPU 列表。

```
extern __host__ cudaError_t CUDARTAPI cudaSetValidDevices(int *device_arr,
int len);
```

5.4.2 存储管理函数

cudaMalloc()：在 GPU 上分配大小为 size 的线性存储空间，起始地址为 *devPtr。

```
extern __host__ cudaError_t CUDARTAPI cudaMalloc(void **devPtr, size_t size);
```

cudaMallocPitch()：在 GPU 上分配大小为 pitch×height 的逻辑 2D 的线性存储空间，首地址为 *devPtr，其中 pitch 是返回的 width 对齐后的存储空间大小。利用该函数分配的 2D 数组每行都对齐，在访存时具有一定优势。下面给出了地址计算方法。

```
extern __host__ cudaError_t CUDARTAPI cudaMallocPitch(void **devPtr, size_t
*pitch, size_t width, size_t height);
devPtr[x]=devPtr[ rowid * pitch+column ];
```

cudaFree()：释放 devPtr 指针指定的 GPU 存储区域。可释放 cudaMalloc() 和 cudaMallocPitch() 分配的 GPU 存储区域。

```
extern __host__ cudaError_t CUDARTAPI cudaFree(void *devPtr);
```

cudaMalloc3D()：在 GPU 上分配逻辑 1D、2D 或 3D 的存储空间，空间大小至少 width×height×depth，首地址为 pitchedDevPtr。利用该函数分配的空间是对齐的，以 pitch 为宽度对齐。xsize 和 ysize 是逻辑宽和高，等于 width 和 height 范围参数。对于 2D 或 3D 数据的存储分配，利用 cudaMallocPitch() 或 cudaMalloc3D() 可以保证对齐访问，有助于提升性能。

```
extern __host__ cudaError_t CUDARTAPI cudaMalloc3D (struct cudaPitchedPtr
*pitchedDevPtr, struct cudaExtent extent);
```

struct __device_builtin__ cudaPitchedPtr
{
  void *ptr;　　/**<Pointer to allocated memory*/
  size_t pitch;　　/**<Pitch of allocated memory in bytes*/
  size_t xsize;　　/**<Logical width of allocation in elements*/
  size_t ysize;　　/**<Logical height of allocation in elements*/
};
	struct __device_builtin__ cudaExtent
	{
	  size_t width;　　/**<Width in elements when referring to array memory, in bytes when referring to linear memory*/
	  size_t height;　　　　/**<Height in elements*/
	  size_t depth;　　　　/**<Depth in elements*/
	};

cudaMallocArray()：在 GPU 中利用 desc 结构体参数分配数组，首地址为*array。下面分别给出了 cudaChannelFormatDesc 结构体的声明和 cudaChannelFormatKind 的值域。Array 传递数组句柄，desc 指定数组元素中成分的数目和类型，width 指定数组采用字节计数的宽度，height 指定数组高度（可选），flags 表示数组使用方式，若用于表面读写操作须声明为 cudaArraySurfaceLoadStore，下面罗列了 flags 的值域范围。

cudaFreeArray()：释放 GPU 中首地址为*array 的数组。

```
extern __host__ cudaError_t CUDARTAPI cudaMallocArray(struct cudaArray **array,
const struct cudaChannelFormatDesc * desc, size_t width, size_t height __dv(0),
unsigned int flags __dv(0));
```

```
extern __host__ cudaError_t CUDARTAPI cudaFreeArray(struct cudaArray * array);
```

struct __device_builtin__ cudaChannel-FormatDesc
{
  int　　　　x; /**<x*/
  int　　　　y; /**<y*/
  int　　　　z; /**<z*/
  int　　　　w; /**<w*/
  enum cudaChannelFormatKind f;
/**<Channel format kind*/
};
	enum __device_builtin__ cudaChannel-FormatKind
	{
	  cudaChannelFormatKindSigned=0,
	    /**<Signed channel format*/
	  cudaChannelFormatKindUnsigned=1,
	    /**<Unsigned channel format*/
	  cudaChannelFormatKindFloat=2,
	    /**<Float channel format*/
	  cudaChannelFormatKindNone=3
	    /**<No channel format*/
	};

```
//flags 值域
#define cudaArrayDefault           0x00   /**<Default CUDA array allocation flag*/
#define cudaArrayLayered           0x01   /**<Must be set in cudaMalloc3DArray to create a layered CUDA array*/
#define cudaArraySurfaceLoadStore  0x02   /**<Must be set in cudaMallocArray or cudaMalloc3DArray in order to bind surfaces to the CUDA array*/
#define cudaArrayCubemap           0x04   /**<Must be set in cudaMalloc3DArray to create a cubemap CUDA array*/
#define cudaArrayTextureGather     0x08   /**<Must be set in cudaMallocArray or cudaMalloc3DArray in order to perform texture gather operations on the CUDA array*/
```

cudaMalloc3DArray()：在 GPU 上利用 desc 结构体参数分配 1D、2D 或 3D 数组，首地址为*array。height 和 depth 为 0 时分配 1D 数组，最大尺寸为{8192,0,0}(Tesla 架构是 8192，Fermi 架构是 32768，Kepler 架构是 65536)；depth 为 0 时分配 2D 数组，最大尺寸为{65536,32768,0}(Tesla 和 Fermi 架构，Kepler 架构是 65535×65536)；均不为 0 时分配 3D 数组，最大尺寸为{2048,2048,2048}(Tesla 和 Fermi 架构，Kepler 架构是 4096×4096×4096)。详细的限制信息获取请查阅 16.7.2 节。

```
extern __host__ cudaError_t CUDARTAPI cudaMalloc3DArray(struct cudaArray**
array, const struct cudaChannelFormatDesc *desc, struct cudaExtent extent,
unsigned int flags __dv(0));
```

cudaHostAlloc()：在主机端根据 flags 值分配页锁定存储，大小为 size，首地址为*pHost。其中，flags 有 4 种选择，默认为 cudaHostAllocDefault，分配默认页锁定存储；cudaHostAllocPortable 分配的页锁定存储可被 CUDA 上下文访问；cudaHostAllocMapped 分配的页锁定存储映射到 GPU，可在 kernel 内访问（一般在零复制访存时使用）；cudaHostAllocWriteCombined 表示分配写结合存储。

```
extern __host__ cudaError_t CUDARTAPI cudaHostAlloc(void **pHost, size_t size,
unsigned int flags);
#define cudaHostAllocDefault         0x00   /**<Default page-locked allocation
                                                flag*/
#define cudaHostAllocPortable        0x01   /**<Pinned memory accessible by all
                                                CUDA contexts*/
#define cudaHostAllocMapped          0x02   /**<Map allocation into device space*/
#define cudaHostAllocWriteCombined   0x04   /**<Write-combined memory*/
```

cudaMallocHost()：在主机端分配分页锁定内存，大小为 size，首地址为*ptr。

cudaFreeHost()：释放 * ptr 指定的分页锁定内存。

```
extern __host__ cudaError_t CUDARTAPI cudaMallocHost(void **ptr, size_t size);
extern __host__ cudaError_t CUDARTAPI cudaFreeHost(void *ptr);
```

cudaHostRegister()：将 ptr 指向的可分页存储空间（尺寸为 size）注册为页锁定存储空间，类型根据 flags 设定，flags 参数与 cudaHostAlloc()函数等同。

cudaHostUnregister()：解除 cudaHostRegister()函数注册的空间。

```
extern __host__ cudaError_t CUDARTAPI cudaHostRegister(void *ptr, size_t size,
unsigned int flags);
extern __host__ cudaError_t CUDARTAPI cudaHostUnregister(void *ptr);
```

cudaHostGetDevicePointer()：对于分配在 CPU 端，但映射到 GPU 端的存储，根据

pHost 指针(CPU 端)获取在 GPU 端的访问指针 * pDevice。在零复制存储访问时与 cudaHostAlloc() 搭配使用。

```
extern __host__ cudaError_t CUDARTAPI cudaHostGetDevicePointer(void **pDevice,
void *pHost, unsigned int flags);
```

cudaMemset()：将 GPU 端 devPtr 指针指向的 count 长度的存储空间赋值为 value。类似的赋值函数还有 cudaMemset2D、cudaMemset3D、cudaMemsetAsync、cudaMemset2DAsync 和 cudaMemset3DAsync。

```
extern __host__ cudaError_t CUDARTAPI cudaMemset(void *devPtr, int value, size_t count);
extern __host__ cudaError_t CUDARTAPI cudaMemset2D(void *devPtr, size_t pitch, int value, size_t width, size_t height);
extern __host__ cudaError_t CUDARTAPI cudaMemset3D(struct cudaPitchedPtr pitchedDevPtr, int value, struct cudaExtent extent);
extern __host__ cudaError_t CUDARTAPI cudaMemsetAsync(void *devPtr, int value, size_t count, cudaStream_t stream __dv(0));
extern __host__ cudaError_t CUDARTAPI cudaMemset2DAsync(void *devPtr, size_t pitch, int value, size_t width, size_t height, cudaStream_t stream __dv(0));
extern __host__ cudaError_t CUDARTAPI cudaMemset3DAsync(struct cudaPitchedPtr pitchedDevPtr, int value, struct cudaExtent extent, cudaStream_t stream __dv(0));
```

5.4.3 数据传输函数

cudaMemcpy()：主机与设备间的数据传输函数，源地址为 *src，目标地址为 *dst，传输长度为 count，kind 指定了传输方向，可选值域如下。

```
extern __host__ cudaError_t CUDARTAPI cudaMemcpy(void *dst, const void *src,
size_t count, enum cudaMemcpyKind kind);
enum __device_builtin__ cudaMemcpyKind
{
  cudaMemcpyHostToHost=0,          /**< Host -> Host */
  cudaMemcpyHostToDevice=1,        /**< Host -> Device */
  cudaMemcpyDeviceToHost=2,        /**< Device -> Host */
  cudaMemcpyDeviceToDevice=3,      /**< Device -> Device */
  cudaMemcpyDefault=4              /**< Default based unified virtual address
                                        space */
};
```

cudaMemcpy2D()：2D 矩阵传输函数，源是以 *src 指针为起始位置的一个二维矩阵

(宽为 width,高为 height),目标位置是*dst 指针指定的位置,dpitch 和 spitch 分别指出目标和源 2D 空间的行长度,kind 指定传输方向。

```
extern __host__ cudaError_t CUDARTAPI cudaMemcpy2D(void *dst, size_t dpitch,
const void *src, size_t spitch, size_t width, size_t height, enum cudaMemcpyKind
kind);
```

cudaMemcpyPeer():GPU 间数据传输函数,源地址为*src,源 GPU 索引为 srcDevice;目标地址为*dst,目标 GPU 索引为 dstDevice,传输数据长度为 count。

```
extern __host__ cudaError_t CUDARTAPI cudaMemcpyPeer(void *dst, int dstDevice,
const void *src, int srcDevice, size_t count);
```

cudaMemcpyToArray():数据传输到数组的传输函数,源地址为*src,数据尺寸为 count;目标是 CUDA 数组 dst,目标起始位置是数组 dst 的左上角位置(wOffset, HOffset),kind 指定传输方向。

```
extern __host__ cudaError_t CUDARTAPI cudaMemcpyToArray(struct cudaArray
*dst, size_t wOffset, size_t hOffset, const void *src, size_t count, enum
cudaMemcpyKind kind);
```

cudaMemcpyFromArray():数据从数组传出的传输函数,源是*src 指向的数组,起始位置为左上角位置(wOffset,HOffset),目标地址为*dst,传输的数据尺寸为 count, kind 指定了传输方向。

```
extern __host__ cudaError_t CUDARTAPI cudaMemcpyFromArray(void *dst, const
struct cudaArray *src, size_t wOffset, size_t hOffset, size_t count, enum
cudaMemcpyKind kind);
```

cudaMemcpy2DToArray():2D 矩阵到 CUDA 数组的传输函数,源是*src 指向的 2D 矩阵,该 2D 矩阵的行长为 spitch,目标是*dst 指向的 CUDA 数组,起始位置为该数组左上角(wOffset,hOffset),传输的数据为宽 width、高 height 的矩阵,kind 指定了传输方向。

```
extern __host__ cudaError_t CUDARTAPI cudaMemcpy2DToArray(struct cudaArray *dst,
size_t wOffset, size_t hOffset, const void *src, size_t spitch, size_t width,
size_t height, enum cudaMemcpyKind kind);
```

cudaMemcpy2DFromArray():从 CUDA 数组到 2D 矩阵的传输函数,源是*src 指向的 CUDA 数组,起始位置为该数组左上角(wOffset,hOffset),目标是*dst 指向的 2D 矩阵,该 2D 矩阵的行长为 spitch,传输的数据为宽 width、高 height 的矩阵,kind 指定了传输方向。

```
extern __host__ cudaError_t CUDARTAPI cudaMemcpy2DFromArray(void *dst, size_t
dpitch, const struct cudaArray *src, size_t wOffset, size_t hOffset, size_t
width, size_t height, enum cudaMemcpyKind kind);
```

cudaMemcpyArrayToArray()：从数组到数组的数据传输函数，源为*src 数组，起始位置为(wOffsetSrc, hOffsetSrc)，目标为*dst 数组，目标起始位置为(wOffsetDst, hOffsetDst)，数据长度为 count，传输方向是 kind。

cudaMemcpy2DArrayToArray()：从 2D 数组到 2D 数组的数据传输函数，源为*src 数组，起始位置为(wOffsetSrc, hOffsetSrc)，目标为*dst 数组，目标起始位置为(wOffsetDst, hOffsetDst)，传输的数据尺寸为宽 width、高 height 的矩阵，kind 指定了传输方向。

```
extern __host__ cudaError_t CUDARTAPI cudaMemcpyArrayToArray ( struct
cudaArray *dst, size_t wOffsetDst, size_t hOffsetDst, const struct cudaArray
*src, size_t wOffsetSrc, size_t hOffsetSrc, size_t count, enum cudaMemcpyKind
kind __dv(cudaMemcpyDeviceToDevice));
extern __host__ cudaError_t CUDARTAPI cudaMemcpy2DArrayToArray ( struct
cudaArray *dst, size_t wOffsetDst, size_t hOffsetDst, const struct cudaArray
*src, size_t wOffsetSrc, size_t hOffsetSrc, size_t width, size_t height, enum
cudaMemcpyKind kind __dv(cudaMemcpyDeviceToDevice));
```

cudaMemcpyToSymbol()：将主机端数据传输到 GPU 存储（全局存储或常量存储）中，源为*src 指针指向的主机端数据，目标为*symbol 指向的偏移为 offset 的存储，传输数据长度为 count，kind 指明了传输方向。

```
extern __host__ cudaError_t CUDARTAPI cudaMemcpyToSymbol(const char *symbol,
const void *src, size_t count, size_t offset __dv(0), enum cudaMemcpyKind kind __dv
(cudaMemcpyHostToDevice));
```

cudaMemcpyFromSymbol()：将 GPU 端数据传输到主机端存储中，源是*symbol 指向的 GPU 端偏移为 offset 的数据，目标是*dst 指向的主机端数据，传输数据长度为 count，kind 是传输方向。

```
extern __host__ cudaError_t CUDARTAPI cudaMemcpyFromSymbol(void *dst, const
char *symbol, size_t count, size_t offset __dv(0), enum cudaMemcpyKind kind __dv
(cudaMemcpyDeviceToHost));
```

cudaGetSymbolSize：查找与符号*symbol 关联的存储空间大小，返回给*size。Symbol 可以是用限定符__device__、__constant__声明的变量（数组）。

```
extern __host__ cudaError_t CUDARTAPI cudaGetSymbolSize(size_t *size, const
char *symbol);
```

下面给出了上述传输函数的异步传输版本，异步传输函数涉及的主机端存储必须是页锁定存储，否则将不起作用。异步传输函数是进行计算通信重叠优化的关键，异步传输函数的参数较普通函数增加了一个参量，即流号，流是 GPU 执行的重要概念，多个流可以同时执行。CUDA 流相关函数请参见 5.4.5 节。

```
extern __host__ cudaError_t CUDARTAPI cudaMemcpyAsync(void *dst, const void *src,
size_t count, enum cudaMemcpyKind kind, cudaStream_t stream __dv(0));
extern __host__ cudaError_t CUDARTAPI cudaMemcpy2DAsync(void *dst, size_t
dpitch, const void *src, size_t spitch, size_t width, size_t height, enum
cudaMemcpyKind kind, cudaStream_t stream __dv(0));
extern __host__ cudaError_t CUDARTAPI cudaMemcpyPeerAsync(void *dst, int
dstDevice, const void *src, int srcDevice, size_t count, cudaStream_t stream __
dv(0));
extern __host__ cudaError_t CUDARTAPI cudaMemcpyToArrayAsync(struct
cudaArray *dst, size_t wOffset, size_t hOffset, const void *src, size_t count,
enum cudaMemcpyKind kind, cudaStream_t stream __dv(0));
extern __host__ cudaError_t CUDARTAPI cudaMemcpyFromArrayAsync(void *dst,
const struct cudaArray *src, size_t wOffset, size_t hOffset, size_t count, enum
cudaMemcpyKind kind, cudaStream_t stream __dv(0));
extern __host__ cudaError_t CUDARTAPI cudaMemcpy2DToArrayAsync(struct
cudaArray *dst, size_t wOffset, size_t hOffset, const void *src, size_t spitch,
size_t width, size_t height, enum cudaMemcpyKind kind, cudaStream_t stream __dv(0));
extern __host__ cudaError_t CUDARTAPI cudaMemcpy2DFromArrayAsync(void *dst,
size_t dpitch, const struct cudaArray *src, size_t wOffset, size_t hOffset, size_t
width, size_t height, enum cudaMemcpyKind kind, cudaStream_t stream __dv(0));
extern __host__ cudaError_t CUDARTAPI cudaMemcpyToSymbolAsync(const char
*symbol, const void *src, size_t count, size_t offset, enum cudaMemcpyKind kind,
cudaStream_t stream __dv(0));
extern __host__ cudaError_t CUDARTAPI cudaMemcpyFromSymbolAsync(void *dst,
const char *symbol, size_t count, size_t offset, enum cudaMemcpyKind kind,
cudaStream_t stream __dv(0));
```

5.4.4 线程管理函数

cudaThreadSynchronize()：CPU 与 GPU 间的同步函数，保证该函数前的 CPU 和 GPU 上的任务均执行完成，并在该函数位置汇合。一般是 CPU 在该函数处等待 GPU 函数执行完成。

```
extern __host__ cudaError_t CUDARTAPI cudaThreadSynchronize(void);
```

cudaThreadExit()：退出 CUDA 并清除资源。显式清除与线程相关的 CUDA 运行时资源，后续的运行时 API 调用需要重新初始化。但程序结束时一般会隐含函数 cudaThreadExit() 调用。

```
extern __host__ cudaError_t CUDARTAPI cudaThreadExit(void);
```

5.4.5 流管理函数

cudaStreamCreate()：CUDA 流创建函数。

```
extern __host__ cudaError_t CUDARTAPI cudaStreamCreate(cudaStream_t *pStream);
```

cudaStreamDestroy()：CUDA 流销毁函数。

```
extern __host__ cudaError_t CUDARTAPI cudaStreamDestroy(cudaStream_t stream);
```

cudaStreamSynchronize()：流同步函数，等待流 stream 中之前的任务全部完成。

```
extern __host__ cudaError_t CUDARTAPI cudaStreamSynchronize(cudaStream_t stream);
```

cudaStreamWaitEvent()：CUDA 流等待 CUDA 事件函数。事件可以与流处于不同上下文，即该函数可以实现 GPU 设备间的同步。

```
extern __host__ cudaError_t CUDARTAPI cudaStreamWaitEvent(cudaStream_t stream, cudaEvent_t event, unsigned int flags);
```

cudaStreamQuery()：流完成状态查询函数，若已完成则返回 cudaSuccess，否则返回 cudaErrorNotReady。

```
extern __host__ cudaError_t CUDARTAPI cudaStreamQuery(cudaStream_t stream);
```

5.4.6 事件管理函数

cudaEventCreate()：事件创建函数。

```
extern __host__ cudaError_t CUDARTAPI cudaEventCreate(cudaEvent_t *event);
```

cudaEventCreateWithFlags()：根据 flags 创建事件函数，flags 的值域如下。

```
extern __host__ cudaError_t CUDARTAPI cudaEventCreateWithFlags(cudaEvent_t
*event, unsigned int flags);
```
```
#define cudaEventDefault        0x00  /**<Default event flag*/
#define cudaEventBlockingSync
                                0x01  /**<Event uses blocking synchronization*/
#define cudaEventDisableTiming  0x02  /**<Event will not record timing data*/
#define cudaEventInterprocess   0x04  /**<Event is suitable for interprocess use.
                                         cudaEventDisableTiming must be set*/
```

cudaEventRecord()：事件记录函数。

```
extern __host__ cudaError_t CUDARTAPI cudaEventRecord(cudaEvent_t event,
cudaStream_t stream __dv(0));
```

cudaEventQuery()：查询事件是否已被记录。若已记录则返回 cudaSuccess，若未记录则返回 cudaErrorNotReady，若未调用事件记录函数则返回 cudaErrorInvalidValue。

```
extern __host__ cudaError_t CUDARTAPI cudaEventQuery(cudaEvent_t event);
```

cudaEventSynchronize()：等待事件被记录。若未调用事件记录函数则返回 cudaError-InvalidValue。

```
extern __host__ cudaError_t CUDARTAPI cudaEventSynchronize(cudaEvent_t event);
```

cudaEventDestroy()：销毁事件。

```
extern __host__ cudaError_t CUDARTAPI cudaEventDestroy(cudaEvent_t event);
```

cudaEventElapsedTime()：计算两次事件之间的时间差值，返回给*ms。

```
extern __host__ cudaError_t CUDARTAPI cudaEventElapsedTime(float *ms,
cudaEvent_t start, cudaEvent_t end);
```

5.4.7 纹理管理函数

cudaCreateChannelDesc()：返回通道描述符，x、y、z、w 是成分，f 是通道格式。

```
extern __host__ struct cudaChannelFormatDesc CUDARTAPI cudaCreateChannelDesc
(int x, int y, int z, int w, enum cudaChannelFormatKind f);
enum __device_builtin__ cudaChannelFormatKind
{
```

```
    cudaChannelFormatKindSigned=0,        /**<Signed channel format */
    cudaChannelFormatKindUnsigned=1,      /**<Unsigned channel format */
    cudaChannelFormatKindFloat=2,         /**<Float channel format */
    cudaChannelFormatKindNone=3           /**<No channel format */
};
struct __device_builtin__ cudaChannelFormatDesc
{
    int x; /**<x */
    int y; /**<y */
    int z; /**<z */
    int w; /**<w */
    enum cudaChannelFormatKind f;         /**<Channel format kind */
};
```

cudaBindTexture()：绑定存储器到纹理。将*devPtr指向的长度为size的存储器绑定到纹理*texref，desc描述了从纹理取值时如何解释存储器，offset是偏移量。绑定纹理到*texref时，之前绑定的存储器将自动解除绑定。

```
extern __host__ cudaError_t CUDARTAPI cudaBindTexture(size_t *offset, const
struct textureReference *texref, const void *devPtr, const struct cudaChannel-
FormatDesc *desc, size_t size __dv(UINT_MAX));
```

cudaBindTexture2D()：绑定2D存储到纹理。将*devPtr指向的宽为width、高为height、行长为pitch的2D存储空间绑定到*texref纹理，desc描述了从纹理取值时如何解释存储器，offset是偏移量。

```
extern __host__ cudaError_t CUDARTAPI cudaBindTexture2D(size_t *offset, const
struct textureReference *texref, const void *devPtr, const struct cudaChannel-
FormatDesc *desc, size_t width, size_t height, size_t pitch);
```

cudaBindTextureToArray()：绑定数组到纹理。将*array指向的数组绑定到纹理*texref。

```
extern __host__ cudaError_t CUDARTAPI cudaBindTextureToArray(const struct
textureReference * texref, const struct cudaArray * array, const struct
cudaChannelFormatDesc * desc);
```

cudaUnbindTexture()：解除纹理绑定。

```
extern __host__ cudaError_t CUDARTAPI cudaUnbindTexture (const struct
textureReference *texref);
```

cudaGetTextureAlignmentOffset()：获得纹理的对齐偏移，返回给 *offset。

```
extern __host__ cudaError_t CUDARTAPI cudaGetTextureAlignmentOffset(size_t
*offset, const struct textureReference *texref);
```

cudaGetTextureReference()：返回关联到符号 *symbol 的纹理引用 *texref。

```
extern __host__ cudaError_t CUDARTAPI cudaGetTextureReference(const struct
textureReference **texref, const char *symbol);
```

cudaBindSurfaceToArray()：绑定数组 * array 到表面 *surfref。

```
extern __host__ cudaError_t CUDARTAPI cudaBindSurfaceToArray(const struct
surfaceReference * surfref, const struct cudaArray * array, const struct
cudaChannelFormatDesc *desc);
```

cudaGetSurfaceReference()：返回关联到符号 *symbol 的表面引用 *surfref。

```
extern __host__ cudaError_t CUDARTAPI cudaGetSurfaceReference(const struct
surfaceReference **surfref, const char *symbol);
```

5.4.8 执行控制函数

cudaConfigureCall()：配置设备启动函数，为要执行的设备指定 grid 和 block 维度、共享存储尺寸（可选）和 CUDA 流（可选），类似于启动 kernel 函数时的 <<<>>>。

```
extern __host__ cudaError_t CUDARTAPI cudaConfigureCall(dim3 gridDim, dim3
blockDim, size_t sharedMem __dv(0), cudaStream_t stream __dv(0));
```

cudaLaunch()：启动函数 *entry。Entry 是 kernel 函数。

```
extern __host__ cudaError_t CUDARTAPI cudaLaunch(const char *entry);
```

其他执行控制函数如 cudaSetupArgument()、cudaFuncSetCacheConfig()、cudaFuncSetSharedMemConfig()、cudaFuncGetAttributes() 请参考 cuda_runtime_api.h 文件。

5.4.9 错误处理函数

cudaGetLastError()：返回运行时调用的最新错误。返回值有 66 个，如下所示。

```
extern __host__ cudaError_t CUDARTAPI cudaGetLastError(void);
enum __device_builtin__ cudaError
{
```

```
cudaSuccess=0,
cudaErrorMissingConfiguration=1,
cudaErrorMemoryAllocation=2,
cudaErrorInitializationError=3,
cudaErrorLaunchFailure=4,
cudaErrorPriorLaunchFailure=5,
cudaErrorLaunchTimeout=6,
cudaErrorLaunchOutOfResources=7,
cudaErrorInvalidDeviceFunction=8,
cudaErrorInvalidConfiguration=9,
cudaErrorInvalidDevice=10,
cudaErrorInvalidValue=11,
cudaErrorInvalidPitchValue=12,
cudaErrorInvalidSymbol=13,
cudaErrorMapBufferObjectFailed=14,
cudaErrorUnmapBufferObjectFailed=15,
cudaErrorInvalidHostPointer=16,
cudaErrorInvalidDevicePointer=17,
cudaErrorInvalidTexture=18,
cudaErrorInvalidTextureBinding=19,
cudaErrorInvalidChannelDescriptor=20,
cudaErrorInvalidMemcpyDirection=21,
cudaErrorAddressOfConstant=22,
cudaErrorTextureFetchFailed=23,
cudaErrorTextureNotBound=24,
cudaErrorSynchronizationError=25,
cudaErrorInvalidFilterSetting=26,
cudaErrorInvalidNormSetting=27,
cudaErrorMixedDeviceExecution=28,
cudaErrorCudartUnloading=29,
cudaErrorUnknown=30,
cudaErrorNotYetImplemented=31,
cudaErrorMemoryValueTooLarge =32,
cudaErrorInvalidResourceHandle=33,
cudaErrorNotReady=34,
cudaErrorInsufficientDriver=35,
cudaErrorSetOnActiveProcess=36,
cudaErrorInvalidSurface=37,
cudaErrorNoDevice=38,
cudaErrorECCUncorrectable=39,
cudaErrorSharedObjectSymbolNotFound=40,
cudaErrorSharedObjectInitFailed=41,
cudaErrorUnsupportedLimit=42,
```

```
    cudaErrorDuplicateVariableName=43,
    cudaErrorDuplicateTextureName=44,
    cudaErrorDuplicateSurfaceName=45,
    cudaErrorDevicesUnavailable=46,
    cudaErrorInvalidKernelImage=47,
    cudaErrorNoKernelImageForDevice=48,
    cudaErrorIncompatibleDriverContext=49,
    cudaErrorPeerAccessAlreadyEnabled=50,
    cudaErrorPeerAccessNotEnabled=51,
    cudaErrorDeviceAlreadyInUse=54,
    cudaErrorProfilerDisabled=55,
    cudaErrorProfilerNotInitialized=56,
    cudaErrorProfilerAlreadyStarted=57,
    cudaErrorProfilerAlreadyStopped=58,
    cudaErrorAssert=59,
    cudaErrorTooManyPeers=60,
    cudaErrorHostMemoryAlreadyRegistered=61,
    cudaErrorHostMemoryNotRegistered=62,
    cudaErrorOperatingSystem=63,
    cudaErrorStartupFailure=0x7f,
    cudaErrorApiFailureBase=10000
};
```

cudaGetErrorString()：根据错误码返回错误信息字符串。

```
extern __host__ const char *CUDARTAPI cudaGetErrorString(cudaError_t error);
```

5.4.10 图形学互操作函数

cudaGraphicsUnregisterResource()：注销图形资源。

```
extern __host__ cudaError_t CUDARTAPI cudaGraphicsUnregisterResource
(cudaGraphicsResource_t resource);
```

cudaGraphicsResourceSetMapFlags()：设置映射图形资源的使用标志。

```
extern __host__ cudaError_t CUDARTAPI cudaGraphicsResourceSetMapFlags
(cudaGraphicsResource_t resource, unsigned int flags);
```

cudaGraphicsMapResources()：映射图形资源以供 CUDA 访问。映射 resources 中的 count 图形资源。

```
extern __host__ cudaError_t CUDARTAPI cudaGraphicsMapResources(int count,
cudaGraphicsResource_t *resources, cudaStream_t stream __dv(0));
```

cudaGraphicsUnmapResources()：解除映射图形资源。

```
extern __host__ cudaError_t CUDARTAPI cudaGraphicsUnmapResources(int count,
cudaGraphicsResource_t *resources, cudaStream_t stream __dv(0));
```

cudaGraphicsResourceGetMappedPointer()：获取已映射图形资源 resource 对应的设备指针*devPtr，从设备指针*devPtr 开始可访问的内存大小返回给*size。

```
extern __host__ cudaError_t CUDARTAPI cudaGraphicsResourceGetMappedPointer
(void **devPtr, size_t *size, cudaGraphicsResource_t resource);
```

cudaGraphicsSubResourceGetMappedArray()：获得已映射图形资源的子资源对应的数组。将对应于数组索引 arrayIndex 和 mipmap 等级 mipLevel 的图形资源 resource 的子资源数组返回给*array。

```
extern __host__ cudaError_t CUDARTAPI cudaGraphicsSubResourceGetMappedArray
(struct cudaArray **array, cudaGraphicsResource_t resource, unsigned int
arrayIndex, unsigned int mipLevel);
```

5.4.11 OpenGL 互操作函数

与 OpenGL 互操作需要添加头文件 cuda_gl_interop.h。

cudaGLGetDevices()：获得与当前 OpenGL 上下文关联的 CUDA 设备。

```
extern __host__ cudaError_t CUDARTAPI cudaGLGetDevices(unsigned int *pCuda-
DeviceCount, int *pCudaDevices, unsigned int cudaDeviceCount, enum cudaGL-
DeviceList deviceList);
```

cudaGLSetGLDevice()：设置一个 CUDA 设备与 OpenGL 进行互操作。

```
extern __host__ cudaError_t CUDARTAPI cudaGLSetGLDevice(int device);
```

cudaGraphicsGLRegisterImage()：注册一个 OpenGL 纹理或渲染缓冲区对象。

```
extern __host__ cudaError_t CUDARTAPI cudaGraphicsGLRegisterImage(struct
cudaGraphicsResource **resource, GLuint image, GLenum target, unsigned int
flags);
```

cudaGraphicsGLRegisterBuffer()：注册一个 OpenGL 缓冲区对象。

```
extern __host__ cudaError_t CUDARTAPI cudaGraphicsGLRegisterBuffer (struct
cudaGraphicsResource **resource, GLuint buffer, unsigned int flags);
```

cudaWGLGetDevice()：获得与 hGpu 相关的 CUDA 设备。

```
extern __host__ cudaError_t CUDARTAPI cudaWGLGetDevice (int *device, HGPUNV
hGpu);
```

5.4.12 Direct3D 互操作函数

与 Direct3D 互操作需要添加 cuda_d3d9_interop.h、cuda_d3d10_interop.h 或 cuda_d3d11_interop.h 头文件。本节以 cuda_d3d10_interop.h 为例进行阐述。

cudaD3D10GetDirect3DDevice()：获得与当前 CUDA 上下文相关的 Direct3D 设备。

```
extern __host__ cudaError_t CUDARTAPI cudaD3D10GetDirect3DDevice (ID3D10Device
**ppD3D10Device);
```

cudaGraphicsD3D10RegisterResource()：注册 Direct3D 10 资源以供 CUDA 访问。

```
extern __host__ cudaError_t CUDARTAPI cudaGraphicsD3D10RegisterResource
( struct cudaGraphicsResource **resource, ID3D10Resource *pD3DResource,
unsigned int flags);
```

cudaD3D10SetDirect3DDevice()：设置 Direct3D 10 设备来与 CUDA 设备互操作。

```
extern __host__ cudaError_t CUDARTAPI cudaD3D10SetDirect3DDevice (ID3D10Device
*pD3D10Device, int device __dv(-1));
```

cudaD3D10GetDevice()：获得配置器*pAdapter 的设备号,返回给 device。

```
extern __host__ cudaError_t CUDARTAPI cudaD3D10GetDevice(int *device, IDXGIAdapter
*pAdapter);
```

cudaD3D10GetDevices()：获得对应与 direct3D 10 设备的 CUDA 设备。

```
extern __host__ cudaError_t CUDARTAPI cudaD3D10GetDevices (unsigned int
*pCudaDeviceCount, int *pCudaDevices, unsigned int cudaDeviceCount, ID3D10Device
*pD3D10Device, enum cudaD3D10DeviceList deviceList);
```

5.5 CUDA C 语言扩展

CUDA C 语言对 C 语言的扩展主要包含以下几个方面。

1. 函数限定符

__device__：声明在设备上执行的函数。该函数无法被 CPU 端调用,只能由 GPU 端程序调用,即只能被__device__或__global__声明的函数调用。

__global__：声明的函数称为 kernel 函数。该函数只能被 CPU 端调用,执行在 GPU 上。Kernel 函数类型必须是 void,即返回类型必须为空。

__host__：声明在主机执行的函数,仅可在 CPU 端调用。一般情况可省略,只有该函数同时存在被设备端和主机端同时调用的情况需要添加该限定符,且无法与__global__联用。

2. 变量限定符

__device__：声明在设备上的变量,该变量位于 global memory(参见 3.3 节 GPU 存储体系),只能在设备端使用,是全局变量,无须也不能在函数参数表中出现。

__constant__：声明在常量存储中的变量,只能在设备端使用。此变量一般情况下是只读的,只能通过特定方式进行修改(详见 16.6 节)。该变量是全局变量,无须也不能在函数参数表中出现。

__shared__：声明在共享存储中的变量,仅供 block 内所有 thread 共享访存,退出 kernel 函数后失效。该变量无法初始化,一般声明在 kernel 函数中。

3. 内置数组变量

dim3 类型:整型数组,等同于 uint3,用于指定线程(块)维度,未赋值时初始化各项值为 1。下面展示了 dim3 结构体的声明。

```
struct __device_builtin__ dim3
{
    unsigned int x, y, z;
#if defined(__cplusplus)
    __host__ __device__ dim3(unsigned int vx=1, unsigned int vy=1, unsigned int vz=1) : x(vx), y(vy), z(vz) {}
    __host__ __device__ dim3(uint3 v) : x(v.x), y(v.y), z(v.z) {}
    __host__ __device__ operator uint3(void) { uint3 t; t.x=x; t.y=y; t.z=z; return t; }
#endif /* __cplusplus */};
```

char2、uchar2、char3、uchar3、char4、uchar4、short2、ushort2、short3、ushort3、short4、ushort4、int2、uint2、int3、uint3、int4、uint4、long2、ulong2、long3、ulong3、long4、ulong4、float2、float3、float4、double2:结构体数据,定义方式类似,以 uchar4 的定义为例。

```
struct __device_builtin__ __align__(4) char4
{
    signed char x, y, z, w;
};
```

4. 内建变量

gridDim：指定 grid 维度，类型为 dim3。
blockDim：指定 block 维度，类型为 dim3。
blockIdx：指定 grid 内 block 索引号，类型为 uint3。
threadIdx：指定 block 内 thread 索引号，类型为 uint3。
warpsize：指定 warp 内 thread 数量，类型为 int。

```
uint3 __device_builtin__ __STORAGE__ threadIdx;
uint3 __device_builtin__ __STORAGE__ blockIdx;
dim3  __device_builtin__ __STORAGE__ blockDim;
dim3  __device_builtin__ __STORAGE__ gridDim;
int   __device_builtin__ __STORAGE__ warpSize;
```

5. kernel 调用

kernel 函数调用就是调用声明为 __global__ 的函数，必须在主机端调用。调用时需要指定线程维度，指定方式为<<<Bs,Ts,Ss,Si>>>，其中 Bs 指定 grid 内 block 维度，类型为 dim3；Ts 指定 block 内 thread 维度，类型为 dim3；Ss 指定共享存储空间大小，类型为 size_t；Si 指定流索引号；其中 Ss 和 Si 可选。下面给出了一个实例（详细代码见第 6 章）进行说明，其中，blocknum 为 grid 内的 block 数量，threadnum 为 block 内的 thread 数量。

```
__global__ void vector_add_gpu_3(DATATYPE *a, DATATYPE *b, DATATYPE *c, int n)
{…}
vector_add_gpu_3<<<blocknum,threadnum>>>(d_a,d_b,d_c,n);
```

6. 特殊函数和内建函数

CUDA C 中引入了大量专用的特殊函数和内建函数以实现特定功能，比如同步函数、数学函数、纹理函数、测时函数、原子函数、存储栅栏函数等。

5.6 grid-block-thread 三维模型

3.2 节提及了 grid-block-thread 到 GPU 硬件的映射关系。本节将重点阐述 grid-block-thread 三维模型。图 5.2 简明地展示了三者之间的关系：1 个 kernel 函数对应 1 个 grid，grid 由大量 block 组成，block 由大量 thread 构建，其维度均为 dim3 类型。下面分别针对 1D 线程维度和 2D 线程维度展开探讨，3D 情况请自行补充。

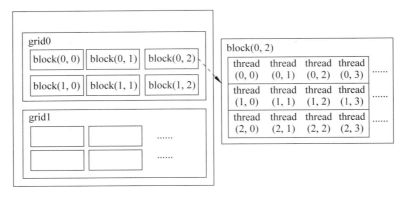

图 5.2 grid-block-thread 三维模型

1D 线程维度：可将 1 个 grid 看成 1 个二维矩阵，1 个 block 视为矩阵 1 行，thread 为行内单个元素（见图 5.3）。这是最简单、最容易掌握的情况，后文中有大量实例。常用的索引计算方法为

```
int id=blockIdx.x * blockDim.x+threadIdx.x;
```

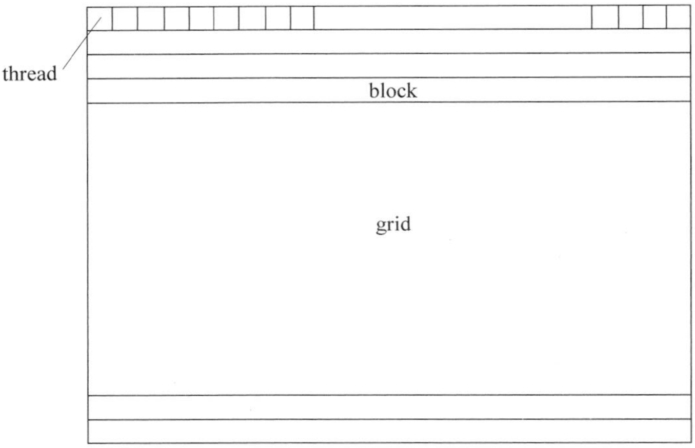

图 5.3 1D 线程维度

2D 线程维度：1 个 grid 是 1 个大矩形，由很多小矩形/正方形（block）构成，每个小矩形中的元素即为 thread（见图 5.4 所示）。这种设计的好处是可以基于数据局部性来充分利用共享存储资源，以减少全局存储的使用，第 8 章矩阵乘法棋盘阵列和第 10 章 2D 卷积都使用了该方法。常用的索引计算方法如下。

```
int j=blockIdx.x *blockDim.x+threadIdx.x;
int i=blockIdx.y *blockDim.y+threadIdx.y;
```

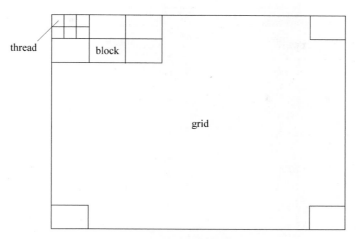

图 5.4　2D 线程维度

第二篇 入门篇

> 第 6 章 向量加法
> 第 7 章 归约：向量内积
> 第 8 章 矩阵乘法
> 第 9 章 矩阵转置

自本篇开始，正式展开循序渐进的实例优化。入门篇以 4 类基础的向量和矩阵运算为实例介绍 GPU 编程和优化的基本方法。

笔者将向量加法作为高性能计算领域的"hello world!"程序，从这个程序开始，介绍如何使用 GPU，如何利用 GPU 进行并行计算，如何进行 CPU 与 GPU 间的数据传输，如何使用 CUBLAS 库进行 GPU 计算等。

接着以归约运算中的向量内积为蓝本，进行循序渐进的代码调优，包括分散归约、低线程归约、CPU 与 GPU 的权衡、原子操作、利用原子操作的计数法实现全局同步、CUBLAS 库、线程（块）配置、算术运算优化、减少同步开销、循环展开等。

矩阵乘法是高性能计算领域的永恒经典，第 8 章将矩阵乘法移植到了 GPU，并介绍了大量的性能优化研究，包括 grid 线程循环、block 线程循环、共享存储优化、最大化共享存储复用、判断移除、CUBLAS 库、共享存储的深入优化等。

本篇最后研究了矩阵转置的 GPU 移植和优化，分别完成了 1D（线程维度）和 2D 的矩阵转置，探索了共享存储优化和全局存储的 diagonal 优化，结合实验结果分析优化的效果，针对其中的不足（共享存储 2D 矩阵转置）开展了进一步的优化研究。

第 6 章 向 量 加 法

6.1 向量加法及其串行代码

向量加法运算是高性能计算领域中最简单的运算,笔者视其为 HPC 领域的"hello world!"。向量加法或矩阵加法的数学表达式如下:

$$C = A + B$$

图 6.1 形象地展示了向量加法过程,即 $C[i]=A[i]+B[i]$,$i=0,1,2,\cdots,n-1$。

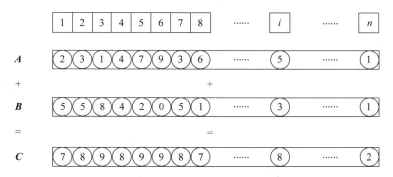

图 6.1 向量加法示意图

下面是串行向量加法的 C 语言源代码:

```
#define DATATYPE float
void vector_add_serial(DATATYPE * a, DATATYPE * b, DATATYPE * c, int n)
{
    for(int i=0;i<n;i++)
    {
        c[i]=a[i]+b[i];
    }
}
```

6.2 单 block 单 thread 向量加

GPU 是协处理器，GPU 程序开发与传统 CPU 程序开发最大的区别是协处理器无法识别 CPU 端数据和代码。如何才能让 GPU 跑起来？这是 GPU 并行程序开发的基础。本节以向量加法为例，首先介绍如何让 GPU 运行向量加法，得到正确结果。

就 GPU 单个线程而言，向量加法运算与 CPU 类似，只是 kernel 函数需要用 __global__ 限定符标识，而 kernel 函数调用时需要用 <<<X,X>>> 配置 grid 和 block 的维度。下面的 CUDA 代码展示了 5.2 节描述的简化的 CUDA 编程五步曲。

```
__global__ void vector_add_gpu_1(DATATYPE *a, DATATYPE *b, DATATYPE *c, int n)
{
    for(int i=0;i<n;i++)
    {
        c[i]=a[i]+b[i];
    }
}

//GPU memory alloc
DATATYPE *d_a, *d_b, *d_c;
cudaMalloc((void**) &d_a, sizeof(DATATYPE) * n);
cudaMalloc((void**) &d_b, sizeof(DATATYPE) * n);
cudaMalloc((void**) &d_c, sizeof(DATATYPE) * n);
//data a and b copy to GPU;
cudaMemcpy(d_a, a, sizeof(DATATYPE) * n, cudaMemcpyHostToDevice);
cudaMemcpy(d_b, b, sizeof(DATATYPE) * n, cudaMemcpyHostToDevice);
//c=a+b
vector_add_gpu_1<<<1,1>>>(d_a,d_b,d_c,n);
//result copy back to CPU
cudaMemcpy(c, d_c, sizeof(DATATYPE) * n, cudaMemcpyDeviceToHost);
//GPU memory free
cudaFree(d_a);
cudaFree(d_b);
cudaFree(d_c);
```

6.3 单 block 多 thread 向量加

通过 6.2 节的努力，向量加法已经成功地在 GPU 运行起来了，但运算依然是串行的。接下来的工作是令 GPU 并行运算向量加法。由 5.1 节可知，kernel 函数启动时，1 个 grid 可拥有多个 block，而 1 个 block 又有多个 thread。本节首先尝试利用 block 内的多个 thread 并行运算向量加法。

针对向量加法过程，kernel 函数中利用 threadIdx.x 获得线程索引号，通过 <<<1,threadnum>>> 指定 1 个 block 内 threadnum 个 thread 同时运算，每个线程完成 1 次向量加法后索引 tid 根据线程总数（blockDim.x）进行跳步，循环处理所有数组元素的加法运算。

```
__global__ void vector_add_gpu_2(DATATYPE *a, DATATYPE *b, DATATYPE *c, int n)
{
    int tid=threadIdx.x;
    const int t_n=blockDim.x;
    while(tid<n)
    {
        c[tid]=a[tid]+b[tid];
        tid+=t_n;
    }
}
vector_add_gpu_2<<<1,threadnum>>>(d_a,d_b,d_c,n);
```

6.4 多 block 多 thread 向量加

6.3 节的代码修改实现了 1 个 block 内多 thread 并行计算向量加法。但 GPU 中不只 1 个 block，此时其他 block 是空闲的，要充分利用 GPU 资源就要利用多个 block 进行计算。本节试图通过代码修改，实现利用 GPU 的多 block 和多 thread 并行运算向量加法。

针对向量加法过程，进行多 block 和多 thread 改造，关键是对线程索引 tid、跳步步长进行重新构造。此时线程索引不再是 threadIdx.x 获得的 block 内部索引，而需要全局线程索引，可利用公式（tid=blockIdx.x * blockDim.x+threadIdx.x）计算。多 block 和多 thread 情况下的跳步也不再是 block 尺寸，而是 grid 内所有 thread 数量（gridDim.x * blockDim.x）。修改后的 CUDA C 代码如下：

```
__global__ void vector_add_gpu_3(DATATYPE *a, DATATYPE *b, DATATYPE *c, int n)
{
    const int tidx=threadIdx.x;
    const int bidx=blockIdx.x;
    const int t_n=gridDim.x*blockDim.x;
    int tid=bidx*blockDim.x+tidx;
    while(tid<n)
    {
        c[tid]=a[tid]+b[tid];
        tid+=t_n;
    }
}
vector_add_gpu_3<<<blocknum,threadnum>>>(d_a,d_b,d_c,n);
```

6.5 CUBLAS 库向量加法

优化的极限在哪儿？NVIDIA 公司花巨资打造的 CUDA 函数库可以看作 GPU 程序员的优化参考极限。学会使用 CUDA 库函数能有效帮助 GPU 程序员快速开发高性能 CUDA 程序。本书针对不同实例，除了阐述笔者的编程和优化方法外，还给出了相应的 CUDA 库函数实现代码，一方面作为性能对比基准，另一方面帮助读者学习库函数的使用方法。库函数看似简单，但真正应用时会遇到各种各样的问题，目前指导新手学习使用库函数的参考书比较匮乏，本书也在试图弥补该方向上的空白。

针对向量加法运算，笔者发现 CUBLAS 库函数中的 cublas<t>axpy() 函数可以实现向量加法功能，在选择 float 类型前提下，选择 cublasSaxpy() 函数。v2 版本 CUBLAS 库函数的调用和 v1 版本库函数的调用略有差别（这里 v2 版本的判定依据就是相应的函数名称中有个 v2，且头文件也有 v2），其中 v1 版本库函数的用法将在第 8 章中阐述。CUBLAS 库（v2 版本）函数调用时需要注意以下几点。

(1) 包含头文件 cublas_v2.h。
(2) 定义 cublasHandle_t 句柄。
(3) 主机端到设备端数据传输函数是 cublasSetVector()，设备端到主机端数据传输函数是 cublasGetVector()。
(4) cublasSaxpy() 函数调用时为 cublasSaxpy_v2()。

下面展示利用 CUBLAS 库（v2 版本）实现的向量加法程序主体部分代码。

```
#include<cublas_v2.h>
    ⋮
  DATATYPE *d_a,*d_b;
  cublasHandle_t handle;
  cublasCreate(&handle);
  cudaMalloc((void**)&d_a, sizeof(DATATYPE) * n);
  cudaMalloc((void**)&d_b, sizeof(DATATYPE) * n);
  float alpha=1.0;
  cublasSetVector(n,sizeof(DATATYPE),a,1,d_a,1);
  cublasSetVector(n,sizeof(DATATYPE),b,1,d_b,1);
  cublasSaxpy_v2(handle,n,&alpha,d_a,1,d_b,1);
  cublasGetVector(n,sizeof(DATATYPE),d_b,1,c1,1);
  cudaFree(d_a);
  cudaFree(d_b);
  cublasDestroy(handle);
```

上述代码编译时需要加入 CUBLAS 库选项-lcublas，在 Linux 下编译指令如下：

```
$nvcc vectoradd_cublas.cpp -O3 -o vectoradd_cublas -lcublas
```

6.6 实验结果分析与结论

6.6.1 本书实验平台

本书实验使用的硬件平台为 2 个 8 核 Intel Xeon CPU E5-2670,内存为 64GB,2 个 NVIDIA Tesla K20C GPU。实验平台配置的软件环境为 ICC version 13.0.0.079、CUDA 5.5。

注意:本书所有串行代码均已经选择最佳优化开关(-O2 或-O3)并且保证最内层循环向量化,即保证向量化(若串行代码不能向量化,性能对比将毫无意义)。

本书部分章节实验的编译环境是 ICC version 11.1。本书实验平台是大集群中的一个节点,有专人管理维护,笔者不知道 ICC 编译器是啥时候升级的,而本书代码编写的时间跨度较大,因此笔者也不确定具体哪个章节用的哪套环境,不过这并不影响实验结果分析。

6.6.2 实验结果

执行 1024×1024 个 float 元素的向量加法时,分别测试统计了计算时间、传输时间(H2D/D2H)和总时间,表 6.1 统计了测试结果。其中,serial 表示串行程序,gpu_1 表示 6.2 节中的单 block 单 thread 的 GPU 向量加法程序,gpu_2 表示 6.3 节开发的单 block 多 thread 的 GPU 向量加法程序,gpu_3 表示 6.4 节开发的多 block 多 thread 的 GPU 向量加法程序,cublas 表示 6.5 节开发的 CUBLAS 库实现的向量加法程序。

注意:首个 kernel 函数会有额外开销,且随着测试次数不同无规律变化,表 6.1 数据是利用 nvprof 测得,关于 nvprof 的用法请查阅 17.8 节。

表 6.1 向量加法执行时间　　　　　　　　　　　　单位:ms

1024×1024	计算时间	传输(H2D)	传输(D2H)	总时间
serial	1.36			1.36
gpu_1	423.97	2.31	1.42	427.70
gpu_2	1.16	2.32	1.43	4.91
gpu_3	0.08	2.32	1.45	3.85
cublas	0.08	2.32	1.53	3.94

6.6.3 结论

分析表 6.1 中的数据,有以下结论。

(1) GPU 单核的计算性能极差,计算时间为 CPU 的 300 多倍。(对比 serial 和 gpu_1 的计算时间)

(2) 满负载时的 1 个 block 性能跟 1 个向量运算的 CPU 核性能相近(稍强)。(对比 gpu_2 和 serial 的计算时间)

(3) GPU 满负载计算时，gpu_3 和 cublas 性能接近；而当数据放大 128 倍时，gpu_3 的计算时间是 10.232ms，cublas 的计算时间是 9.879ms，说明当数据量大时，CUBLAS 库的性能更好。

(4) 单论计算时间，GPU 并行计算时间明显比 CPU 串行计算时间短很多，说明 GPU 并行计算性能高。

(5) 在向量加法运算中，CPU 与 GPU 的通信时间比串行计算时间还长，说明该类问题（向量加法）的 GPU 移植瓶颈在数据传输，在当前平台上若数据位于 CPU 端情况下不适合移植到 GPU 计算；当然，本节代码的通信时间还能进一步优化（利用页锁定存储），但依然无法匹配串行计算时间。在 NVIDIA 最新发布的 Tesla P100 GPU 配置了 NVLink 可令通信带宽高达 160GB/s，若使用该产品可能向量加法中的通信瓶颈将大大削弱甚至不再存在。

6.7 知识点总结

CPU 端开辟存储空间尽量使用 malloc，而不使用 new。new 开辟的空间有大小限制，而 malloc 没有此限制，实际应用中更加方便和自由。

```
DATATYPE *a=(DATATYPE *)malloc(sizeof(DATATYPE) * n);
```

cudaMalloc 函数：在 GPU 上分配空间，空间的开始指针是 d_a，空间大小为 sizeof(DATATYPE) * n。注意，初始接触异构编程时，一定要理清主机端和设备端（GPU）的存储区别。在 GPU 上分配空间、存储数据，必须使用指针进行操作。

cudaFree 函数：释放指针 d_a 对应的位于 GPU 上的已分配存储空间。

```
cudaMalloc((void**) &d_a, sizeof(DATATYPE) * n);
cudaFree(d_a);
```

cudaMemcpy 函数：实现 CPU 与 GPU 之间的数据传输，第一个参数 d_a 是目标地址，第二个参数 a 是源地址，第三个参数是传输数据长度（单位为 B），第四个参数指明传输方向。传输方向值域有多个选择，常用的有 cudaMemcpyHostToDevice 表示数据从 CPU 端传输到 GPU 端，cudaMemcpyDeviceToHost 表示数据从 GPU 端传输到 CPU 端，cudaMemcpyDeviceToDevice 表示数据 GPU 端传输到 GPU 端等。

```
cudaMemcpy(d_a, a, sizeof(DATATYPE) * n, cudaMemcpyHostToDevice);
cudaMemcpy(c, d_c, sizeof(DATATYPE) * n, cudaMemcpyDeviceToHost);
```

__global__ 限定符：要在 GPU 上执行的 kernel 函数必须用该符号声明，才能被 CPU 端程序调用。类似的限定符还有__device__、__host__，其中__device__声明的是 GPU 端函数,且仅能被 GPU 上执行的代码调用;__host__声明的是普通 CPU 端函数,一般情况下可省略。

<<<**blocknum,threadnum**>>>:调用__global__声明的 kernel 函数时配置的线程维度,其中第 1 个参数 blocknum 表示 grid 里 block 的数量,threadnum 表示 block 里 thread 的数量。这里两个参数都是 dim3 类型数据,即(x,y,z),分别表示 x、y、z 三个维度的数值。

```
vector_add_gpu_3<<<blocknum,threadnum>>>(d_a,d_b,d_c,n);
```

其实,<<<>>>符号内共有 4 个参数,前两个分别表示 block 维度和 thread 维度,第 3 个参数表示共享存储大小,第 4 个参数表示 CUDA 流索引号。

GPU 线程索引方法:利用(threadIdx.x,threadIdx.y,threadIdx.z)或(blockIdx.x,blockIdx.y,blockIdx.z)分别对 thread 和 block 进行索引。在使用 1D 索引时(即只有 threaIdx.x 和 blockIdx.x 大于 1),以下方法可以唯一地标识线程索引号。2D 索引计算请参见 5.6 节。

```
const int tidx=threadIdx.x;
const int bidx=blockIdx.x;
int tid=bidx*blockDim.x+tidx;
```

Kernel 函数中的循环:kernel 函数运行时,所有的 block 并行执行,每个 block 内所有线程也是并发运算的(实际上一个 warp 的 32 个线程同时执行同一条指令),所以 kernel 循环处理不同于 CPU 的 for 循环累加,而采用了固定跳步的方法,步长一般是 grid 内所有的 thread 数量,即

```
const int t_n=gridDim.x*blockDim.x;
```

用 for 或 while 均可实现 kernel 函数的循环处理过程,下面展示了实现 kernel 函数循环的两种方法:

| ```
const int t_n=gridDim.x*blockDim.x;
int tid=blockIdx.x*blockDim.x+threadIdx.x;
while(tid<n){
    c[tid]=a[tid]+b[tid];
    tid+=t_n;
}
``` | ```
const int t_n=gridDim.x*blockDim.x;
int tid=blockIdx.x*blockDim.x+threadIdx.x;
for(;tid<n;tid+=t_n)
{
    c[tid]=a[tid]+b[tid];
}
``` |
| --- | --- |

注意:有说法称 for 循环的性能会更好,笔者经过实验测试,发现两者性能没有差异。

GPU 运算五步曲:分配空间、传输数据、运算、数据传出、释放空间。

```
cudaMalloc((void**) &d_a, sizeof(float) * n);        //分配显存空间
cudaMemcpy(d_a,a,sizeof(float) * n,cudaMemcpyHostToDevice);    //数据传输(H2D)
gpu_kernel<<<blocks,threads>>>(***);                 //kernel 函数
cudaMemcpy(a,d_a,sizeof(float) * n,cudaMemcpyDeviceToHost);    //数据传输(D2H)
cudaFree(d_a);                                       //释放显存空间
```

CUDA 函数错误检测：本书为了方便读者阅读，删除了所有的函数调用状态判断（即 CUDA 函数错误检测），建议读者在实践时加上。添加函数调用状态判断，有利于 GPU 程序的调试，代码出错位置的分析等。CUDA 函数运行状态错误判断有很多种方法，比如 cutilSafeCall、CUDA_SAFE_CALL、cudaError_t 状态等（下面展示了详细代码）。笔者建议使用 cudaError_t 状态判断方法来检测 CUDA 函数错误。原因是其他两种方法需要引用 cutil 库，但程序移植到其他平台时，目标平台可能没有 cutil 库，比如 CUDA 5.0 及之后的版本默认不提供 cutil 库。

```
#include<cutil_inline.h>
⋮
cutilSafeCall(cudaMalloc((void**) &d_a, sizeof(DATATYPE) * n));
```
```
#include<cuda_runtime.h>
#include<cutil.h>
⋮
CUDA_SAFE_CALL(cudaMalloc((void**)&d_a,sizeof(float) * n));
```
```
#include<cuda_runtime.h>
⋮
cudaError_t err=cudaSuccess;
err=cudaMalloc((void**)&d_a,size);
if(err != cudaSuccess) { fprintf (stderr, "Failed to allocate device vector A
(error code %s)!\n", cudaGetErrorString(err)); exit(EXIT_FAILURE);}
```

封装的 cudaError_t 状态检测方法：上述 cudaError_t 状态判别方法如果插入程序，将大幅增加程序复杂度，进而影响程序的阅读和理解，因此本书提供了 cudaError_t 状态检测方法的封装技巧，以帮助读者方便快捷地检测和分析 CUDA 函数错误。

```
void hpc6_cudaSafeCall(cudaError_t err){
    if(err !=cudaSuccess){
        printf("Error: CUDA runtime API error %d: %s\n",(int)err,
        cudaGetErrorString(err));
      exit(-1);
    }
}
void hpc6_cudaKernelCheck(void){
    cudaError_t err=cudaGetLastError();
```

```
    if(err !=cudaSuccess){
        printf("Error: CUDA kernel launch error %d: %s\n",(int)err,
        cudaGetErrorString(err));
    exit(-1);
    }
}
hpc6_cudaSafeCall(cudaMalloc((void**) &d_a, sizeof(DATATYPE) * n));
vector_add_gpu_3<<<blocknum,threadnum>>>(d_a,d_b,d_c,n);
hpc6_cudaKernelCheck();
```

CUBLAS 库(v2)的调用方法：本章利用 CUBLAS 库(v2)实现了向量加法，主要流程有：①引用头文件 cublas_v2.h；②利用 cublasHandle_t 声明句柄；③利用 cublasCreate()函数创建句柄；④利用 cublasSetVector()函数实现 H2D(host to device)数据传输；⑤利用 cublasSaxpy_v2()函数实现向量加法运算；⑥通过 cublasGetVector()函数实现 D2H 数据传输；⑦用 cublasDestroy()函数销毁句柄。

Linux 下调用 CUBLAS 库的编译技巧：下面给出了在 Linux 环境下利用 ICC 和 NVCC 两种编译器编译 CUBLAS 库函数调用代码的编译命令。

```
$icc vectoradd_cublas.cpp -o vectoradd_cublas_icc -O3 -I/vol/home/fangmq/gpu_cuda/cuda/cuda_5_5/include -L/vol/home/fangmq/gpu_cuda/cuda/cuda_5_5/lib -L/vol/home/fangmq/gpu_cuda/cuda/cuda_5_5/lib64 -lcudart -lcublas
$nvcc vectoradd_cublas.cpp -O3 -o vectoradd_cublas -lcublas
```

VS 调用 CUBLAS 库的方法：在 VS 开发环境中，要调用 CUBLAS 库函数，首先要添加相关目录，然后指明链接 cublas.lib 库。具体方法如下。

(1) 添加相关目录。CUDA 安装时一般已经添加，如没有添加，添加方法是在"工具"→"选项"→"项目和解决方案"→"VC++目录"中分别选择下拉选项包含文件、库文件添加相应目录。

(2) 指明链接 cublas.lib 库。在工程上右击，选择"属性"→"配置属性"→"链接器"→"输入"命令，在附件依赖项中加入 cublas.lib(一般原来有 cudart.lib，保留)。

(3) 正常编译执行。

6.8 扩展练习

(1) 实现维度为 $m \times n$ 的矩阵加减法的 GPU 并行计算。
(2) 在 GPU 上实现维度为 $m \times n \times k$ 的三维矩阵加减法计算。

第7章 归约：向量内积

7.1 向量内积及其串行代码

相较于各元素计算完全独立的向量加法，向量内积运算中各元素的运算间存在简单联系，即需要累加所有元素乘积结果。向量内积结果的累加过程称为归约（reduction），这也是本章探讨研究的主题。

向量内积运算在数学上的计算公式为

$$c = \mathbf{A} \cdot \mathbf{B}$$

即

$$c = \sum_{i=1}^{n}(\mathbf{A}[i] \times \mathbf{B}[i])$$

$\mathbf{A}$ 和 $\mathbf{B}$ 向量长度相等，相同索引的元素相乘，然后将乘积结果累加，得到向量内积结果 c。

下面是 C 语言实现的向量内积运算源代码。

```
void vector_dot_product_serial(DATATYPE *a, DATATYPE *b,
DATATYPE *c, int n)
{
    double temp=0;
    for(int i=0;i<n;i++)
    {
        temp+=a[i]*b[i];
    }
    *c=temp;
}
```

关于向量内积运算精度的探讨：在上述代码中，为什么中间变量 temp 必须声明为 double？并行计算往往会引入精度损失，而计算科学应用领域对计算精度的要求非常高，因此在编程时需要更加注意计算精度问题。笔者在研究过程中发现，向量内积运算中，中间变量一般采用 double 类型，否则当循环量增大到一定程度，结果会出错（或损失精度），至于循环量多大时会出错，读者可以通过实践测试来体会。8.1 节串行矩阵乘法的运行结果也验证了这一点。

7.2 单 block 分散归约向量内积

本节讨论如何在 GPU 上单个 block 内实现向量内积运算。首先各线程独立完成所有向量元素的乘积运算，然后将乘积结果累加得到向量内积结果。由于寄存器是 thread 私有的，block 内的 threads 要进行数据通信，必须使用共享存储或全局存储，而共享存储访存速度比全局存储快 1 个量级，故采用共享存储器。各 thread 的乘积运算将得到 1 个与 thread 数量相等的乘积结果数组，存储于共享存储中，然后对该数组进行归约处理。GPU 上归约的方法有很多种，常见的有分散归约和低线程归约两种。

图 7.1 展示了分散归约的过程，相邻两个数组元素成对做加法运算，逐层循环直至仅剩下 1 个结果。该方法将归约运算步数从原来的 $n-1$ 步降为 $\log(n)$ 步。通过分散归约，不仅把串行归约计算过程改为并行归约，还减少了运算步数。

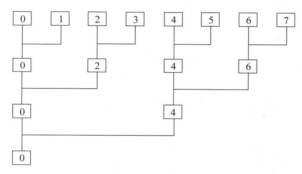

图 7.1 分散归约图

下面展示单 block 分散归约向量内积的 CUDA C 代码（标记为 gpu_1）。同步（__syncthreads();）之前的部分实现两个向量的所有元素两两相乘，同步后的代码实现单个 block 的分散归约。

```
__global__ void vector_dot_product_gpu_1(DATATYPE *a, DATATYPE *b, DATATYPE *c, int n)
{
    __shared__ DATATYPE tmp[threadnum];
    const int tidx=threadIdx.x;
    const int t_n=blockDim.x;
    int tid=tidx;
    double temp=0.0;
    while(tid<n)
    {
        temp+=a[tid]*b[tid];
        tid+=t_n;
    }
```

```
    tmp[tidx]=temp;
    __syncthreads();
    int i=2,j=1;
    while(i<=threadnum)
    {
        if ((tidx%i)==0)
        {
            tmp[tidx]+=tmp[tidx+j];
        }
        __syncthreads();
        i*=2;
        j*=2;
    }
    if (tidx==0)
    {
        c[0]=tmp[0];
    }
}
vector_dot_product_gpu_1<<<1,threadnum>>>(d_a,d_b,d_c,n);
```

分散归约方法在 GPU 上存在以下 3 个方面的缺陷：①单个线程访问全局存储的相邻两个单元，违背了访存对齐原则；②共享存储访问可能会引发 bank conflict；③由于 warp 内所有 thread 执行同一指令，因而运算时需要多个 warp，从而导致性能损失。这些缺陷对性能有什么影响？影响多大？将在 7.9 节的实验结果中分析。

7.3 单 block 低线程归约向量内积

低线程归约是对分散归约的改进，其中融入了 GPU 体系结构和线程执行思想。7.2 节分析了分散归约可能存在的访存不对齐、共享存储 bank conflict、需要多 warp 线程执行计算等问题，低线程归约主要针对这些问题进行改进。低线程归约流程如图 7.2 所示，每次归约运算位于当前所有线程的前一半线程（低线程），循环执行，最终在 0 号线程得到最终结果。下面展示了采用低线程归约的向量内积 CUDA C 代码（标记为 gpu_2）。

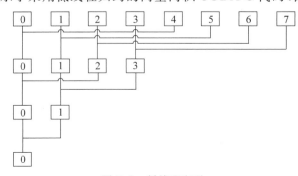

图 7.2 低线程归约

```
__global__ void vector_dot_product_gpu_2(DATATYPE *a, DATATYPE *b, DATATYPE
*c, int n)
{
    __shared__ DATATYPE tmp[threadnum];
    const int tidx=threadIdx.x;
    const int t_n=blockDim.x;
    int tid=tidx;
    double temp=0.0;
    while(tid<n)
    {
        temp+=a[tid]*b[tid];
        tid+=t_n;
    }
    tmp[tidx]=temp;
    __syncthreads();
    int i=threadnum/2;
    while(i!=0)
    {
        if (tidx<i)
        {
            tmp[tidx]+=tmp[tidx+i];
        }
        __syncthreads();
        i/=2;
    }
    if(tidx==0)
    {
        c[0]=tmp[0];
    }
}
vector_dot_product_gpu_2<<<1,threadnum>>>(d_a,d_b,d_c,n);
```

7.4 多 block 向量内积（CPU 二次归约）

最大化线程数量是使用 GPU 的一个重要优化策略，仅 1 个 block 显然无法发挥 GPU 的真实性能。因此，本节探讨多个 block 的向量内积运算。归约方法取效率较高的低线程归约（详见 7.9 节实验结果）。每个 block 计算得到一个中间结果，所有 block 的中间结果构成一个中间结果数组，该数组共有 blocknum（block 数量）个元素，对该数组进行二次归约即可得到最终结果。需要中间结果数组的原因是 GPU 中不同的 blocks 间无法保证同步。中间结果的二次归约有两种实现方案：CPU 实现和 GPU 实现。

考虑 CPU 二次归约方案（标记为 gpu_3），首先利用 GPU 上多个 block 归约，得到等同 block 数量的中间结果，即下面 CUDA 代码中 vector_dot_product_gpu_3() 函数；接着将中间结果数组传回主机端；再由主机端完成串行二次归约（主机端也是可以并行的，但是这里运算量较少，向量化的串行运算依然能够完全满足需求，并行反而可能引入并行开销）。这种方法的思想是中间结果数组元素数量有限，一般不大于 1024，传输开销也有限，CPU 端代码经向量化优化后性能理想。

```
__global__ void vector_dot_product_gpu_3(DATATYPE *a, DATATYPE *b, DATATYPE *c_tmp, int n)
{
    __shared__ DATATYPE tmp[threadnum];
    const int tidx=threadIdx.x;
    const int bidx=blockIdx.x;
    const int t_n=blockDim.x * gridDim.x;
    int tid=bidx * blockDim.x+tidx;
    double temp=0.0;
    while(tid<n)
    {
        temp+=a[tid] * b[tid];
        tid+=t_n;
    }
    tmp[tidx]=temp;
    __syncthreads();
    int i=threadnum/2;
    while (i!=0)
    {
        if (tidx<i)
        {
            tmp[tidx]+=tmp[tidx+i];
        }
        __syncthreads();
        i/=2;
    }
    if(tidx==0)
    {
        c_tmp[bidx]=tmp[0];
    }
}
```

```
vector_dot_product_gpu_3<<<blocknum,threadnum>>>(d_a,d_b,d_c_tmp,n);
cudaMemcpy(c_tmp, d_c_tmp, sizeof(DATATYPE) * blocknum,
cudaMemcpyDeviceToHost);
for (int i=0;i<blocknum;i++)
{
    temp+=c_tmp[i];
}
c=temp;
```

7.5 多 block 向量内积（GPU 二次归约）

7.4 节讨论了多 block 向量内积运算中的 CPU 实现二次归约方案，本节讨论 GPU 实现二次归约方案。

利用 GPU 实现二次归约：在利用 vector_dot_product_gpu_3() 函数计算得到中间结果数组后，另外启动一个 kernel 函数 vector_dot_product_gpu_4() 对中间结果进行二次归约，得到最终向量内积结果。该方法的主要优势是减少通信数据量、利用 GPU 高运算性能。GPU 实现二次归约的向量内积 CUDA C 代码（记为 gpu_4）如下：

```
__global__ void vector_dot_product_gpu_4(float *result_tmp,float *result)
{
    __shared__ float temp[blocknum];
    const int tidx=threadIdx.x;
    temp[tidx]=result_tmp[tidx];
    __syncthreads();
    int i=blocknum/2;
    while (i!=0)
    {
        if(tidx<i)
        {
            temp[tidx]+=temp[tidx+i];
        }
        __syncthreads();
        i/=2;
    }
    if(tidx==0)
    {
        result[0]=temp[0];
    }
}
vector_dot_product_gpu_3<<<blocknum,threadnum>>>(d_a,d_b,d_c_tmp,n);
vector_dot_product_gpu_4<<<1,blocknum>>>(d_c_tmp,d_c);
```

7.6 基于原子操作的多 block 向量内积

原子操作（ATOM）能保证每个线程互斥地访问全局存储或共享存储上的某一数据，即同一数据在某一时刻只能由一个线程访问，避免了并行程序中多个线程同时修改某一变量出现访存冲突问题。原子操作函数种类丰富，包括加法（atomicAdd）、减（atomicSub）、与（atomicAnd）、或（atomicOr）、异或（atomicXor）、最大（atomicMax）、最小

(atomicMin)、增量(atomicInc)、减量(atomicDec)、比较交换(atomicCAS)、交换(atomicExch)等。需要注意的是,原子操作的结果是直接写回存储的,返回值是旧值,所以要获得新值需重新从存储中读取数据。此外,原子操作可能会导致不可控的死锁。

在向量内积运算中,需要用到原子操作函数是原子加法函数。GPU 的向量内积运算包含两个累加部分可采用原子操作替换,分别是 block 内 thread 间归约和 block 间归约,当然此两者亦可合并为一次原子操作。

原子操作亦分为全局存储原子操作和共享存储原子操作,恰好能对应向量内积中的两个累加部分。

利用原子操作替换归约过程的排列组合有很多种组合,本节将探讨两种组合,分别是两次归约替换一次原子操作、block 内归约 block 间原子操作,另外,在 16.4.4 节探讨了两部分累加归约分别用共享存储原子操作和全局存储原子操作替换的组合。

两次归约替换一次原子操作:采用原子加法操作对所有线程的乘积结果进行归约。这种方法的 kernel 函数相对比较简单,每个线程求得乘积结果 temp 后,采用原子加法运算将其累加到最终值 c 即可,其 CUDA C 代码如下。由于原子加法完成的是累加运算,因此,结果变量必须赋初值 0,否则运行结果将出错。

```
__global__ void vector_dot_product_gpu_5_0(DATATYPE *a, DATATYPE *b, DATATYPE *c, int n)
{
    if ((threadIdx.x==0)&&(blockIdx.x==0))
    {
        c[0]=0.0;
    }
    const int tidx=threadIdx.x;
    const int bidx=blockIdx.x;
    const int t_n=blockDim.x*gridDim.x;
    int tid=bidx*blockDim.x+tidx;
    double temp=0.0;
    while(tid<n)
    {
        temp+=a[tid]*b[tid];
        tid+=t_n;
    }
    atomicAdd(c,temp);
}
```

block 内归约 block 间原子操作:原子操作对数据的访问是串行的,特别是当数据位于全局存储中时,频繁的原子操作将会非常耗时。为了尽量减少全局存储的原子操作数量,在上述全局原子操作代码中加入 block 内低线程归约运算,最后对每个 block 的归约结果进行原子加法运算,相应的 CUDA C 代码(标记为 gpu_5)如下:

```
__global__ void vector_dot_product_gpu_5(DATATYPE *a, DATATYPE *b, DATATYPE
*c, int n)
{
    if((threadIdx.x==0)&&(blockIdx.x==0))
    {
        c[0]=0.0;
    }
    __shared__ DATATYPE tmp[threadnum];
    const int tidx=threadIdx.x;
    const int bidx=blockIdx.x;
    const int t_n=blockDim.x * gridDim.x;
    int tid=bidx * blockDim.x+tidx;
    double temp=0.0;
    while(tid<n)
    {
        temp+=a[tid] * b[tid];
        tid+=t_n;
    }
    tmp[tidx]=temp;
    __syncthreads();
    int i=blockDim.x/2;
    while(i!=0)
    {
        if(tidx<i)
        {
            tmp[tidx]+=tmp[tidx+i];
        }
        __syncthreads();
        i/=2;
    }
    if(tidx==0)
    {
        atomicAdd(c,tmp[0]);
    }
}
```

启动<<<512,512>>>线程执行 1024×1024 个元素的向量内积运算,测试执行时间,全局原子操作的 kernel 执行时间为 1121.6μs,块内归约块间原子操作的 kernel 函数执行时间为 84.384μs,性能提升明显。

注意:原子操作是 CUDA 计算能力 1.1 版本之后加入的内容,nvcc 编译默认采用 1.0 版本,因此在编译包含原子操作的 CUDA 代码时,必须指定其计算能力高于 1.1。例如,在本书测试平台(计算能力 3.5 的 Tesla K20c GPU),编译命令如下:

```
$nvcc vector_dot_product_gpu_5.cu -O3 -o vector_dot_product_gpu_5 -arch
'compute_35' -code 'compute_35'
```

7.7 计数法实现多 block 向量内积

原子加法操作除了能对结果实现累加运算外,还能实现栅栏计数,并以此来突破 block 间无法同步的难题,在一个 kernel 函数中实现多个 block 内数据的归约运算。下面是利用计数法实现向量内积运算 CUDA C 代码(gpu_6):①在语句"__threadfence();"前,完成了 block 内归约,结果保存在中间结果 c_tmp 中,其功能等同 7.4 节的 kernel 函数;②其后利用原子加法运算实现 block 间的栅栏同步,实现方法是每到 1 个 block,记录 lockcount 加 1,判断是否最后一个 block;③所有 block 抵达栅栏后,最后一个抵达的 block(最后一个 block 抵达时,lock 为真)读取中间数组 c_tmp 完成归约运算(同 7.5 节中 GPU 二次归约)。

```
__device__ void vector_dot(DATATYPE *out,volatile DATATYPE *tmp)
{
    const int tidx=threadIdx.x;
    int i=blockDim.x/2;
    while (i!=0)
    {
        if(tidx<i)
        {
            tmp[tidx]+=tmp[tidx+i];
        }
        __syncthreads();
        i/=2;
    }
    if (tidx==0)
    {
        out[0]=tmp[0];
    }
}
```

```
__device__ unsigned int lockcount=0;
__global__ void vector_dot_product_gpu_6(DATATYPE *a, DATATYPE *b, DATATYPE *c
_tmp, DATATYPE *c, int n)
{
    __shared__ DATATYPE tmp[threadnum];
    const int tidx=threadIdx.x;
    const int bidx=blockIdx.x;
    const int t_n=blockDim.x*gridDim.x;
```

```
        int tid=bidx*blockDim.x+tidx;
        double temp=0.0;
        while(tid<n)
        {
            temp+=a[tid]*b[tid];
            tid+=t_n;
        }
        tmp[tidx]=temp;
        __syncthreads();
        vector_dot(&c_tmp[blockIdx.x],tmp);
        __shared__ bool lock;
        __threadfence();
        if (tidx==0)
        {
            unsigned int lockiii=atomicAdd(&lockcount,1);
            lock=(lockcount==gridDim.x);
        }
        __syncthreads();
        if (lock)
        {
            tmp[tidx]=c_tmp[tidx];
            __syncthreads();
            vector_dot(c,tmp);
            lockcount=0;
        }
}
```

其中，__threadfence()函数与__syncthreads()函数类似，均为同步函数，能够保证后面读到的是前面代码写到共享存储器或全局存储器上的值。类似的同步函数还有__threadfence_block()，同步函数间的辨析参见15.5.1节。

注意："lock=(lockcount==gridDim.x);"中 lockcount 不能是 lockiii，否则结果错误。原因是原子操作的结果是直接写回存储的，返回值是旧值，所以要获得新值需重新从存储中读取数据。

7.8 CUBLAS 库向量内积

CUBLAS 函数库提供了向量内积运算函数 cublas<t>dot()，选择 v2 版本，float 类型函数 cublasSdot_v2()进行运算，CUBLAS 库(v2)函数的使用过程在第 6 章详细阐述，本节不再赘述。下面是关键代码(标记为 cublas)。

```
#include<cublas_v2.h>
    ⋮
    cublasSdot_v2(handle,n,d_a,1,d_b,1,&c1);
```

7.9 实验结果与结论

7.9.1 实验结果

对 1024×1024 长度的 float 数组进行向量内积运算,统一设置 block 和 thread 为 512,测试各项时间,测试结果统计如表 7.1 所示。对比输出运行的结果,serial、gpu_1、gpu_3、cublas 的执行结果为 4399734194176;而 gpu_2、gpu_4 和 gpu_6 的执行结果为 4399733669888,存在精度损失;而原子操作的 gpu_5 的结果不固定,每次运行都会发生变化。

表 7.1 向量内积执行时间　　　　　　　　　　单位:μs

1024×1024	计算 1	计算 2	传输(H2D)	传输(D2H)	总时间
serial	522.85				522.85
gpu_1	1133.40		2328.40	2.75	3464.55
gpu_2	1120.00		2306.80	2.82	3429.62
gpu_3	84.61	0.95	2321.80	3.10	2410.47
gpu_4	83.68	6.37	2318.60	2.75	2411.40
gpu_5	84.38		2317.80	2.56	2404.74
gpu_6	120.86		2325.00	3.07	2448.93
cublas	54.72	4.32	2432.49	2.59	2494.12

7.9.2 结论

从实验结果数据中可以总结出以下结论。

(1) gpu_2、gpu_4 和 gpu_6 的执行结果有精度损失,而 gpu_3 没有精度损失。可得出结论:**低线程归约做最终归约计算会有精度损失;而低线程归约用于中间计算则无精度损失**。

(2) gpu_5 结果不定,结果随执行变化。说明**原子操作结果不稳定,存在精度损失**。

(3) 对比 gpu_1、gpu_2 和 serial 三组数据,**单 block 归约性能还不如 ICC 编译器向量化后的 CPU 性能**,还存在通信开销,读者使用时需要慎重考虑。这一点在结论(5)中也可得到验证。

(4) 对比 gpu_1 和 gpu_2 的计算时间,发现分散归约和低线程归约在 Tesla K20C GPU 上性能差距不大;此时由于 kernel 函数中不仅仅是归约运算,其时间更多地花费在固定跳步的循环乘加运算中,为了更好地进行对比,将数据规模设置为与线程数量相等,即 512,再次测试,发现分散归约耗时 9.1520μs,而低线程归约耗时 6.6240μs,说明低维度归约性能较好。

(5) 对比 gpu_3 和 gpu_4 中计算 2 和传输(D2H)的时间,得出结论:**CPU 二次归约比 GPU 二次归约性能更好**。

(6) 对比 gpu_3 和 gpu_4 中传输(D2H)的时间,还能得到一个结论:**1 个 float 数据和上百个 float 数据的传输时间差距极小**,还不到 $1\mu s$,传输启动时间(即 1 个 float 数据传输时间)约为 $2\mu s$。关于通信时间和速率的探讨详见 17.7 节。

(7) 从 cublas 时间分布中可以看出,CUBLAS 库在实现向量内积计算时采用的是 GPU 二次归约计算,其**将 cublasSdot_v2()函数拆分为两个 kernel 函数实现**,分别是 dot_kernel 和 reduce_1Block_kernel 函数(通过 nvprof 分析可知,具体用法参考本章知识点总结)。

(8) 对比 gpu_3、gpu_4 和 cublas 中的计算 1 时间,说明笔者书写的归约代码与库函数还有巨大差距,在 7.10 节将进一步探讨归约的优化及效果。

(9) 对比 gpu_3 和 cublas 的计算 2 和传输(D2H)时间,又能得出一个非常有意义的结论,NVIDIA 的 CUBLAS 库还有提升空间,若用 CPU 实现二次归约是否性能更好呢?

(10) 对比串行计算时间和传输(H2D)时间,发现传输时间比串行计算时间还长(即便采用页锁定内存依然如此),且两者都与数据量线性相关,因此无论数据量多大,只把向量内积运算从 CPU 端移到 GPU 端是毫无意义的。本书分别将数据放大 128、512 倍,测试使用页锁定内存的通信开销和串行计算时间,结果与推测一致。

(11) 结合向量加法和向量内积,在当前计算平台上,单独把这两个计算过程从 CPU 移植到 GPU 是没有意义的,因为传输时间比 CPU 串行计算时间还长,此时 GPU 计算再快也毫无价值。这就涉及数据使用次数的问题,如果在应用中**数据使用次数与数据量是线性关系时,数据传输开销可能超过串行运算时间**。那么当两者是什么关系时,GPU 并行计算才有意义呢?有待进一步研究讨论。

开始两个章节实例的结论都如此悲观,不知道是否打击到大家的 GPU 编程积极性呢?第 8 章的矩阵乘法中,将会获得巨大的加速效果,敬请期待!

7.10 归约的深入优化探讨

7.9 节通过实验得出低线程归约比分散归约性能更好,本节以低线程归约为对象,进一步深入研究归约的优化策略及其效果。

影响 kernel 函数在 GPU 上执行性能的因素有很多,主要包括 block 数量和 thread 数量、算术运算(采用内置运算)、映射方法设计、判读消除、减少同步开销等。下面逐一尝试,实验采用向量长度为 1024×1024。

7.10.1 block 数量和 thread 数量对归约性能的影响

以 gpu_3(7.4 节中的 CUDA 代码)为基准(记为 dot_0),测试不同 block 数量和 thread 数量情况下向量内积执行时间(执行时间刨除了主机端到设备端的数据传输(H2D)的时间,仅统计 GPU 计算时间、CPU 计算时间和中间结果传输(D2H)时间),时间数据见表 7.2。

表 7.2　block 数量和 thread 数量对归约性能影响　　　　　　　　单位：ms

gridDim.x	blockDim.x				
	1024	512	256	128	64
1024	0.457	0.306	0.258	0.247	0.262
512	0.179	0.108	0.089	0.084	0.097
256	0.134	0.088	0.084	0.084	0.108
128	0.097	0.083	0.086	0.105	0.101
64	0.093	0.089	0.078	0.099	0.165
32	0.092	0.080	0.102	0.162	0.300

表 7.2 中数据显示，block 和 thread 维度为<<<64,256>>>时执行时间最小，利用 nvprof 测得其 kernel 函数执行时间为 58.400μs，已经与 CUBLAS 库的 54.72μs 非常接近了。

7.10.2　算术运算优化

修改 kernel 函数，用向右移位替代除法运算，用内置函数__fmul_rn()和__fadd_rn()替代乘法和加法运算。下面是修改后的 CUDA C 代码，标记为 dot_1。

```
__global__ void vector_dot_product_gpu_3_1(DATATYPE *a, DATATYPE *b, DATATYPE *c_tmp, int n)
{
    extern __shared__ DATATYPE tmp[];
    const int tidx=threadIdx.x;
    const int bidx=blockIdx.x;
    const int t_n=blockDim.x*gridDim.x;
    int tid=bidx*blockDim.x+tidx;
    double temp=0.0;
    while(tid<n)
    {
        temp=__fadd_rn(temp,__fmul_rn(a[tid],b[tid]));
        tid+=t_n;
    }
    tmp[tidx]=temp;
    __syncthreads();
    int i=blockDim.x>>1;
    while(i!=0)
    {
        if(tidx<i)
        {
            tmp[tidx]=__fadd_rn(tmp[tidx],tmp[tidx+i]);
        }
```

```
        __syncthreads();
        i>>=1;
    }
    if (tidx==0)
    {
        c_tmp[bidx]=tmp[0];
    }
}
vector_dot_product_gpu_3_1<<<blocknum,threadnum,threadnum*sizeof(DATATYPE)
>>>(d_a,d_b,d_c_tmp,n);
```

执行并测试修改前后 kernel 函数的执行时间,修改前(7.4 节代码)的 kernel 函数(dot_0)执行时间是 58.400μs,内置函数和移位运算修改后 kernel 函数(dot_1)的执行时间为 57.440μs,性能有微小提升,但由于提升太小,甚至可视为误差。

那么问题又来了,如果数据量提高时,这种微小提升能否累积呢?即在应用中采用内置函数的方法是否有意义?针对向量内积问题,将数据量增加 128 倍,再次测试,得到 dot_0 的 kernel 函数执行时间为 6.5688ms,dot_1 的 kernel 函数执行时间为 6.5653ms,两者差距几乎可以忽略,故得出结论:**向量内积中运算改为内置函数和移位运算意义不大,反而会增加 kernel 函数的代码复杂度**。

7.10.3 减少同步开销

GPU 线程执行时以 warp 为单位进行调度,每个 warp(32 个线程)执行同一条指令,存在着天然的同步机制,无须再调用__syncthreads()函数。因此当归约到 32 线程之后,可删除__syncthreads()函数和判断直接计算。下面展示了相应的 CUDA C 代码(记为 dot_2),其中,"volatile float * tmp_1=tmp;"非常重要,如果去掉,将导致执行结果错误;若去掉又要保证结果正确,每次累加都需要加入判断,又将增加分支判断。

```
__global__ void vector_dot_product_gpu_3_2(DATATYPE *a, DATATYPE *b, DATATYPE
*c_tmp, int n)
{
    extern __shared__ DATATYPE tmp[];
    const int tidx=threadIdx.x;
    const int bidx=blockIdx.x;
    const int t_n=blockDim.x * gridDim.x;
    int tid=bidx * blockDim.x+tidx;
    double temp=0.0;
    while(tid<n)
    {
        temp=__fadd_rn(temp,__fmul_rn(a[tid],b[tid]));
        tid+=t_n;
    }
```

```
    tmp[tidx]=temp;
    __syncthreads();
    int i=blockDim.x>>1;
    while(i!=32)
    {
        if(tidx<i)
        {
            tmp[tidx]=__fadd_rn(tmp[tidx],tmp[tidx+i]);
        }
        __syncthreads();
        i>>=1;
    }
    if (tidx<32)
    {
        volatile float * tmp_1=tmp;
        tmp_1[tidx]=__fadd_rn(tmp_1[tidx],tmp_1[tidx+32]);
        tmp_1[tidx]=__fadd_rn(tmp_1[tidx],tmp_1[tidx+16]);
        tmp_1[tidx]=__fadd_rn(tmp_1[tidx],tmp_1[tidx+8]);
        tmp_1[tidx]=__fadd_rn(tmp_1[tidx],tmp_1[tidx+4]);
        tmp_1[tidx]=__fadd_rn(tmp_1[tidx],tmp_1[tidx+2]);
        tmp_1[tidx]=__fadd_rn(tmp_1[tidx],tmp_1[tidx+1]);
        if(tidx==0)
        {
            c_tmp[bidx]=tmp_1[0];
        }
    }
}
vector_dot_product_gpu_3_2<<<blocknum,threadnum,threadnum * sizeof(DATATYPE)
>>>(d_a,d_b,d_c_tmp,n);
```

通过实验已经初步了解 volatile 关键字的效果，volatile 关键字将全局存储和共享存储中的变量声明为敏感变量，默认其他线程随时可能会修改该值，每次运算都会取读到新值。在向量内积问题中，volatile 关键字能保证 warp 内每次归约用的都是上一次归约的结果，否则可能读取到同步之前的结果。

测试后发现，dot_2 的 kernel 函数执行时间为 $56.416\mu s$，对比 dot_1 的 kernel 函数执行时间 $57.440\mu s$ 又有细微提升。

7.10.4 循环展开

将归约中的 while 循环展开，由于启动 kernel 函数的线程维度是<<<64,256>>>，将 while 部分代码改写成如下形式，再次测试，发现 kernel 函数的执行时间为 $55.168 \sim 56.576\mu s$，性能提升不明显。

```
if (tidx<128){tmp[tidx]=__fadd_rn(tmp[tidx],tmp[tidx+128]);}
__syncthreads();
if (tidx<64){tmp[tidx]=__fadd_rn(tmp[tidx],tmp[tidx+64]);}
__syncthreads();
```

7.10.5　总结

以 gpu_5 为基准,同样采取上述优化策略,实现相应的 CUDA 代码。然后运行 gpu_3(优化后)、gpu_5(优化后)和 cublas 三个版本,统计 kernel 函数执行时间和通信(D2H)时间(见表 7.3)。

表 7.3　优化后的向量内积时间　　　　　　　　　　单位:μs

64×256	计算 1	计算 2	通信(D2H)	总时间
gpu_3	56.29	0.95	2.82	60.06
gpu_5	57.50		2.56	60.06
cublas	54.72	4.32	2.59	61.63

从计算 1 的时间列可以看出,采取本节阐述的多种优化策略后,kernel 函数与 CUBLAS 库函数依然存在少许差距。从总时间上看,本节设计的 CUDA 代码性能已经略微超越了 CUBLAS 库性能,同时证实了 7.9 节的结论"CPU 二次归约性能更佳"和"CUBLAS 库仍存在优化空间"。

本节主要针对归约过程进行了 GPU 性能优化研究,其中线程(块)维度配置比较重要,合理的线程(块)维度对性能提升效果显著,而其他几种优化策略的效果并不明显,读者在优化真实应用程序时可以酌情考虑。本节的研究更多的是提供一些 GPU 优化的思路,供读者借鉴。

7.11　知识点总结

高精度中间变量:中间变量采用的数据类型精度要尽可能高,一般采用 double 类型,否则当循环次数过多时会出错。

归约的 GPU 映射方法:归约运算在 GPU 中映射有两种方法:分散归约和低线程归约。通过这两种方法,归约运算的计算步数从串行的 $n-1$ 减少为 $\log(n)$。具体映射思想参见 7.2 节和 7.3 节。

低线程归约效率更优:低线程归约有 3 大优势,分别是最小化执行 warp 数量、数据对齐访问和避免 bank conflict。7.9 节对比了低线程归约和分散归约的性能。

__shared__:变量限定符,用该限定符声明存储在共享存储的变量。共享存储器比全局存储器访存速度快了一个量级,使用共享存储可以加快数据访问时间,其定义方法如下:

```
__shared__ DATATYPE tmp[threadnum];
```

  __syncthreads()：block 内所有 thread 同步。线程执行时以 warp 为单位，不同 warp 间不同步，通过该函数可以实现 block 内所有线程的同步。一般该函数用于先写后读的情况，后来读到的内容是前面写的内容。

  __threadfence()、__threadfence_block()：线程同步函数，类似于__syncthreads()。

  GPU 不同 block 间无法通信或同步：GPU 上不同 block 之间无法通信或同步，当不同 block 需要数据交互时，7.4 节~7.7 节阐述了 4 种解决方法。

  CPU 二次归约破解 block 间无法同步：由于 GPU 不同 block 间无法通信和同步，在 7.4 节提出 CPU 二次归约的方法，即各 block 将中间结果保存到全局存储中，再把中间结果数组传回 CPU 完成第二次归约运算，得到最终结果。

  GPU 二次归约破解 block 间无法同步：由于 GPU 不同 block 间无法通信和同步，在 7.5 节提出 GPU 二次归约的方法，即各 block 将中间结果保存到全局存储中，再启动另一个单 block 的 kernel 函数，归约中间结果得到最终结果。

  原子操作破解 block 间无法同步：由于 GPU 不同 block 间无法通信和同步，在 7.6 节提出利用原子操作的方法来解决该问题，即利用原子操作互斥访问数据的特点实现。

  计数法破解 block 间无法同步：由于 GPU 不同 block 间无法通信和同步，在 7.7 节提出计数法，即在各 block 计算得到中间结果后，设置栅栏并利用原子加法运算来做计数，记录每个到达栅栏的 block，当最后一个 block 到达后，最后一个 block 完成后续中间结果的归约计算。这里的栅栏本身是无法拦住 block 的，仅仅是做了个计数工作，判断是否所有的 block 都已到达栅栏。

  原子操作：能保证线程互斥地访问全局存储或共享存储上的同一数据，即同一数据在某一时刻只能被一个线程访问或修改，避免了并行程序中多个线程同时修改某一变量出现的冲突。原子操作函数包括加法(atomicAdd)、减(atomicSub)、与(atomicAnd)、或(atomicOr)、异或(atomicXor)、最大(atomicMax)、最小(atomicMin)、增量(atomicInc)、减量(atomicDec)、比较交换(atomicCAS)、交换(atomicExch)等。具体用法详见 7.6 节。

  原子操作值的初始化：原子操作前，需要对原子操作修改的数据赋值，否则将导致结果错误。

  尽量少用原子操作：尽管原子操作功能强大，但其本质是串行处理，大量使用会对性能产生较大影响，故应尽量少用原子操作。在 7.6 节进行了相关测试和探讨。

  原子操作编译：原子操作是 CUDA 计算能力 1.1 以后加入的功能(详见 3.4 节)，NVCC 编译器默认采用 1.0 版本进行编译，因此编译包含原子操作的代码时，需要明确指定更高级的计算能力。

```
$nvcc vector_dot_product_gpu_5.cu -O3 -o vector_dot_product_gpu_5 -arch 'compute_35' -code 'compute_35'
```

  nvprof 命令行性能测试工具：通过该命令行性能测试工具，可以得到与 GPU 相关的函数(包括 kernel 函数和通信函数)的耗时信息,包括时间比率、时间、启动次数、平均时间、最小时间和最大时间等信息。下面给出 nvprof 命令的使用实例。

```
[fangmq@cn18%yhstar vectordot]$nvprof ./vector_dot_product_cublas
==20100== NVPROF is profiling process 20100, command: ./vector_dot_product
_cublas
cublas vector dot:n=1048576 ; timeused=3.723860 (ms)
4399734194176.000000
==20100==Profiling application: ./vector_dot_product_cublas
==20100==Profiling result:
Time(%)      Time     Calls      Avg       Min       Max    Name
97.56%    2.4159ms     3     805.30μs   1.4080μs  1.2485ms  [CUDA memcpy HtoD]
 2.16%   53.376μs     1     53.376μs   53.376μs  53.376μs  void dot_kernel<float, int
=128, int=0, int=0>(cublasDotParams<float>)
 0.17%   4.3200μs     1     4.3200μs   4.3200μs  4.3200μs  void reduce_1Block_kernel
<float, int=128, int=7>(float *, int, float *)
 0.11%   2.6240μs     1     2.6240μs   2.6240μs  2.6240μs  [CUDA memcpy DtoH]
```

低线程归约用于最终结果计算时有精度损失,而用于中间结果计算时没有精度损失。原子操作计算结果不稳定,有精度损失。

CPU 二次归约性能更好:单 block 的归约性能不及 ICC 编译器向量化后的 CPU 串行归约性能。

传输函数启动时间:大约 2~3μs 之间,实验中 1 个 float 数据(4B)和数百个 float 数据传输时间差距极小。当数据量逐步增大时,传输时间如何变化? 详见 17.7 节。

CUBLAS 库函数的实现可能不止一个 kernel 函数:在 cublasSdot 函数执行时,nvprof 结果显示其真正执行的 kernel 函数有两个,分别是 dot_kernel 和 reduce_1Block_kernel 函数。

CUBLAS 库仍有提升空间:CUBLAS 库函数中的 cublasSdot 函数拆分为两个 kernel 函数执行,应该是采取了 GPU 二次归约的方法,经本章实验得出结论是 CPU 二次归约方法性能更优,那么如果将 CUBLAS 库函数的二次归约方案改成 CPU 二次归约,是否能获得更好的性能呢? 关于这点,在 7.10 节对归约过程进行了深入优化,并对比优化后的代码和 CUBLAS 库的性能,结果显示优化后总性能比 CUBLAS 库还好,说明 CUBLAS 库确实仍有优化空间。

线性关系的计算量和通信量(数据量)不适合移植到 GPU:向量加法和向量内积的实验结果显示,当计算量和数据量成线性关系时,通信量与数据量一致,通信时间比 CPU 串行计算时间还长。

线程维度对性能影响较大:7.10.1 节探讨了不同 block 数量和 thread 数量对向量内积运算性能的影响,其执行时间 0.078~0.457ms,差别极大,因此在 GPU 并行程序优化时一定要合理配置线程维度。

warp 内无须同步:GPU 中单个 warp 的所有线程同时执行同一指令,因此 warp 内的线程无须同步就能获得新更新的数据。但编程实现时必须使用 volatile 关键字定义敏感变量。7.10.3 节给出了相应的实现方法。

volatile 关键字：volatile 关键字可以将全局存储和共享存储中的变量声明为敏感变量，默认其他线程可能随时会修改其值，每次运算都会取读新值。

7.12 扩展练习

（1）除了向量内积外，归约运算还有很多，比如最大最小值、求和、求平方和、逻辑与、逻辑或等，实现其他运算的归约操作，并分析其性能。

（2）本章的归约运算代码仅适用于 2 的幂次线程数量，设计能够处理任意线程数量的归约运算代码。

（3）设计实验探索 7.9 节得出的结论 11："数据使用次数与数据量是线性关系时，数据传输开销可能超过串行运算时间。那么当两者是什么关系时，GPU 并行计算才有意义呢？"

第 8 章 矩阵乘法

8.1 矩阵乘法及其 3 种串行代码

与向量加法和向量内积相比,矩阵乘法在运算复杂性上有所提升,并且运算量和数据量的关系也不再是线性关系,其中 $O(n^2)$ 的数据量对应 $O(n^3)$ 的运算量。矩阵乘法的数学表达式如下:

$$C = AB$$

$$C_{ij} = \sum_{k=0}^{n} A_{ik} \times B_{kj}$$

从本质上看,矩阵乘法是向量内积的集合。C 矩阵的每一个元素都是 A 矩阵一行和 B 矩阵一列的向量内积。

参与内积运算的是 B 矩阵的一列,而在 C 语言中,数据是按行存储的,因此 B 矩阵访存不连续,导致 cache 命中失效。本节除了给出一般的矩阵乘法代码外,还阐述了两种解决访存不连续问题的方法,并分别给出相应的 C 代码。

8.1.1 一般矩阵乘法

一般矩阵乘法即 C 矩阵的一个元素等于 A 矩阵的一行和 B 矩阵一列的向量内积,如图 8.1 所示。

图 8.1 一般矩阵乘法

根据图 8.1 和矩阵乘法的数学公式,可以写出 C 语言版本的矩阵乘法代码,其中 A 矩阵大小为 m 行 l 列,B 矩阵大小为 l 行 n 列,相乘后得到 C 矩阵大小为 m 行 n 列。这种方法的思路是一次完成 C 矩阵一个元素的运算,然后通过循环迭代的方法完成 **C** 矩阵所有元素的运算。

```
//a:m行1列,b:1行n列,c:m行n列
void matrix_multiplication_serial_1(DATATYPE *a, DATATYPE *b, DATATYPE *c,int
m, int n, int l)
{
    int i,j,k;
    double temp=0.0;
    for(i=0; i<m; i++) {
        for(j=0; j<n; j++) {
            temp=0.0;
            for(k=0; k<l; k++) {
                temp+=a[i*l+k]*b[k*n+j];
            }
            c[i*n+j]=temp;
        }
    }
}
```

注意：代码中最内层循环变量 k，对 B 矩阵的访问 b[k＊n＋j]是不连续的。这种访存上的不连续将导致代码在 CPU 上编译无法向量化，下面是 ICC(v11.1)编译上述代码的向量化报告，报告指出存在向量依赖、不支持数据类型等原因无法向量化。向量化是 CPU 性能发挥的关键，其重要性不言而喻。

```
$icc matrix_multiplication_serial.cpp -O3 -o matrix_multi_serial -vec-report2
 :
matrix_multiplication_serial.cpp(34): (col. 4) remark: loop was not vectorized:
insufficient computational work.
matrix_multiplication_serial.cpp(34): (col. 4) remark: loop was not vectorized:
unsupported data type.
```

笔者在做代码的结果正确性和性能准确性验证的时候发现了一个现象,在此进行扩展说明。结果验证时发现执行时间比最初的测试时间快了 1 倍,笔者发现这段代码已被向量化,再深入研究,原来最初笔者写本章时(一年前)的实验环境是 ICC 11.1 版本编译器,现在的环境改成了 ICC 13.0.0.079。下面给出了利用 ICC 13.0.0.079 编译的向量化报告。说明利用较新版本的编译器能够更好地达到向量化的目的。

```
$icc matrix_multiplication_serial.cpp -O3 -o matrix_multi_serial -vec-report2
 :
matrix_multiplication_serial.cpp(34): (col. 4) remark: LOOP WAS VECTORIZED.
```

8.1.2 循环交换矩阵乘法

针对一般矩阵乘法中存在的访存不连续、无法向量化的问题,一种改进方法是交换内部两层循环,使得数据访问连续,以达到向量化的目的。这种矩阵乘法的思想是每次读取 **A** 矩阵的一个元素,与 **B** 矩阵对应的一行做乘积运算,累加乘积结果到 **C** 矩阵的对应行,如图 8.2 所示。

图 8.2 循环交换矩阵乘法

下面是循环交换矩阵乘法的源代码。

```
//a:m行1列,b:1行n列,c:m行n列
void matrix_multiplication_serial_2(DATATYPE *a, DATATYPE *b, DATATYPE *c,int m, int n, int l)
{
    int i,j,k;
    double temp=0.0;
    for(i=0;i<m*n;i++)
    {
        c[i]=0.0;
    }
    for(i=0; i<m; i++) {
        for(k=0; k<l; k++) {
            temp=a[i*l+k];
            for(j=0; j<n; j++) {
                c[i*n+j]+=temp*b[k*n+j];
            }
        }
    }
}
```

代码经循环交换后,使用 ICC 编译器编译,并查看向量化报告,发现最内层循环已经向量化。由于循环交换后要对 C 矩阵进行累加运算,因此必须赋初值,赋值过程将引入额外开销,向量化报告指出编译赋值循环亦能向量化。

```
matrix_multiplication_serial.cpp(46): (col. 2) remark: LOOP WAS VECTORIZED.
matrix_multiplication_serial.cpp(53): (col. 4) remark: LOOP WAS VECTORIZED.
```

8.1.3 转置矩阵乘法

要解决矩阵乘法中访存不连续、无法向量化的问题,除了内部两层循环交换的方法,还有矩阵转置的方法,即对 **B** 矩阵转置。**B** 矩阵转置后,原来 **A** 矩阵一行和 **B** 矩阵一列的向量内积运算就变成了 **A** 矩阵一行和 **B** 矩阵一行的向量内积运算(见图 8.3),此时 **B** 矩阵是按行访问的,因此具有良好的 cache 命中率、数据连续访存、可向量化等优点。

图 8.3 转置矩阵乘法

下面给出了转置矩阵乘法的 C 语言代码。

```
void matrix_multiplication_serial_3(DATATYPE *a, DATATYPE *b, DATATYPE *c,int m, int n, int l)
{
    int i,j,k;
    double temp=0.0;
    DATATYPE *b1;
    b1=(DATATYPE *) malloc(sizeof(DATATYPE) * l * n);
    for (int i=0;i<l;i++)
    {
        for (int j=0;j<n;j++)
        {
            b1[i*l+j]=b[j*n+i];
        }
    }
    for(i=0; i<m; i++) {
        for(j=0; j<n; j++) {
            temp=0.0;
            for(k=0; k<l; k++) {
                temp+=a[i*l+k] * b1[j*n+k];
            }
            c[i*n+j]=temp;
        }
    }
    free(b1);
}
```

使用 ICC 编译转置矩阵乘法代码,向量化报告显示矩阵相乘的最内层循环已经向量化，但矩阵转置过程的内层循环由于对 **B** 矩阵的访存不连续而无法向量化。因此，这种方法将会引入可观的额外转置开销。

```
matrix_multiplication_serial.cpp(69): (col. 3) remark: loop was not vectorized:
unsupported data type.
matrix_multiplication_serial.cpp(77): (col. 4) remark: LOOP WAS VECTORIZED.
```

注意：上述代码不能使用 ICC 编译器的 O3 优化开关，否则运行结果出错。笔者认为是由于 O3 优化开关优化比较激进，导致程序运行结果出错。使用默认的 O2 优化开关时运行结果正确，且经测试上述代码的 O2 开关和 O3 开关性能相当（这段话写在最初，当时用的 v11.1 版本的 ICC 编译器，但在做结果验证时发现：仍然用 v11.1 版本的 ICC 编译器开 O3 优化，结果出错无法复现；O2 和 O3 性能相当。由于本段描述的现象在笔者的编程开发过程中也曾遇到，故保留于此作为经验分享给读者）。

8.1.4 实验结果与最优串行矩阵乘

在 CPU 上编译运行 3 种串行矩阵乘代码，测试不同规模数据的耗时信息，统计在表 8.1 中。其中，改进方法的耗时已包含引入的额外开销，比如内部两层循环交换中赋初值，转置矩阵乘法中矩阵转置、额外矩阵存储分配和释放等。表中数据显示，**B** 矩阵转置的改进方法性能最佳，是最优串行矩阵乘法，选为基准用于衡量 GPU 矩阵乘法性能。

由于结果验证时发现了与原来的差距，笔者在表 8.1 同时罗列了使用两个编译器版本的执行时间结果。对比不同编译器版本的执行结果，发现新版本编译器的性能明显好过旧版本，特别是其中串行矩阵乘法 1 中，新版本能够向量化，加速了近 1 倍。横向对比 3 个串行矩阵乘法，不同版本的表现基本相同，v11.1 版本的数据也能反映情况，故本章后续分析时仍然采用 v11.1 版本的矩阵乘法 3 作为对比依据。

表 8.1 串行矩阵乘法时间 单位：ms

ICC v11.1	matrix_multi_1	matrix_multi_2	matrix_multi_3
256	33.22	17.78	8.31
512	322.14	125.06	64.76
1024	2455.54	999.16	480.21
2048	19642.04	7952.84	3816.80
4096	716571.48	69411.94	40812.96
ICC v13.0	matrix_multi_1	matrix_multi_2	matrix_multi_3
256	18.35	15.75	8.05
512	181.83	126.04	57.18
1024	1358.98	996.53	440.77
2048	10953.12	7947.39	3495.49
4096	356152.21	66718.97	32921.03

对比 3 个程序的矩阵乘法结果，发现采用高精度中间变量的矩阵乘法 1 和矩阵乘法 3 的结果是一致的，矩阵乘法 2 由于直接累加到了 **C** 矩阵，而 **C** 矩阵是单精度的，导致了

精度损失。这也验证了 7.1 节关于中间变量计算精度的论述(8.8 节有详细数据对比)。

基于 CPU,矩阵乘法还有很多深入的优化手段,比如 intrinsic 函数、基于 cache 的矩阵切分等,但本书选择最优串行程序时不考虑这些优化措施,主要有以下几点原因:①数据访存连续、向量化后的程序已能发挥一定的 CPU 性能;②这些深入的优化手段获得的性能收益相对有限;③深入优化需要程序员对代码做大幅修改,还需要运用 CPU 底层的体系结构知识,投入与产出不成比例。

8.2 grid 线程循环矩阵乘法

一般矩阵乘法代码的外面两层循环可以合并为一层循环,大幅增加并行度。在 GPU 映射时,grid 内所有的 thread 对应合并后的外层循环(即每个 thread 完成 **A** 一行和 **B** 一列的向量内积运算)。此时,需要定位 thread 在 grid 的索引(idx),并计算线程所属 **C** 矩阵元素所在的行号(row)和列号(column),最后各 thread 完成对应 **C** 矩阵元素的计算。下面给出相应的 CUDA C 代码(标记为 gpu_1)。

```
//A:n行 lda列;B:lda行 ldb列;C:n行 ldc列,其中 ldb=ldc
__global__ void matrix_multiplication_gpu_1(const DATATYPE *a, size_t lda,
const DATATYPE *b, size_t ldb, DATATYPE *c, size_t ldc, int n)
{
    const int tidx=threadIdx.x;
    const int bidx=blockIdx.x;
    const int idx=bidx * blockDim.x+tidx;
    const int row=idx/n;
    const int column=idx%n;
    if(row<n && column<ldc)
    {
        double tmp=0.0;
        for(int i=0; i<n; i++)
        {
            tmp+=a[row * lda+i] * b[i * ldb+column];
        }
        c[row * ldc+column]=tmp;
    }
}
```

```
int blocks=(n+threadnum-1)/threadnum;
matrix_multiplication_gpu_1<<<blocks * n, threadnum>>>(d_a,n,d_b,n,d_c1,n,
n);
```

grid 内线程循环的矩阵乘法可能存在以下缺陷:运算时,一个 block 内的线程可能需要运算 **C** 矩阵不同行的矩阵元素,block 内 thread 对相应的 **A** 矩阵的访存不一致,导致无法广播和额外的访存开销,最终导致执行时间增加。

8.3　block 线程循环矩阵乘法

本节针对 8.2 节 grid 循环矩阵乘法方案中可能存在的缺陷做了如下改进：每个 block 计算 C 矩阵一行，block 内的 thread 以固定跳步步长 blockDim.x 的方法循环计算 C 矩阵的一行；每一行启动一个 block，共计 n 个 block。下面给出相应的 CUDA C 代码（标记为 gpu_1_0）。

```
//A:n 行 lda 列;B:lda 行 ldb 列;C:n 行 ldc 列,其中 ldb=ldc
__global__ void matrix_multiplication_gpu_1_0(const DATATYPE *a, size_t lda,
const DATATYPE *b, size_t ldb, DATATYPE *c, size_t ldc, int n)
{
    int tidx=threadIdx.x;
    int bidx=blockIdx.x;
    double tmp=0.0;
    int i;
//  for(;bidx<n;bidx+=gridDim.x)
    {
        for(tidx=threadIdx.x;tidx<ldc;tidx+=blockDim.x)
        {
            tmp=0.0;
            for(i=0;i<lda;i++)
            {
                tmp+=a[bidx * lda+i] * b[i * ldb+tidx];
            }
            c[bidx * ldc+tidx]=tmp;
        }
    }
}
matrix_multiplication_gpu_1_0<<<n, threadnum>>>(d_a,n,d_b,n,d_c2,n,n);
```

一般情况下，block 数量很难达到 GPU 规定的上限值（详见 3.2 节），若矩阵行数超过该上限值，只需去掉代码中的注释行，即可令 block 以固定跳步的方法循环计算矩阵所有行。

8.4　行共享存储矩阵乘法

共享存储与 L1 Cache 同级，其访存延迟较全局存储小一个量级。用共享存储替代全局存储是 GPU 最重要的优化手段之一。但共享存储中的数据不是凭空而来，一般读取自全局存储，因此共享存储的访存必然伴随有至少 1 次全局存储的访存。故采取共享存储优化的关键是数据复用，数据复用次数越多，共享存储优化可获得的收益也越高。本节将探讨矩阵乘法中的数据复用情况和利用共享存储优化的尝试。

8.3 节 block 循环矩阵乘法中,1 个 block 内所有 thread 都会用到 **A** 矩阵的一行,此时 **A** 矩阵中该行复用了 **A** 矩阵列数次。故可考虑将 **A** 矩阵一行读入 shared memory,运算时从 shared memory 读取相应数据。根据上述思想,实现相应 CUDA C 代码如下(标记为 gpu_2)。

```
__global__ void matrix_multiplication_gpu_2(const DATATYPE *a, size_t lda,
const DATATYPE *b, size_t ldb, DATATYPE *c, size_t ldc, int n)
{
    extern __shared__ DATATYPE data[];
    const int tid=threadIdx.x;
    const int row=blockIdx.x;
    int i, j;
    for(i=tid; i<n; i+=blockDim.x)
    {
        data[i]=a[row * lda+i];
    }
    __syncthreads();
    double tmp=0.0;
    for(j=tid; j<n; j+=blockDim.x)
    {
        tmp=0.0;
        for(i=0; i<n; i++)
        {
            tmp+=data[i] * b[i * ldb+j];
        }
        c[row * ldc+j]=tmp;
    }
}
matrix_multiplication_gpu_2<<<n, threadnum, sizeof(DATATYPE) * n>>>(d_a,n,d_b,n,d_c2,n,n);
```

对齐访存优化:如果参与运算的矩阵不是 512×512、1024×1024 这种规整的矩阵,而是 1000×1000,甚至任意维度的矩阵,此时矩阵存储不对齐,将影响访存性能。为了解决这个问题,采用对齐访存优化。思想是在矩阵空间分配时每行后添加空白空间,从而保证下一行数据的访存对齐。下面是访存对齐优化后 GPU 矩阵乘法的 CUDA C 代码(标记为 gpu_2_1)。

```
//对齐存储
    size_t pitch_a, pitch_b, pitch_c;
    cudaMallocPitch((void**) &d_a, &pitch_a, sizeof(DATATYPE) * n, n);
```

```
cudaMallocPitch((void**) &d_b, &pitch_b, sizeof(DATATYPE) * n, n);
cudaMallocPitch((void**) &d_c3, &pitch_c, sizeof(DATATYPE) * n, n);
cudaMemcpy2D(d_a, pitch_a, a, sizeof(DATATYPE) * n, sizeof(DATATYPE) * n, n,
cudaMemcpyHostToDevice);
cudaMemcpy2D(d_b, pitch_b, b, sizeof(DATATYPE) * n, sizeof(DATATYPE) * n, n,
cudaMemcpyHostToDevice);
matrix_multiplication_gpu_2<<<n, threadnum, sizeof(DATATYPE) * n>>>(d_a,
pitch_a/sizeof(DATATYPE), d_b, pitch_b/sizeof(DATATYPE), d_c3, pitch_c/
sizeof(DATATYPE),n);
cudaMemcpy2D(c3, sizeof(DATATYPE) * n, d_c3, pitch_c, sizeof(DATATYPE) * n,
n, cudaMemcpyDeviceToHost);
cudaFree(d_a);
cudaFree(d_b);
cudaFree(d_c3);
```

编译执行 gpu_2 和 gpu_2_1 版本矩阵乘法代码，测试两者计算时间，统计在表 8.2 中。表中数据显示，当矩阵规整（2 的幂次）时，对齐访存优化前后基本一致；**而矩阵不规整时（500、1000 等任意大小），对齐访存优化能获得部分性能收益**。

表 8.2 对齐访存优化效果 单位：ms

n	500	1000	2000	4000	512	1024	2048	4096
gpu_2	4.25	28.57	193.18	1767.07	4.16	27.16	192.18	1848.09
gpu_2_1	4.55	26.04	182.96	1752.23	4.15	27.16	193.16	1847.79

上述共享存储优化方案（A 矩阵 1 行的共享存储优化）仍然存在缺陷，读者可以先思考或实践测试。随后 8.5 节将分析其缺陷，8.8 节用实验验证该缺陷，并在 8.9 节给出解决方案。

8.5 棋盘阵列矩阵乘法

8.4 节的 A 矩阵"行共享存储优化"可能存在以下缺陷：①若 A 矩阵 1 行过大，将超出 shared memory 容量；②B 矩阵依然位于全局存储，且访存次数没有变化，仍有巨大的优化空间。

面对上述问题，本节采用一种棋盘阵列分块的矩阵乘法。棋盘阵列分块思想如图 8.4 所示，利用二维 block 和二维 thread 的概念，每个 block 对应 C 矩阵的一个子矩阵（小窗口），block 内所有 thread 完成该子矩阵结果的运算。子矩阵结果计算时，需要的数据是与 C 的子矩阵行号相同的 A 矩阵的整行数据和与 C 子矩阵列号相同的 B 矩阵的整列数据，即图示 A 和 B 矩阵中的两个长方形。每个小窗口（假设小窗口尺寸是 nx×nx）映射到 GPU 中 1 个 block，nx 是 block 内二维 thread 的尺寸，每次将小窗口内的数据存放入共享存储，该数据将会被复用 nx 次，非常划算。同时 shared memory 使用量可以通过 nx

进行调整和控制。

根据上述棋盘阵列矩阵乘法思想,编程实现以下 CUDA C 代码(标记为 gpu_3)。其中,threadnx 是图 8.4 中小窗口的尺寸,声明了两个共享存储矩阵 matA 和 matB,用于存放 **A** 矩阵和 **B** 矩阵中小窗口中的数据。为保证程序的正确性,共享存储数据的加载和读出都要同步。代码中,i 循环实现小窗口内数据的计算,j 循环实现小窗口的移动。kernel 函数启动时采用了 2D 线程维度(详见 5.6 节)。

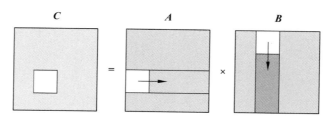

图 8.4 棋盘阵列矩阵乘法

```
__global__ void matrix_multiplication_gpu_3(const DATATYPE *a, size_t lda,
const DATATYPE *b, size_t ldb, DATATYPE *c, size_t ldc, int n)
{
    __shared__ DATATYPE matA[threadnx][threadnx];
    __shared__ DATATYPE matB[threadnx][threadnx];
    const int tidc=threadIdx.x;
    const int tidr=threadIdx.y;
    const int bidc=blockIdx.x * threadnx;
    const int bidr=blockIdx.y * threadnx;
    int i, j;
    double results=0.0;
    for(j=0; j<n; j+=threadnx)
    {
        if(tidr+bidr<n && tidc+j<n)
        {
            matA[tidr][tidc]=a[(tidr+bidr) * lda+tidc+j];
        } else {
            matA[tidr][tidc]=0;
        }
        if(tidr+j<n && tidc+bidc<n)
        {
            matB[tidr][tidc]=b[(tidr+j) * ldb+tidc+bidc];
        } else {
            matB[tidr][tidc]=0;
        }
        __syncthreads();
        for(i=0; i<threadnx; i++)
        {
```

```
            results+=matA[tidr][i] * matB[i][tidc];
        }
        __syncthreads();
    }
    if(tidr+bidr<n && tidc+bidc<n)
    {
        c[(tidr+bidr) * ldc+tidc+bidc]=results;
    }
}
int bx=(n+threadnx-1)/threadnx;
dim3 blocks(bx, bx);
dim3 threads(threadnx, threadnx);
matrix_multiplication_gpu_3<<<blocks, threads>>>(d_a,pitch_a/sizeof(DATATYPE),d_b,pitch_b/sizeof(DATATYPE),d_c1,pitch_c/sizeof(DATATYPE),n);
```

8.6 判断移除

8.5 节中棋盘阵列矩阵乘法仍有性能缺陷,即判断过多,且判断位于循环内,每次循环都需要执行判断语句,而 GPU 不擅长分支处理(15.4 节详细地讨论了 GPU 分支处理)。减少判断亦是 GPU 性能优化的关键。针对棋盘阵列矩阵乘法运算,进行判断移除优化,修改后 CUDA C 代码如下(标记为 gpu_4):

```
__global__ void matrix_multiplication_gpu_4(const DATATYPE *a, size_t lda,
const DATATYPE *b, size_t ldb, DATATYPE *c, size_t ldc, int n){
    __shared__ DATATYPE matA[threadnx][threadnx];
    __shared__ DATATYPE matB[threadnx][threadnx];
    const int tidc=threadIdx.x;
    const int tidr=threadIdx.y;
    const int bidc=blockIdx.x * threadnx;
    const int bidr=blockIdx.y * threadnx;
    int i, j;
    double results=0.0;
    for(j=0; j<n; j+=threadnx)
    {
        matA[tidr][tidc]=a[(tidr+bidr) * lda+tidc+j];
        matB[tidr][tidc]=b[(tidr+j) * ldb+tidc+bidc];
        __syncthreads();
        for(i=0; i<threadnx; i++)
        {
            results+=matA[tidr][i] * matB[i][tidc];
        }
```

```
        __syncthreads();
    }
    if(tidr+bidr<n && tidc+bidc<n)
    {
        c[(tidr+bidr)*ldc+tidc+bidc]=results;
    }
}
matrix_multiplication_gpu_4<<<blocks,threads>>>(d_a,pitch_a/sizeof
(DATATYPE),d_b,pitch_b/sizeof(DATATYPE),d_c2,pitch_c/sizeof(DATATYPE),n);
```

8.7 CUBLAS 矩阵乘法

CUBLAS 库是 NVIDIA 官方提供的线性代数函数库,本书第 6 章和第 7 章已有相关介绍和应用,本节将探讨利用 cublas(默认为 v1)和 cublas_v2 两种版本的 CUBLAS 库来实现矩阵乘法运算,同时提供这两种 CUBLAS 库的使用方法供读者参考。

cublas 版本矩阵乘法:cublas 版本需要包含头文件 cublas.h,GPU 上存储分配和释放、数据传输等有特定的函数,比如存储分配使用 cublasAlloc()函数,释放使用 cublasFree()函数,传输函数与 V2 版本相同(详见 6.5 节)。单精度浮点矩阵乘法运算使用 cublasSgemm()函数。函数功能和使用方法可参考 BLAS 库说明文档。

```
cublasStatus CUBLASAPI cublasAlloc (int n, int elemSize, void **devicePtr);
cublasStatus CUBLASAPI cublasFree (void *devicePtr);
void CUBLASAPI cublasSgemm (char transa, char transb, int m, int n, int k, float
alpha, const float *A, int lda, const float *B, int ldb, float beta, float *C, int
ldc);
```

下面给出了 cublas 版本实现的矩阵乘法,C 语言程序员需要特别注意,由于 BLAS 库最初用 FORTRAN 语言书写,其矩阵数据是按列存储的,而 C 语言中矩阵是按行存储的。因此对于存储空间中的矩阵 A,相当于是其转置矩阵 A^T。因此 $C=A\times B$ 就变成了 $C^T=B^T\times A^T$,细看代码中 cublasSgemm()函数的参数应该就能有所收获了。读者可扩展利用 CUBLAS 库实现 3 个维度尺寸均不相同的矩阵乘法来加强理解。

```
#include<cublas.h>
    :
DATATYPE *d_a, *d_b, *d_c1;
cublasInit();
cublasAlloc(n*n,sizeof(DATATYPE),(void**)&d_a);
cublasAlloc(n*n,sizeof(DATATYPE),(void**)&d_b);
```

```
    cublasAlloc(n*n,sizeof(DATATYPE),(void**)&d_c1);
    cublasSetVector(n*n,sizeof(DATATYPE),a,1,d_a,1);
    cublasSetVector(n*n,sizeof(DATATYPE),b,1,d_b,1);
    cublasSetVector(n*n,sizeof(DATATYPE),c1,1,d_c1,1);
    cublasSgemm('n','n',n,n,n,1.0,d_b,n,d_a,n,0.0,d_c1,n);
    cublasGetVector(n*n,sizeof(DATATYPE),d_c1,1,c1,1);
    cublasFree(d_a);
    cublasFree(d_b);
    cublasFree(d_c1);
    cublasShutdown();
```

下面是利用 cublas_v2 实现的矩阵乘法代码。

```
#include<cublas_v2.h>
    ⋮
    DATATYPE *d_a, *d_b, *d_c1;
    cublasHandle_t handle;
    cublasCreate(&handle);
    cudaMalloc((void**)&d_a, sizeof(DATATYPE)*n*n);
    cudaMalloc((void**)&d_b, sizeof(DATATYPE)*n*n);
    cudaMalloc((void**)&d_c1, sizeof(DATATYPE)*n*n);
    float alpha=1.0;
    float beta=0.0;
    cublasSetVector(n*n,sizeof(DATATYPE),a,1,d_a,1);
    cublasSetVector(n*n,sizeof(DATATYPE),b,1,d_b,1);
    cublasSetVector(n*n,sizeof(DATATYPE),c1,1,d_c1,1);
    cublasSgemm_v2(handle, CUBLAS_OP_N, CUBLAS_OP_N,n,n,n,&alpha,d_b,n,d_a,n,
    &beta,d_c1,n);
    cublasGetVector(n*n,sizeof(DATATYPE),d_c1,1,c1,1);
    cudaFree(d_a);
    cudaFree(d_b);
    cudaFree(d_c1);
    cublasDestroy(handle);
```

经研究发现，v1 版本 CUBLAS 库函数在最终实现时调用了 v2 版本库函数，在 cublas_v2.h 文件中做了如下定义：

```
#define cublasSgemm          cublasSgemm_v2
```

故 v1 版本和 v2 版本的 CUBLAS 库函数性能是一致的，区别只是编程方式不同。cublas_v1 版本的好处之一是能让熟悉 BLAS 库的用户快速上手。

8.8 实验结果分析与结论

8.8.1 矩阵乘精度分析

运算精度在多数科学计算应用中至关重要。并行编程对精度的影响如何？如何才能在并行编程中提高精度呢？8.1节已提及矩阵乘法的精度问题,但未给出详细的实验结果数据。本节将通过实验探讨矩阵乘法中的精度问题。

将输入矩阵中的所有元素赋值为0.1,测试并统计本章代码的输出结果(见表8.3)。其中,serial_3_1是在serial_3的基础上采取Kahan's Summation Formula方法,避免了高精度中间变量(double)的使用;gpu_4_0是将gpu_4的中间变量类型由double型改为float型,而gpu_4_1利用Kahan's Summation Formula方法来提高gpu_4_0精度。

表 8.3 矩阵乘法结果精度对比

	512	1024	2048	4096	8192
serial_1	5.120 000	10.240 001	20.480 001	40.960 003	81.920 006
serial_2	5.120 022	10.240 139	20.480 373	40.959 133	81.919 090
serial_3	5.120 000	10.240 001	20.480 001	40.960 003	81.920 006
serial_3_1	5.120 000	10.240 001	20.480 001	40.960 003	81.920 006
mkl	5.119 996	10.239 992	20.479 982	40.959 965	81.919 930
gpu_4	5.120 000	10.240 001	20.480 001	40.960 003	81.920 006
gpu_4_0	5.120 022	10.240 139	20.480 373	40.959 133	81.919 090
gpu_4_1	5.120 000	10.240 001	20.480 001	40.960 003	81.920 006
cublas	5.120 022	10.240 139	20.480 373	40.959 133	81.919 090

表8.3中的数据显示serial_2、mkl、gpu_4_0、cublas都有较大精度损失,其中mkl相对精度损失较小。MKL库(math kernel library,数学核心库)和CUBLAS库在单精度运算精度上仍有不足之处,有进一步提升的空间。

采用了高精度中间变量方法的serial_1、serial_3、gpu_4和采取了Kahan's Summation Formula方法的serial_3_1、gpu_4_1精度损失较小。说明两种方法都能有效提升计算精度。

本节涉及了两种精度提升方法,其中高精度中间变量方法已在7.1节提及,即用高精度数据类型(double)存储中间累加变量,以避免精度损失。Kahan's Summation Formula方法在改进矩阵乘法时,将原来的加法运算改为减法运算。下面分别展示了改进后的串行矩阵乘代码和GPU矩阵乘代码。

```
//serial_3_1
    for(i=0; i<m; i++) {
        for(j=0; j<n; j++) {
            temp=0.0;
```

```
                comp=0.0;
                for(k=0; k<l; k++) {
                    comp-=a[i * l+k] * b1[j * l+k];
                    t=temp-comp;
                    comp=(t-temp)+comp;
                    temp=t;
                }
                c[i * n+j]=temp;
        }
    }
//gpu_4_1
__global__ void matrix_multiplication_gpu_4_1(const DATATYPE *a, size_t lda,
const DATATYPE *b, size_t ldb, DATATYPE *c, size_t ldc, int n){
    __shared__ DATATYPE matA[threadnx][threadnx];
    __shared__ DATATYPE matB[threadnx][threadnx];
    const int tidc=threadIdx.x;
    const int tidr=threadIdx.y;
    const int bidc=blockIdx.x * threadnx;
    const int bidr=blockIdx.y * threadnx;
    int i, j;
    float results=0.0;
    float comp=0.0;
    float t;
    for(j=0; j<n; j+=threadnx)
    {
        matA[tidr][tidc]=a[(tidr+bidr) * lda+tidc+j];
        matB[tidr][tidc]=b[(tidr+j) * ldb+tidc+bidc];
        __syncthreads();
        for(i=0; i<threadnx; i++)
        {
            comp-=matA[tidr][i] * matB[i][tidc];
            t=results-comp;
            comp=(t-results)+comp;
            results=t;
//          results+=matA[tidr][i] * matB[i][tidc];
        }
        __syncthreads();
    }
    if(tidr+bidr<n && tidc+bidc<n)
    {
        c[(tidr+bidr) * ldc+tidc+bidc]=results;
    }
}
```

8.8.2 实验结果分析

编译本章代码,分别执行 n 为 512、1024、2048、4096 的 $n\times n$ 维度的矩阵乘法,统计执行时间(包括数据传输时间)如表 8.4 所示。

表 8.4　矩阵乘法执行时间　　　　　　　　　　单位:ms

n	serial_3	mkl	gpu_1	gpu_1_0	gpu_2	gpu_3	gpu_4	gpu_4_1	cublas_v1	cublas_v2
512	64.76	2.15	4.38	4.34	4.16	3.08	2.83	2.71	1.86	1.86
1024	480.21	6.69	27.76	26.80	27.16	13.73	13.18	12.71	5.14	5.12
2048	3816.80	35.02	195.04	190.70	192.18	92.17	85.68	82.10	23.94	23.88
4096	40812.96	986.95	1921.77	1432.82	1848.09	667.17	620.30	588.90	131.75	133.58

表 8.4 中数据显示:

(1) 对比 serial_3 与 gpu_1 两行数据,即使是没有使用任何优化手段的 gpu_1 也能加速 14.8~21.2 倍,说明 GPU 并行矩阵乘法相对 CPU 串行矩阵乘有绝对的性能优势。

(2) 对比 mkl 和 gpu_4 两组数据,发现当矩阵规模小于 4096 时,MKL 库函数性能更好,而当矩阵尺寸达到 4096 后,优化后的 GPU 并行矩阵乘法比 CPU 并行矩阵乘法(MKL 库)更快。本节 MKL 库矩阵乘法运行平台配置两个 8 核 Intel Xeon E5-2670 CPU。

(3) 对比 gpu_1、gpu_1_0、gpu_2,有意思的是,使用了共享存储的 gpu_2 没有性能提升,反而采用 block 内线程循环方法的矩阵乘 gpu_1_0 取得了更好的性能。说明共享存储必须注意使用方法,盲目使用将导致负效果。那么面临这种情况怎么办呢?在 8.9 节中将给出解决方案。

(4) 对比 gpu_1、gpu_1_0 两组数据,gpu_1_0 的性能更好,说明 block 行循环比 grid 全局循环性能更好。

(5) gpu_3 和 gpu_4 相对 gpu_1、gpu_1_0 和 gpu_2 有明显的性能提升,说明棋盘阵列的设计对矩阵乘法性能提升正效果明显。在该问题中,**二维线程网格能更好地复用 shared memory,减少 global memory 的访问开销**。笔者认为这里性能提升的本质是更好地复用共享存储,至于二维线程网格设计是否有助于性能提升?留待读者实践探索。

(6) gpu_4 比 gpu_3 快,说明 GPU 中判断处理比较耗时,移除判断是提升 GPU 程序性能的重要优化方法。

(7) 对比 gpu_4_1 和 gpu_4 的实验结果和代码,又有一个有意思的现象:在访存量相同的情况下,增加了额外的运算量,程序性能反而有所提升。原因何在?

下面通过控制变量法探索其原因,这两个代码的主要区别有两点:gpu_4 用的是 double 双精度中间变量,而 gpu_4_1 用的都是 float 单精度运算;gpu_4_1 采用了 Kahan's Summation Formula 方法优化计算精度,增加了计算量。此时引入前文提及的单精度中间变量的版本 gpu_4_0,在表 8.5 中统计测试结果。表 8.5 中数据显示,gpu_4_0 的性能与 gpu_4_1 几乎一致,增加了同类型数据计算量而访存不变情况下 GPU 性能基本不变,说明该问题优化到现在仍是访存受限。gpu_4 比其他两组数据执行时间长,说明 **GPU 计算中 float 类型运算比 double 类型运算快。**

表 8.5 数据类型和计算量对 GPU 性能的影响 单位:ms

n	gpu_4	gpu_4_0	gpu_4_1
512	2.83	2.77	2.71
1024	13.18	12.68	12.71
2048	85.68	82.01	82.10
4096	620.30	589.58	588.90

(8) **CUBLAS 库的矩阵乘法运算性能最好**,比笔者编写的矩阵乘法快了近 10 倍(见表 8.6)。遗憾的是其计算精度不是很好。

表 8.6 矩阵乘法中的浮点运算能力

n	计算时间/ms			Gflops		
	serial_3	gpu_4_1	cublas_v2	serial_3	gpu_4_1	cublas_v2
512	64.76	1.16	0.34	4.14	231.62	796.25
1024	480.21	8.31	1.25	4.47	258.33	1712.39
2048	3816.80	64.23	7.64	4.50	267.45	2248.99
4096	40 812.96	511.78	52.62	3.37	268.55	2611.97
8192	317 056.58	4123.03	413.39	3.47	266.68	2659.76

8.8.3 浮点运算能力分析

为了更直观地了解 GPU 的性能,分别测试 serial_3、gpu_4_1 和 cublas_v2 这 3 个版本的矩阵乘法,仅统计计算时间,排除通信时间。然后计算相应的浮点运算能力 Gflops,如表 8.6 所示。其中,矩阵乘法的浮点运算次数为 $n\times n\times n\times 2$。

表 8.6 中数据显示:

(1) Xeon E5-2670 CPU 执行"最优串行程序"的浮点运算性能约为 4 ± 0.5 Gflops,过低说明串行代码没写好,过高则采用了深入优化。本书设定的"最优串行程序"仅包含访存对齐、向量化、最佳优化开关等优化,不包含深入体系结构的优化。

(2) CUBLAS 库的矩阵乘法运算几乎达到了峰值性能,比笔者用 CUDA-Z 工具测出来的计算性能还高(详见 17.1 节)。

(3) 优化后的 GPU 并行程序,在访存和计算量相当时(在矩阵乘法中,取两个数据运算两次),性能大约为 200~300Gflops。当然,该值也受 GPU 型号的限制。当程序优化至此,尽管与 NVIDIA 函数库的性能差距仍然很大,但已能基本满足一般计算应用需求。

8.9 行共享存储矩阵乘法改进

8.8 节结果显示 A 矩阵一行存放到共享存储后并没有给程序带来性能提升,反而导致部分性能损失。从理论上看,共享存储的访存延迟比全局存储小一个量级,加上前文分析的数据复用情况,该优化不应该导致负效果。那么是什么原因导致 8.4 节代码性能不好?又如何才能获得性能收益呢?本节试图解决这两个问题。

分析 8.4 节共享存储优化性能不好的原因,8.8 节数据中,当矩阵规模小时,性能并没有变差,而随矩阵规模的增大,性能差距才越来越大。一种可能原因是当矩阵增大时,每行数据量增加,而共享存储容量有限,尽管 block 内每行的数据量依然没有超过共享存储容量,但共享存储的使用量却会影响活跃 warp 数量,影响 occupancy,最终影响 GPU 程序执行性能。

基于上述猜测,解决方案是将原来 8.4 节中由 kernel 函数动态分配的 shared memory 固定下来,通过循环的方法完成矩阵乘法运算。下面 CUDA 代码基于 8.4 节矩阵乘法代码进行修改,额外加入一层循环层,即共享存储循环(A 的行循环),共享存储一次读取 A 矩阵一行中 sharesize 尺寸的数据,后利用这些数据与 B 矩阵对应行的所有列元素运算,中间结果暂存于 C 矩阵中。然后共享存储循环直接从 C 矩阵取相应元素值赋给中间变量,进行下一轮计算,因此 C 矩阵需要初始化。

```
#define sharesize 512
__global__ void matrix_multiplication_gpu_2_2(const DATATYPE *a, size_t lda,
const DATATYPE *b, size_t ldb, DATATYPE *c, size_t ldc, int n)
{
    __shared__ DATATYPE data[sharesize];
    const int tid=threadIdx.x;
    const int row=blockIdx.x;
    int i, j;
    int p,bidx;
    for(bidx=row;bidx<n;bidx+=gridDim.x)
    {
        for(j=tid; j<n; j+=blockDim.x)
        {
            c[bidx * ldc+j]=0.0;
        }
    }
    __syncthreads();
    double tmp=0.0;
    for(bidx=row;bidx<n;bidx+=gridDim.x)
    {
        for (p=tid;p<n;p+=sharesize)
        {
            data[tid]=a[bidx * lda+p];
            for(j=tid; j<n; j+=blockDim.x)
            {
                tmp=c[bidx * ldc+j];
                for(i=0; i<sharesize; i++)
```

```
            {
                tmp+=data[i]*b[i*ldb+j];
            }
            c[bidx*ldc+j]=tmp;
        }
    }
}
matrix_multiplication_gpu_2_2<<<sharesize, sharesize>>>(d_a,n,d_b,n,d_c3,n,n);
```

编译执行上述代码,统计时间信息如表8.7所示。表8.7中数据显示,经本节代码修改后,共享存储优化起到了正效果,说明笔者对该问题的原因和解决方案的分析是正确的。本节验证了共享存储的一个使用原则,即**少量多次使用共享存储以减少硬件负载**。

表 8.7　A 一行共享存储改进效果

n	gpu_1_0	gpu_2	gpu_2_2
512	4.34	4.21	3.89
1024	26.80	26.28	24.33
2048	190.70	192.54	177.25
4096	1432.82	1846.78	1341.51

从某种意义上说,本节研究是多余的,因为棋盘阵列矩阵乘法中 2D 共享存储的优化效果远超 A 矩阵 1 行的共享存储优化效果。但从另一种角度看又是非常有意义的,本节给出一个正确的共享存储使用原则,同时为读者提供一条面临性能瓶颈的解决思路。

8.10　知识点总结

串行程序必须最优化,并行程序加速比才有意义。本书的串行"最优化"主要指连续访存、向量化、release 模式(优化开关 O2 或 O3),而不涉及深入 CPU 体系结构的优化。Xeon E5-2670 CPU 的最优串行浮点计算能力约为 4 ± 0.5 Gflops。

-vec-report<t>:向量化报告开关,其中,t 的值域为 0~6,指示向量化报告从无到详细的详细程度。

数据连续访存:数据连续访存是最基本、最重要的优化方法之一,是合并访存和数据预取的前提,无论是 CPU 还是 GPU。若数据访存不连续,可选的优化策略包括循环交换、矩阵转置、数据结构重构等。

O3 优化开关可能导致结果出错,使用时需要注意,最好验证下使用前后的结果。

GPU 线程循环方案:GPU 的线程循环有多种方案,本章涉及 grid 内线程循环和 block 内线程循环。

1. grid 内线程循环

以整个 grid 为单位步长跳步，线程索引 idx 和跳步 nn 计算公式为

```
int idx=bidx*blockDim.x+tidx;
const int nn=blockIdx.x*blockDim.x+threadIdx.x;
```

2. block 内线程循环

以 block 为单位步长跳步，线程索引 tidx 和跳步 nn 计算公式为

```
int tidx=threadIdx.x;
const int nn=blockDim.x;
```

关于这两种线程循环方案，从本章实验数据可知，**block 内线程循环方法性能较好**，特别是针对二维空间循环问题，详细分析请查阅前文相关章节。

数据对齐有助于提高不规整矩阵的运算性能。数据对齐方法及其效果见 8.4 节。

2D 线程维度在某些应用中能更好地复用共享存储，从而提升程序性能。

棋盘阵列矩阵乘法：详见 8.5 节。

移除判断是 GPU 常用且有效的优化方法之一。

CUBLAS 库的两种用法：**cublas_v1** 和 **cublas_v2**，详见 8.7 节。

CUBLAS 库使用时的注意事项：涉及矩阵运算时，库函数采用的是 FORTRAN 标准，即按列存储，在使用 C 语言调用库函数时需要特别注意。

CUBLAS 库的矩阵乘法运算性能极高。在处理 8192×8192 矩阵乘法运算时，浮点计算性能可达 2.6Tflops，几乎达到了 GPU 的峰值运算性能。

两种提升计算精度方法如下。

（1）高精度数据类型（double）中间变量，详见 7.1 节。

（2）Kahan's Summation Formula 方法，详见 8.8 节。

对比 MKL 库和 CUBLAS 库函数精度，**MKL 库函数精度较高，而 CUBLAS 库函数精度较低**。

GPU 并行矩阵乘法对比 CPU 矩阵乘有绝对性能优势，无论是 CPU 串行还是并行。这是本书第 1 个真正意义上取得正加速效果的实例（第 6 章和第 7 章，向量加法和向量内积运算的计算量和数据量相当，仅传输时间就超过了 CPU 串行计算时间）。

shared memory 不能盲目使用，否则不仅无法加速，还会导致性能下降。

一种 **shared memory 使用原则是少量多次循环使用**，以减少 shared memory 硬件资源负载，增加活跃 warp 数量，最大化 occupancy，详见 8.9 节。

一种 **shared memory 的使用原则是尽量多的数据复用**，共享存储中的数据无法凭空生产，一般来自全局存储，因此只有在多次数据复用的情况下使用共享存储才能获得收益。复用次数越多，性能收益越大。

一种 **shared memory 的设计原则是尽量减少 global memory 访问**。

冗余计算不会影响 GPU 的性能：在 8.8.2 节分析时发现，增加同类型数据计算量而访存不变情况下 GPU 性能基本不变。当然前提是计算总量未超出峰值浮点运算性能。

参与计算的数据类型影响 GPU 性能：8.8.2 节对比了中间变量数据类型分别是 double 和 float 时程序的性能，显然 float 类型运算更快，说明参与运算的数据类型直接影响程序性能。

8.11 扩展练习

（1）本章最终矩阵乘程序与 CUBLAS 库性能差异巨大，有兴趣的读者可以进一步研究矩阵乘法优化方法，争取接近 CUBLAS 库函数的水平。

（2）实现其他矩阵运算的 GPU 移植（可从 BLAS 库 3-level 中挑选）。

（3）实现矩阵长、高不等的矩阵乘法，特别是 CUBLAS 库的调用。

第9章 矩阵转置

9.1 矩阵转置及其串行代码

矩阵转置应用广泛,是最常见的运算之一。比如,8.1节改进的转置矩阵乘法就包含矩阵转置运算,通过矩阵转置实现矩阵乘法中矩阵数据的连续访存,令矩阵乘法运算向量化。

矩阵转置的数学表达:对于矩阵 $\boldsymbol{X}$,其转置矩阵 $\boldsymbol{X}^\mathrm{T}$,有

$$\boldsymbol{X} = \begin{bmatrix} X_{11} & X_{12} & \cdots & X_{1n} \\ X_{21} & X_{22} & \cdots & X_{2n} \\ \vdots & \vdots & \ddots & \vdots \\ X_{n1} & X_{n2} & \cdots & X_{nn} \end{bmatrix}, \quad \boldsymbol{X}^\mathrm{T} = \begin{bmatrix} X_{11} & X_{21} & \cdots & X_{n1} \\ X_{12} & X_{22} & \cdots & X_{n2} \\ \vdots & \vdots & \ddots & \vdots \\ X_{1n} & X_{2n} & \cdots & X_{nn} \end{bmatrix}$$

矩阵转置非常简单,但在编程实现时有两种版本,分别是读连续和写连续。下面分别阐述这两种矩阵转置运算。

读连续矩阵转置:$\boldsymbol{A}$ 矩阵读取连续(按行读),而 $\boldsymbol{C}$ 矩阵写入不连续(按列写)。下面展示了相应的 C 语言矩阵转置代码(标记为 serial_1)。

```
void matrix_transposition_serial_1(DATATYPE *a,DATATYPE *c,int m,int n)//a:m行n列,c:n行m列
{
    int i,j;
    for(i=0;i<m;i++)
    {
        for(j=0;j<n;j++)
        {
            c[j*m+i]=a[i*n+j];
        }
    }
}
matrix_transposition_serial_1(a,c1,n,n);
```

写连续矩阵转置:$\boldsymbol{A}$ 矩阵读取不连续,$\boldsymbol{C}$ 矩阵写入连续。下面是写连续矩阵转置的 C 语言代码(标记为 serial_2)。

```
void matrix_transposition_serial_2(DATATYPE *a,DATATYPE *c,int m,int n)//a:m
行 n 列,c:n 行 m 列
{
    int i,j;
    for(i=0;i<n;i++)
    {
        for(j=0;j<m;j++)
        {
            c[i*n+j]=a[j*m+i];
        }
    }
}
matrix_transposition_serial_2(a,c2,n,n);
```

9.2　1D 矩阵转置

9.1 节在 CPU 上实现了矩阵转置运算,本节的目标是将矩阵转置过程移植到 GPU 上。本节采用 block 内循环的方法来实现矩阵转置的 GPU 映射,即在 thread 维度,1 个 block 转置矩阵 1 行(或 1 列),由于线程数量限制,采用固定跳步的方式循环处理,block 维度亦是如此(参考 5.6 节的 1D 线程维度)。与 9.1 节类似,笔者分别设计了读连续和写连续两种矩阵转置实现方式。

GPU 上读连续 1D 矩阵转置:保证 **A** 矩阵的读取是对齐可合并的,但不保证 **C** 矩阵存储模式。下面展示了读连续 1D 矩阵转置 CUDA C 代码(标记为 gpu_1_0)。

```
__global__
void matrix_transposition_gpu_1d_1(DATATYPE *a,DATATYPE *c,int m,int n)//a:m
行 n 列,c:n 行 m 列
{
    const int tidx=threadIdx.x;
    const int bidx=blockIdx.x;
    int tid=tidx;
    int bid=bidx;
    while(bid<m)
    {
        while(tid<n)
        {
            c[tid*m+bid]=a[bid*n+tid];
            tid+=blockDim.x;
        }
        bid+=gridDim.x;
    }
}
matrix_transposition_gpu_1d_1<<<512,512>>>(d_a,d_c1,n,n);
```

GPU 上写连续 1D 矩阵转置:保证 **C** 矩阵的对齐合并存储,而不保证 **A** 矩阵的访问模式。下面给出了写连续 1D 矩阵转置 CUDA C 代码(标记为 gpu_1_1)。

```
__global__
void matrix_transposition_gpu_1d_2(DATATYPE *a,DATATYPE *c,int m,int n)//a:m
行 n 列,c:n 行 m 列
{
    const int tidx=threadIdx.x;
    const int bidx=blockIdx.x;
    int tid=tidx;
    int bid=bidx;
    while(bid<m)
    {
        while(tid<n)
        {
            c[bid*n+tid]=a[tid*m+bid];
            tid+=blockDim.x;
        }
        bid+=gridDim.x;
    }
}
matrix_transposition_gpu_1d_2<<<512,512>>>(d_a,d_c2,n,n);
```

9.3 2D 矩阵转置

9.2 节使用了 1D 线程维度做矩阵转置,本节讨论在 GPU 上实现 2D 矩阵转置。2D 线程维度概念请查阅 5.6 节。2D 矩阵转置的实现并不复杂,线程索引定位数据单元,然后复制赋值即可,下面给出相应的 2D 矩阵转置的 CUDA C 代码(标记为 gpu_2_0)。

```
__global__
void matrix_transposition_gpu_2d_1(DATATYPE *a,DATATYPE *c,int m,int n)//a:m
行 n 列,c:n 行 m 列
{
    const unsigned int xIndex=blockDim.x*blockIdx.x+threadIdx.x;
    const unsigned int yIndex=blockDim.y*blockIdx.y+threadIdx.y;

    if((xIndex<n)&&(yIndex<m))
    {
        unsigned int index_in=xIndex+n*yIndex;
        unsigned int index_out=yIndex+m*xIndex;
        c[index_out]=a[index_in];
```

```
    }
}
    dim3 threads(BLOCK_DIM,BLOCK_DIM,1);
    dim3 blocks((n+BLOCK_DIM-1)/BLOCK_DIM,(n+BLOCK_DIM-1)/BLOCK_DIM,1);
    matrix_transposition_gpu_2d_1<<<blocks,threads>>>(d_a,d_c1,n,n);
```

9.4 共享存储 2D 矩阵转置

9.3 节代码中，相邻线程的 **C** 矩阵写操作是不连续的（即不对齐写，或不可写合并），而这种不对齐、不可合并的写操作将导致较大的访存开销。能否将这种不对齐、不可合并访存操作转移到访存更快的存储单元，以达到减少开销的目的呢？首先考虑共享存储，因为只有它才能同时兼备可读、可写和线程间共享功能。

图 9.1 展示了共享存储 2D 矩阵转置过程。图中小方格代表 block，也代表了矩阵子块。在共享存储 2D 矩阵转置过程中，首先 block 内线程将矩阵子块的数据读取到共享存储中，在共享存储完成转置，然后将转置后的数据写入相应的转置结果矩阵子块。

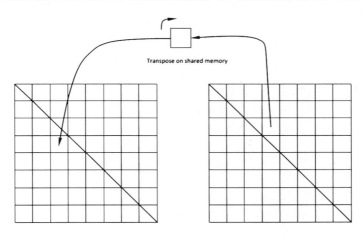

图 9.1 共享存储优化的 2D 矩阵转置

下面给出共享存储 2D 矩阵转置的 CUDA C 代码（标记为 gpu_2_1）。首先建立输入矩阵 **A** 的索引，以对齐合并访问的方式读取 **A** 矩阵；共享存储写后读之间需要同步；同步后建立输出矩阵 **C** 的索引，转置共享存储中的矩阵并输出到 **C** 矩阵。

```
#define BLOCK_DIM 16
__global__
void matrix_transposition_gpu_2d_2(DATATYPE *a,DATATYPE *c,int m,int n)//a:m
行 n 列,c:n 行 m 列
{
```

```
        __shared__ DATATYPE tmp[BLOCK_DIM][BLOCK_DIM];
        unsigned int xIndex=blockDim.x*blockIdx.x+threadIdx.x;
        unsigned int yIndex=blockDim.y*blockIdx.y+threadIdx.y;

        if((xIndex<n)&&(yIndex<m))
        {
            unsigned int index_in=xIndex+n*yIndex;
            tmp[threadIdx.y][threadIdx.x]=a[index_in];
        }
        __syncthreads();
        xIndex=blockIdx.y*BLOCK_DIM+threadIdx.x;
        yIndex=blockIdx.x*BLOCK_DIM+threadIdx.y;
        if((xIndex<m)&&(yIndex<n))
        {
            unsigned int index_out=yIndex*m+xIndex;
            c[index_out]=tmp[threadIdx.x][threadIdx.y];
        }
    }
matrix_transposition_gpu_2d_2<<<blocks,threads>>>(d_a,d_c2,n,n);
```

上述代码中,存储于 shared memory 中的中间矩阵 tmp 读取数据是按列访问的,存在 bank conflict(详见 16.4.2 节)。bank conflict 的解决方法之一是在定义矩阵时增加 1 列,这样矩阵中同一列的数据就错开位于不同的 bank 中,从而避免 bank conflict。CUDA C 代码的修改部分如下(标记为 gpu_2_2)。

```
    ...
        __shared__ DATATYPE tmp[BLOCK_DIM][BLOCK_DIM+1];
    ...
matrix_transposition_gpu_2d_3<<<blocks,threads>>>(d_a,d_c3,n,n);
```

9.5 共享存储 2D 矩阵转置 diagonal 优化

矩阵转置的 diagonal 优化是 CUDA SDK 中的示例代码,其思想是解决矩阵转置过程中全局存储访问的分区冲突问题。CUDA SDK 中认为在读数据过程中 block 对全局存储的不同分区的访问是均衡的,而写数据的过程都发生在分区 0 中,是不均衡的。因此做了移位,即通过将 block 的 x 和 y 索引相加,实现输入数据块的增量(可参考《GPU 高性能运算之 CUDA》[2] 第 187 页和 CUDA Toolkit Samples 中 transpose 实例的 MatrixTranspose.pdf 文件)。下面展示了共享存储 2D 矩阵转置 diagonal 优化的 CUDA C 代码(标记为 gpu_3)。

```
__global__
void matrix_transposition_gpu_diagonal(DATATYPE *a,DATATYPE *c,int m,int n)
//a:m行n列,c:n*m
{
    __shared__ float tile[BLOCK_DIM][BLOCK_DIM+1];
    int blockIdx_x, blockIdx_y;
    // do diagonal reordering
    if(n==m) {
        blockIdx_y=blockIdx.x;
        blockIdx_x=(blockIdx.x+blockIdx.y)%gridDim.x;
    }else {
        int bid=blockIdx.x+gridDim.x*blockIdx.y;
        blockIdx_y=bid%gridDim.y;
        blockIdx_x=((bid/gridDim.y)+blockIdx_y)%gridDim.x;
    }
    int xIndex=blockIdx_x * BLOCK_DIM+threadIdx.x;
    int yIndex=blockIdx_y * BLOCK_DIM+threadIdx.y;
    int index_in=xIndex+ (yIndex) * n;
    xIndex=blockIdx_y * BLOCK_DIM+threadIdx.x;
    yIndex=blockIdx_x * BLOCK_DIM+threadIdx.y;
    int index_out=xIndex+ (yIndex) * m;
    tile[threadIdx.y][threadIdx.x]=a[index_in];
    __syncthreads();
    c[index_out]=tile[threadIdx.x][threadIdx.y];
}
matrix_transposition_gpu_diagonal<<<blocks,threads>>>(d_a,d_c1,n,n);
```

但笔者认为上述分析存在缺陷,代码中所有线程同时对所有的矩阵元素进行访存(读取和输出),而且对共享存储的读取和存储过程都已对齐,对全局存储的访问可合并,因此笔者认为上述全局存储访问的分区冲突问题是不存在的,没有必要进行 diagonal 优化。

9.6 节的实验结果显示,矩阵转置在 diagonal 优化前比 diagonal 优化后执行更快,说明笔者的分析是正确的。

9.6 实验结果分析与结论

编译上述代码,对 $n \times n$ 的矩阵进行转置变换,测试并统计执行时间,如表 9.1 所示。表 9.1 中统计了各版本代码的转置执行时间和 CPU/GPU 通信时间(包括输入和输出)。再依据表 9.1 的数据,计算访存性能(见表 9.2)。对于 gpu_2_1、gpu_2_2 和 gpu_3 这 3 个版本,BLOCK_DIM 值的设定将直接影响程序执行性能,有兴趣的读者可进行实践探索,本书(本书结果已取最优值)不再展开论述。

表 9.1　矩阵转置时间对比　　　　　　　　　　　　　　　单位：ms

n	serial_1	serial_2	gpu_1_0	gpu_1_1	gpu_2_0	gpu_2_1	gpu_2_2	gpu_3	memcpy
256	0.129	0.126	0.065	0.071	0.062	0.059	0.058	0.060	0.220
512	0.638	0.600	0.217	0.244	0.203	0.197	0.197	0.199	1.118
1024	2.006	2.105	0.252	0.307	0.273	0.254	0.254	0.258	3.048
2048	12.517	12.745	0.337	0.435	0.554	0.473	0.474	0.483	12.630
4096	106.007	106.227	1.210	0.701	1.645	1.347	1.348	1.401	50.978
8192	472.448	472.812	3.380	2.009	6.332	5.757	5.860	6.275	203.989

表 9.2　矩阵转置访存性能对比　　　　　　　　　　　　　单位：GB/s

n	serial_1	serial_2	gpu_1_0	gpu_1_1	gpu_2_0	gpu_2_1	gpu_2_2	gpu_3	memcpy
256	1.94	1.99	3.84	3.52	4.03	4.25	4.32	4.16	1.14
512	1.57	1.67	4.61	4.10	4.92	5.07	5.08	5.02	0.89
1024	1.99	1.90	15.87	13.03	14.65	15.75	15.74	15.51	1.31
2048	1.28	1.26	47.46	36.79	28.88	33.83	33.76	33.12	1.27
4096	0.60	0.60	52.89	91.30	38.90	47.51	47.48	45.68	1.26
8192	0.54	0.54	75.74	127.43	40.43	44.47	43.69	40.80	1.25

通过分析表 9.1 和表 9.2 中的数据,可以得出以下结论。

(1) 比较 serial_1 和 serial_2 两列数据,读连续和写连续的转置时间基本相等,得出结论是单核 CPU 对内存的读写性能相当。

(2) 纵向分析表 9.2 中 serial_1 列,矩阵越大性能越差,原因可能是矩阵增大后 cache 命中率下降。

(3) 总体而言,GPU 上转置比 CPU 上转置快得多,特别是矩阵越大优势越明显。

(4) 矩阵规模较小时(256,512),2D 矩阵转置比 1D 矩阵转置速度快。

(5) 对比 gpu_2_0 和 gpu_2_1,共享存储优化后程序性能获得了提升。

(6) 对比 gpu_2_1 和 gpu_2_2,矩阵较小时,避免 bank conflict 优化后有细微性能提升;矩阵增大后,bank conflict 对性能的影响几乎可以忽略。

(7) 比较 gpu_2_2 和 gpu_3 两列,diagonal 优化后性能不升反降,这一结果验证了 9.5 节的分析,即共享存储 2D 矩阵转置中不存在全局存储分区冲突,不需要 diagonal 优化。

(8) 当矩阵规模增大时(1024 以上),1D 矩阵转置比 2D 矩阵转置速度快。当规模为 1024、2048 时,读连续性能较好;而数据规模增大到 4096 以上时,写连续性能更优,且性能提升明显。

9.7　共享存储 2D 矩阵转置的深入优化

9.6 节的实验结果显示,利用了共享存储的 2D 矩阵转置的 gpu_2_2 版本在矩阵小时比 1D 版本快,但矩阵增大后性能反而不如 1D 矩阵转置。

可能原因之一是共享存储使用不当。当矩阵增大时,9.3 节和 9.4 节中的 block 启动数量增加，需要的总共享存储空间也增加，而硬件中共享存储空间是有限的。为了验证这种可能性并试图解决该问题，可对代码进行如下修改：申请固定 block 数量（通过测试获得最佳 block 数量，实验结果显示矩阵转置中 32×32 的 block 数量性能最佳），通过循环方式完成矩阵所有元素的转置。下面是修改后的共享存储 2D 矩阵转置 CUDA C 代码（标记为 gpu_2_2_1）。

```
#define BLOCK_DIM 32
__global__
void matrix_transposition_gpu_2d_3_1(DATATYPE *a,DATATYPE *c,int m,int n)
//a:m 行 n 列,c:n 行 m 列
{
    __shared__ DATATYPE tmp[BLOCK_DIM][BLOCK_DIM+1];
    unsigned int xIndex;
    unsigned int yIndex;
    unsigned int bidx,bidy;
    unsigned int index_in,index_out;
    const int mm= (m+BLOCK_DIM-1)/BLOCK_DIM;
    const int nn= (n+BLOCK_DIM-1)/BLOCK_DIM;
    for(bidy=blockIdx.y;bidy<mm;bidy+=gridDim.y)
    {
        for(bidx=blockIdx.x;bidx<nn;bidx+=gridDim.x)
        {
            xIndex=blockDim.x * bidx+threadIdx.x;
            yIndex=blockDim.y * bidy+threadIdx.y;
            if((xIndex<n)&&(yIndex<m))
            {
                index_in=xIndex+n * yIndex;
                tmp[threadIdx.y][threadIdx.x]=a[index_in];
            }
            __syncthreads();
            xIndex=bidy * BLOCK_DIM+threadIdx.x;
            yIndex=bidx * BLOCK_DIM+threadIdx.y;
            if((xIndex<m)&&(yIndex<n))
            {
                index_out=yIndex * m+xIndex;
                c[index_out]=tmp[threadIdx.x][threadIdx.y];
            }
        }
    }
}
dim3 blocks1(32,32,1);
matrix_transposition_gpu_2d_3_1<<<blocks1,threads>>>(d_a,d_c2,n,n);
```

测试上述代码的执行时间,与 gpu_2_2 版本对比(见表9.3)。表9.3中数据显示,深入优化后的共享存储 2D 矩阵转置,在原基础上进一步提升了性能。该实验结果也进一步验证了第 8 章矩阵乘法中得出 shared memory 使用原则的结论,即少量多次循环使用共享存储。

表 9.3 共享存储继续优化效果 单位:ms

n	gpu_2_2	gpu_2_2_1	n	gpu_2_2	gpu_2_2_1
256	0.059	0.056	2048	0.535	0.519
512	0.202	0.190	4096	1.593	1.275
1024	0.269	0.203	8192	5.901	4.368

9.8 知识点总结

GPU 线程维度(grid-block-thread):kernel 函数启动时配置的线程(块)维度,一般有 1D 线程维度和 2D 线程维度等,3D 线程维度比较少见。5.6 节阐述了 1D 线程维度和 2D 线程维度。

1D 固定线程维度的双向循环遍历:由于共享存储需求或启动的线程网格尺寸随问题规模扩大而增大,一定程度后往往超出 GPU 极限或导致性能下降,这就需要在 kernel 函数内以循环遍历的方式来减少启动的线程(块)数量,此时启动的线程网格是固定的。1D 线程维度情况下 block 和 thread 双向 while 循环遍历方式如下。同理读者可补充 for 循环遍历方式和 2D 线程网格的循环遍历方式。

```
int tid=threadIdx.x;
int bid=blockIdx.x;
while(bid<m)
{
    while(tid<n)
    {
        ⋮
        tid+=blockDim.x;
    }
    bid+=gridDim.x;
}
```

bank conflict:共享存储包含 32 个 bank(Tesla 架构是 16 个 bank,与 warp 内同时执行的 thread 数量相等),不同 bank 上的数据可以同时访问,相同 bank 上不同数据则只能串行访问,当 warp 内多个 thread 同时访问同一 bank 上的不同数据时就会导致 bank conflict,影响程序性能。

一种避免 bank conflict 的方法:对于 2D 共享存储数组,增加 1 列,即可使相同列的元素错开布置在不同的 bank,避免 bank conflict。

矩阵转置（共享存储优化后）不存在分区冲突，因此 diagonal 优化后性能没有提升。

CPU 读连续和写连续矩阵转置时间基本相等，说明 CPU 的读写性能相当。

矩阵增大时，cache 命中率下降，CPU 串行矩阵转置性能下降。

GPU 矩阵转置明显比 CPU 快得多，说明 GPU 访存带宽是 GPU 的一大优势。

矩阵转置避免 bank conflict 后性能提升较少，也说明 Kepler 架构中 bank conflict 对性能的影响在某种程度上可以忽略。

共享存储的关键优化原则之一：少量多次循环使用。

9.9 扩展练习

（1）阅读文献、NVIDIA 官方资料，进一步研究矩阵转置的性能优化。

（2）矩阵继续增大时，性能会否发生其他变化？可否进一步挖掘性能？

第三篇　提　高　篇

> 第 10 章　卷积
> 第 11 章　曼德博罗特集
> 第 12 章　扫描：前缀求和
> 第 13 章　排序
> 第 14 章　几种简单图像处理

　　本篇仍然以循序渐进的实例优化研究为主线，选取了并行计算中的一些经典实例作为研究对象，比如卷积、曼德博罗特集、前缀求和、排序、图像直方图和图像滤波等。

　　第 10 章在 GPU 上分别实现了 1D 和 2D 卷积(分别进行了判断法和扩展法的边界处理)，并基于这两种卷积运算展开性能优化研究，如常量存储优化、共享存储优化等。在 2D 卷积的共享存储优化中，由于线程块内的线程和需要加载的数据不完全匹配，需要关注共享存储的数据加载过程。

　　曼德博罗特集通过迭代运算作图获得绚丽的图像，这类计算密集型应用在 GPU 上性能尤佳。笔者在将该过程移植到 GPU 上的基础上，进行了一系列的优化尝试和效果分析，特别是实现了 4 种不同零拷贝版本的代码，从结果分析中总结出了真实有效的零拷贝使用技巧(目前在其他文档或书籍中均未涉及)。该章节还进行了计算和通信重叠的研究，阐述了突破 kernel 函数执行时间限制的方法。

　　GPU 上前缀求和的运算是一类特殊的并行问题，第 12 章分别利用 Kogge-Stone 和 Brent-Kung 两种并行方法求解前缀和，并实现两层的扇入、扇出的完整前缀求和。在使用 Kogge-Stone 方法并行时发现"同步也无法保证共享存储的正确读写"的问题，提出了交错读写的解决方案。Brent-Kung

方法进行了两倍数据优化和 bank conflict 避免等优化研究。接着以 Kogge-Stone 为例进行了 warp 级分段的优化研究。

第 13 章首先分析了几种经典串行排序算法的性能,选择性能较好的 4 种排序(基数排序、双调排序网络、快速排序和合并排序)开展 GPU 并行移植和优化研究。对于基数排序,探讨了几种不同合并方法的性能,利用 Thrust 库函数实现了完整的基数排序。

最后对几种简单的图像处理算法进行了 GPU 移植和优化,包括图像直方图统计、中值滤波和均值滤波。涉及原子操作、共享存储优化、最大化共享存储复用次数、扩展法优化共享存储数据加载等优化策略。对 2D 共享存储数据加载时遇到的特殊问题,针对性地提出并实现了两种解决方案。

第 10 章 卷 积

10.1 卷积及其串行实现

卷积是一种常见的数组运算,在信号处理、图像处理、视频处理、计算机视觉和数字录音等领域有广泛应用。

数学上的卷积是一种数组运算,每个输出元素等于其周围输入元素的加权总和。加权计算中的权重由掩码数组 M 决定,称为卷积核或卷积掩码。一个数组中所有元素的卷积掩码一般是相同的。

常见的卷积运算有一维卷积和二维卷积两种,下面分别阐述这两种卷积运算。

10.1.1 一维卷积

一维卷积中,若要计算卷积数组 P 的第 i 个元素值,分别取掩码数组 M 的全部元素(设 M 数组长为 L)和 N 数组中以第 i 个元素为中心的 L 个元素(在后文代码中用 m 变量表示 L),两两相乘并累加得到目标值。其计算公式如下:

$$P[i] = \sum_{j=0}^{L} N[i+j-L/2]M[j]$$

图 10.1 是 1D 卷积的示意图。图示运算为 $P[2]=N[0]\times M[0]+N[1]\times M[1]+N[2]\times M[2]+N[3]\times M[3]+N[4]\times M[4]$。

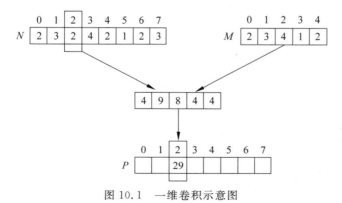

图 10.1 一维卷积示意图

一维卷积边界处理：当需要求解的 P 数组中元素位于数组的前 $L/2$ 和后 $L/2$ 时，以该索引为中心取 N 数组的 L 个值必定越界，此时就涉及边界处理。1D 卷积的边界处理方法有判断法和扩展法两种。

判断法通过判定当前索引是否发生越界，若未越界即正常计算，若越界则进行边界计算。下面是判断法实现的一维卷积代码，其中有明确的边界判断和边界计算。

```
void convolution_cpu1(float *array_m,float *array_n,float *array_p,int n)
{
    int i,j,k,l,kk,ll;
    int m_2=m/2;
    double temp;
    for(j=0;j<n;j++)
    {
        temp=0.0;
        if(j<m_2)
        {
            for(l=m_2-j;l<m;l++)
            {
                temp+=array_m[l] * array_n[j-m_2+l];
            }
        }
        else if(j>=(n-m_2))
        {
            for(l=0;l<((n-j)+m_2);l++)
            {
                temp+=array_m[l] * array_n[j-m_2+l];
            }
        }
        else
        {
            for(l=0;l<m;l++)
            {
                temp+=array_m[l] * array_n[j-m_2+l];
            }
        }
        array_p[j]=temp;
    }
}
    convolution_cpu1(array_m,array_n,array_p1,n);
```

扩展法是通过扩展边界来进行边界处理，即在数组前后均增加 $L/2$ 个元素并赋值为 0，此时卷积计算无须考虑边界条件，但要注意正确计算索引。下面是扩展法实现的一维卷积计算代码。

```
void convolution_cpu2(float *array_m,float *array_n1,float *array_p,int n,int
n1)
{
    int i,j,k,l,kk,ll;
    int m_2=m/2;
    double temp;
    for(j=0;j<n;j++)
    {
        temp=0.0;
        for(l=0;l<m;l++)
        {
            ll=j+l;
            temp+=array_m[l] * array_n1[ll];
        }
        array_p[j]=temp;
    }
}
```

```
    int n1=n+(m-1);
    for(j=0;j<n1;j++)
    {
        if((j<m_2)||(j>=(n+m_2)))
        {
            array_n1[j]=0.0;
        }
        else
        {
            array_n1[j]=array_n[(j-m_2)];
        }
    }
    convolution_cpu2(array_m,array_n1,array_p2,n,n1);
```

10.1.2 二维卷积

二维卷积计算与一维卷积类似,不同的是将一维卷积运算扩展到二维空间。2D 卷积中,要计算卷积 $P[m][n]$,需要的数据包括掩码矩阵 $M[][]$ 和 N 矩阵中以 $[m][n]$ 为中心与 M 矩阵大小一致的二维子矩阵,计算卷积时 M 和 N 矩阵对应的元素相乘并累加,即可算出目标卷积结果。相应的 2D 卷积计算公式为

$$P[m][n] = \sum_{i=0}^{L}\sum_{j=0}^{L} N[m+i-L/2][n+j-L/2]M[i][j]$$

图 10.2 展示了一个 2D 卷积实例,图中计算 $P[2,2]$ 过程如下:

$$P[2,2] = N[1,1] \times M[1,1] + N[1,2] \times M[1,2] + N[1,3] \times M[1,3]$$
$$+ N[2,1] \times M[2,1] + N[2,2] \times M[2,2] + N[2,3] \times M[2,3]$$

$$+ N[3,1] \times M[3,1] + N[3,2] \times M[3,2] + N[3,3] \times M[3,3]$$

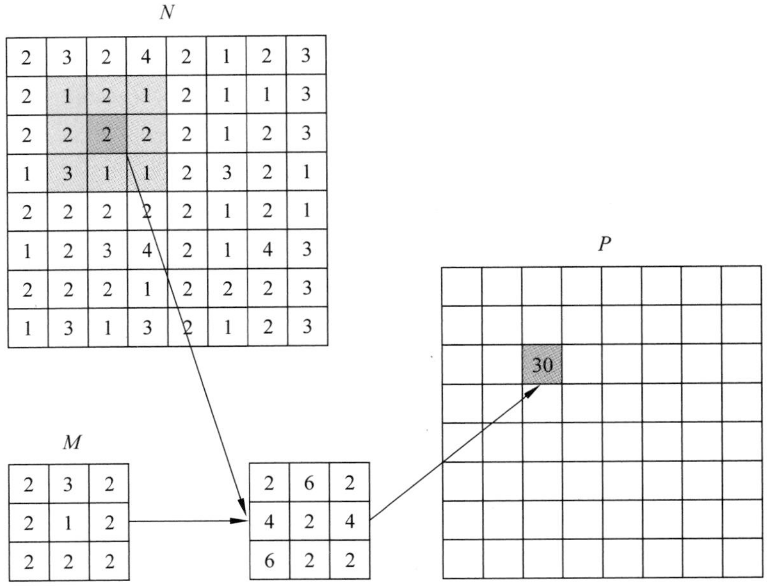

图 10.2 二维卷积示意图

二维卷积也会面临边界问题,与 1D 卷积类似,二维边界处理也有判断法和扩展法两种边界处理方法。

判断法处理二维卷积边界时,通过边界判定来辨别边界和非边界,非边界的元素参与正常计算,边界元素计算采用特定的方法,下面给出了使用判断法的二维卷积 C 代码,其中包含边界元素判断和处理方法。

```
void convolution_2d_cpu1(float *array_m,float *array_n,float *array_p,int n)
{
    int i,j,k,l,kk,ll;
    int m_2=m/2;
    double temp;
    int l_0,l_1,l_2,l_3;
    for(i=0;i<n;i++)
    {
        if(i<m_2)
        {
            l_0=m_2-i;l_1=m;
        }
        else if(i>=(n-m_2))
        {
            l_0=0;l_1=(n-i)+m_2;
        }
```

```
            else
            {
                l_0=0;l_1=m;
            }
            for(j=0;j<n;j++)
            {
                if(j<m_2)
                {
                    l_2=m_2-j;l_3=m;
                }
                else if(j>=(n-m_2))
                {
                    l_2=0;l_3=(n-j)+m_2;
                }
                else
                {
                    l_2=0;l_3=m;
                }
                temp=0.0;
                for(k=l_0;k<l_1;k++)
                {
                    for(l=l_2;l<l_3;l++)
                    {
                        temp+=array_m[k*m+l] * array_n[(i-m_2+k) * n+(j-m_2+l)];
                    }
                }
                array_p[i*n+j]=temp;
            }
        }
    }
    convolution_2d_cpu1(array_m,array_n,array_p1,n);
```

扩展法计算二维卷积时,矩阵行列的前后均扩展 $L/2$,并赋值为 0。下面是扩展法实现的二维卷积计算 C 代码,其中需要特别注意索引的计算。

```
void convolution_2d_cpu2(float *array_m,float *array_n1,float *array_p,int n,
int n1)
{
    int i,j,k,l,kk,ll;
    int m_2=m/2;
    double temp;
    for(i=0;i<n;i++)
    {
```

```
        for(j=0;j<n;j++)
        {
            temp=0.0;
            for(k=0;k<m;k++)
            {
                for(l=0;l<m;l++)
                {
                    kk=i+k;
                    ll=j+l;
                    temp+=array_m[k*m+l] * array_n1[kk*n1+ll];
                }
            }
            array_p[i*n+j]=temp;
        }
    }
}
    for(i=0;i<n1;i++)
    {
        for(j=0;j<n1;j++)
        {
            if((i<m_2)||(i>=(n+m_2))||(j<m_2)||(j>=(n+m_2)))
            {
                array_n1[i*n1+j]=0.0;
            }
            else
            {
                array_n1[i*n1+j]=array_n[(i-m_2)*n+(j-m_2)];
            }
        }
    }
    convolution_2d_cpu2(array_m,array_n1,array_p2,n,n1);
```

10.2　GPU 上 1D 卷积

由 10.1 节的卷积运算和串行代码可知,每个元素的卷积计算都是独立的,并不依赖于周围或其他元素的卷积计算,故卷积计算是一种理想的并行问题。在参考文献[4]的第 8 章介绍了判断法 1D 卷积的 GPU 移植和优化,本书主要阐述边界扩展方法 1D 卷积运算的 GPU 移植和优化,同时实现了判断法 1D 卷积 CUDA 代码(见附录 A)并通过实验对两类方法的卷积性能进行对比和分析。

扩展法 1D 卷积的 GPU 移植比较简单,只需将串行代码中 j 层循环拆开,利用 GPU 线程索引号代替 j 的索引,结合固定跳步循环的方法,理论上能够处理 n 无限大的数据规模(若 n 过大导致 GPU 显存不足则另当别论)。下面是扩展法 1D 卷积的 CUDA C 代码(标记为 gpu_1)。

```
__global__ void convolution_gpu1(float *array_m,float *array_n1,float *array_p,int n,int n1)
{
    int tidx=threadIdx.x;
    int bidx=blockIdx.x;
    int j,l;
    float temp=0.0;
    int ll;
    for(j=bidx*blockDim.x+tidx;j<n;j+=gridDim.x*blockDim.x)
    {
        temp=0.0;
        for(l=0;l<m;l++)
        {
            ll=j+l;
            temp+=array_m[l] * array_n1[ll];
        }
        array_p[j]=temp;
    }
}

#define m 25
#define block_width 512
    int grid_width=block_width;
    convolution_gpu1<<<grid_width,block_width>>>(d_array_m,d_array_n1,d_array_p1,n,n1);
```

10.3 M 常量 1D 卷积

分析卷积计算中的掩码数组 M,有以下特征:①卷积计算过程中,掩码数组 M 是不变的(只读性);②M 数组长度不大(数据量小);③所有线程都需要用到 M 数组,且 warp 内所有线程读取 M 数组同一元素(广播性)。GPU 中常量存储和共享存储均适合存储少量共享数据,且都能提供广播功能(参见第 16 章)。合理利用这两种存储单元存储 M 数组可以加速 GPU 上 1D 卷积计算。本节和 10.4 节将分别探讨如何利用这两种存储单元加速 1D 卷积计算。

本节讨论利用常量存储单元存储掩码数组 M 来加速卷积计算。常量数组使用 __constant__ 限定符声明,声明后无法直接修改常量存储,可以通过特殊传输函数 cudaMemcpyToSymbol() 进行复制。下面展示了常量存储优化卷积掩码数组 M 后的 1D 卷积运算 CUDA C 代码(标记为 gpu_2)。

```
__constant__ float d_array_m1[m];
__global__ void convolution_gpu2(float *array_n1,float *array_p,int n,int n1)
{
    int tidx=threadIdx.x;
    int bidx=blockIdx.x;
    int j,l;
    float temp=0.0;
    int ll;
    for(j=bidx*blockDim.x+tidx;j<n;j+=gridDim.x*blockDim.x)
    {
        temp=0.0;
        for(l=0;l<m;l++)
        {
            ll=j+l;
            temp+=d_array_m1[l] * array_n1[ll];
        }
        array_p[j]=temp;
    }
}
cudaMemcpyToSymbol(d_array_m1,array_m,sizeof(float)*m);
convolution_gpu2<<<grid_width,block_width>>>(d_array_n1,d_array_p1,n,n1);
```

10.4 M 共享 1D 卷积

本节探讨如何利用共享存储来存储卷积掩码数组 M 来加速 GPU 上 1D 卷积运算。共享存储数组利用 __shared__ 限定符声明,将 M 数组赋值给共享存储后需要同步操作 __syncthreads() 来保证共享存储数据的正确访问。下面是共享存储优化掩码数组 M 后的 1D 卷积 CUDA C 代码(标记为 gpu_3)。

```
__global__ void convolution_gpu3(float *array_m,float *array_n1,float *array_p,int n,int n1)
{
    __shared__ float array_m2[m];
    int tidx=threadIdx.x;
    if(tidx<(m))
    {
        array_m2[tidx]=array_m[tidx];
    }
    int bidx=blockIdx.x;
    int j,l;
    float temp=0.0;
    int ll;
```

```
    __syncthreads();
    for(j=bidx*blockDim.x+tidx;j<n;j+=gridDim.x*blockDim.x)
    {
        temp=0.0;
        for(l=0;l<m;l++)
        {
            ll=j+l;
            temp+=array_m2[l] * array_n1[ll];
        }
        array_p[j]=temp;
    }
}
convolution_gpu3<<<grid_width,block_width>>>(d_array_m,d_array_n1,d_array_p1,n,n1);
```

10.5 N 共享 1D 卷积

分析数组 N，其中任一元素被其自身和前后 $m/2$（$L/2$，为结合代码阐述并行和优化过程，使用代码中的变量）个元素的卷积运算使用，共使用 m 次（m 次复用）。数据复用情况一般采取共享存储进行优化。m 次全局存储访问开销较大，因此考虑利用访存延迟小的共享存储单元来替代全局存储单元。

1D 卷积计算中，x 个连续的卷积计算需要的数据包括 N 数组中这 x 个元素本身以及其前后的 $m/2$ 个元素。由于 m 值不会特别大，1 个 block 的线程数量是 block_width，block 中 block_width 个卷积计算需要的 N 数组元素个数为（block_width＋m－1），其尺寸远小于共享存储容量，因此可以利用共享存储优化 N 数组的访存。

由于最终加载到共享存储中的 N 数组元素个数比线程数量多 $m-1$ 个，因此需要分两次加载数据。下面是同时利用共享存储存放 N 数组和利用常量存储存放 M 数组优化后的 1D 卷积 CUDA C 代码（标记为 gpu_5）。

```
__constant__ float d_array_m1[m];
#define mm (block_width+m-1)
__global__ void convolution_gpu5(float *array_n1,float *array_p,int n,int n1)
{
    __shared__ float array_ns[mm];
    const int tidx=threadIdx.x;
    const int bidx=blockIdx.x;
    int j,l;

    int tidk;
    float temp=0.0;
```

```
    int ll;
    for(j=bidx*blockDim.x;j<n;j+=gridDim.x*blockDim.x)
    {
        array_ns[tidx]=array_n1[j+tidx];
        if(tidx<(m-1))
        {
            tidk=tidx+block_width;
            array_ns[tidk]=array_n1[tidk+j];
        }

        __syncthreads();
        temp=0.0;
        for(l=0;l<m;l++)
        {
            ll=tidx+l;
            temp+=d_array_m1[l] * array_ns[ll];
        }
        array_p[j+tidx]=temp;
    }
}
convolution_gpu5<<<grid_width,block_width>>>(d_array_n1,d_array_p1,n,n1);
```

此外,实现同时利用共享存储存放掩码数组 M 和 N 数组的 1D 卷积,下面是相应的 CUDA C 代码(标记为 gpu_6)。

```
__global__ void convolution_gpu6(float *array_m,float *array_n1,float *array_p,int n,int n1)
{
    __shared__ float array_m2[m];
    __shared__ float array_ns[mm];
    const int tidx=threadIdx.x;
    if(tidx<(m))
    {
        array_m2[tidx]=array_m[tidx];
    }
    __syncthreads();
    const int bidx=blockIdx.x;
    int j,l;
    j=bidx*blockDim.x;
    int tidk;
    float temp=0.0;
    int ll;
    for(j=bidx*blockDim.x;j<n;j+=gridDim.x*blockDim.x)
```

```
    {
        array_ns[tidx]=array_n1[j+tidx];
        if(tidx<(m-1))
        {
            tidk=tidx+block_width;
            array_ns[tidk]=array_n1[tidk+j];
        }

        __syncthreads();
        temp=0.0;
        for(l=0;l<m;l++)
        {
            ll=tidx+l;
            temp+=array_m2[l] * array_ns[ll];
        }
        array_p[j+tidx]=temp;
    }
}
convolution_gpu6<<<grid_width,block_width>>>(d_array_m,d_array_n1,d_array_p1,n,n1);
```

10.6 实验结果分析

10.6.1 扩展法 1D 卷积实验结果分析

编译并执行上述代码,设定 $m=25$,线程维度为 <<<512,512>>>,测试结果统计如表 10.1 所示,其中,GPU 版本的 1D 卷积计算均采用扩展法。

表 10.1 1D 卷积计算时间 单位:ms

n	判断法	扩展法	gpu_1	gpu_2	gpu_3	gpu_5	gpu_6
1 048 576	21.90	14.70	2.72	0.47	0.49	0.17	0.23
2 097 152	45.10	29.91	3.12	0.92	0.96	0.32	0.44
4 194 304	89.00	60.17	6.07	1.84	1.93	0.62	0.88
8 388 608	178.37	120.91	11.81	3.65	3.78	1.23	1.69
16 777 216	357.12	242.38	23.48	7.25	7.51	2.42	3.37
33 554 432	714.83	484.35	46.75	14.43	14.99	4.84	6.85
67 108 864	1429.93	969.18	93.26	28.88	29.91	9.61	13.37
134 217 728	2859.75	1938.42	185.71	57.67	59.70	19.26	26.70
268 435 456	5721.60	3878.27	375.53	115.32	122.11	38.47	53.39
536 870 912	11 440.01	7752.81	751.71	230.47	244.60	77.31	107.24

表 10.1 中数据显示：

（1）对比判断法和扩展法的两列 CPU 串行执行时间，扩展法比判断法的 1D 卷积运算速度快、性能好，原因是判断法分支处理比较耗时，应尽量避免。

（2）对比扩展法和 gpu_1 两列数据，1D 卷积运算从 CPU 端移植到 GPU 端可加速近 10 倍。

（3）对比 gpu_1、gpu_2 两列数据，gpu_2 的耗时为 gpu_1 的三分之一，说明常量存储优化掩码数组 M 能获得很好的性能提升。

（4）对比 gpu_1 和 gpu_3 两列，gpu_3 比 gpu_1 快很多，说明共享存储优化掩码数组 M 也能获得很好的加速效果。

（5）对比 gpu_2 和 gpu_5 两列数据或对比 gpu_3 和 gpu_6 两列数据，发现采用共享存储优化 N 数组后，性能均获得了大幅度提升，加速了两倍多。

（6）对比 gpu_2 和 gpu_3 两列数据或对比 gpu_5 和 gpu_6 两列数据，发现掩码数组放在常量存储中比在共享存储中加速效果更好。

（7）同时使用共享存储单元存储 N 数组和常量存储单元存放 M 数组的 gpu_5 性能最好。

（8）对比 gpu_5 和 gpu_1 两列数据，优化前后的 GPU 上 1D 卷积计算性能提升近 10 倍，说明优化是 GPU 编程中至关重要的一个环节。

（9）对比 gpu_5 和扩展法两列数据，优化后的 GPU 上 1D 卷积比 CPU 端快了近 100 倍。

10.6.2 判断法与扩展法 1D 卷积对比

笔者实现了参考文献[4]第 8 章的 1D 卷积 GPU 并行代码（见附录 A），测试并统计 kernel 函数执行时间（见表 10.2）。对照表 10.2 和表 10.1，basic（代码见附录 A.1）对应 gpu_1，constant（代码见附录 A.2）对应 gpu_2，shared（代码见附录 A.3）对应 gpu_5。cache 版本（代码见附录 A.4）是共享存储未扩展版本，即 shared 版本中共享存储除了加载线程本身对应的元素外，还需要前后 $(m-1)/2$ 个数据；而 cache 版本中仅仅加载对应数据而忽略前后数据，认为前后数据会被 cache 缓存。

表 10.2 判断法边界处理 1D 卷积计算时间　　　　　　单位：ms

n	basic	constant	shared	cache
1 048 576	1.48	0.54	0.18	0.50
2 097 152	2.96	1.06	0.36	0.98
4 194 304	5.91	2.10	0.70	1.95
8 388 608	11.81	4.19	1.38	3.87
16 777 216	23.60	8.36	2.76	7.70

对比表 10.1 和表 10.2 中相应的数据，显然扩展法边界处理的 1D 卷积比判断法边界处理的 1D 卷积性能要好。另外，参考文献[4]中提及的利用 L2 cache 优化方法的性能并不好，且从结果数据中亦无法判断 cache 是否起到正效果。

值得一提的是,当问题规模 n 大于或等于 33 554 432 时,判断法 1D 卷积程序无法执行,主要原因是数据规模过大,超出 grid 维度限制(即 block 数量不够),解决方法是在 kernel 函数中加入循环处理。而本书设计扩展法 1D 卷积融入了循环处理的思想,理论上能进行任意大小规模的卷积运算。

10.6.3 加速比分析

10.6.2 节通过实验对比了 CPU 和 GPU 上 1D 卷积计算时间,但未考虑 GPU 异构计算引入的通信开销等额外开销。本节对比完整 GPU 计算时间(设定线程维度为 <<<512,512>>>)和 CPU 计算时间。时间数据统计如表 10.3 所示,其中主机端内存为页锁定内存。表中数据显示,即便加上通信开销,GPU 上卷积运算比 CPU 串行卷积快了 10 倍多,若仅考虑计算时间则可加速 100 多倍。

表 10.3 1D 卷积执行时间对比 单位:ms

n	扩展法	仅计算	计算+通信
1 048 576	14.70	0.17	1.67
2 097 152	29.91	0.32	3.27
4 194 304	60.17	0.62	6.19
8 388 608	120.91	1.23	11.97
16 777 216	242.38	2.42	23.14
33 554 432	484.35	4.84	46.46
67 108 864	969.18	9.61	92.48
134 217 728	1938.42	19.26	184.39
268 435 456	3878.27	38.47	368.44
536 870 912	7752.81	77.31	736.92

10.6.4 线程维度对性能的影响

本节分析线程维度对 1D 卷积 kernel 函数执行性能的影响。设计一组实验,以性能最好的 gpu_5 代码为对象,令 $n=67\ 108\ 864, m=25$,控制线程块和线程数量在 256、512 和 1024 间组合变化。测试并统计执行时间,如表 10.4 所示,表中数据显示线程维度为 <<<512,256>>> 时 GPU 上 1D 卷积的性能最佳。但总体来看,在该实例中,线程维度配置对 kernel 函数执行性能的影响不大。

表 10.4 线程维度对性能的影响 单位:ms

grid_width	block_width	gpu_5	grid_width	block_width	gpu_5
256	256	10.883	512	1024	9.857
256	512	9.676	1024	256	9.518
256	1024	9.861	1024	512	9.517
512	256	9.450	1024	1024	9.804
512	512	9.551			

10.7　2D 卷积的 GPU 移植与优化

前文阐述了 1D 卷积的 GPU 移植和优化工作，本节将着重介绍扩展法 2D 卷积的移植和优化，并通过实验分析其移植和优化效果。

10.7.1　GPU 上 2D 卷积

在 10.1.2 节已给出了扩展法 2D 卷积的串行实现代码，要将其移植到 GPU 上执行，需要注意其 2D 索引的计算。下面展示了具体的扩展法 2D 卷积的 CUDA C 代码。

```
__global__ void convolution_2d_gpu1(float *array_m,float *array_n1,float *array_p,int n,int n1)
{
    int tidx=threadIdx.x;
    int tidy=threadIdx.y;
    int bidx=blockIdx.x;
    int bidy=blockIdx.y;
    int i,j,k,l;
    j=bidx*blockDim.x+tidx;
    i=bidy*blockDim.y+tidy;
    float temp=0.0;
    int kk,ll;
    for(k=0;k<m;k++)
    {
        for(l=0;l<m;l++)
        {
            kk=i+k;
            ll=j+l;
            temp+=array_m[k*m+l] * array_n1[kk*n1+ll];
        }
    }
    array_p[i*n+j]=temp;
}
```

```
#define block_width 16
    dim3 blockdimnum(block_width,block_width,1);
    int grid_width=n/block_width;
    dim3 griddimnum(grid_width,grid_width,1);
    convolution_2d_gpu1<<<griddimnum,blockdimnum>>>(d_array_m,d_array_n1,d_array_p1,n,n1);//global memory
```

10.7.2 M 常量 2D 卷积

1D 卷积的实验结果表明采用常量存储单元存储掩码矩阵 **M** 可获得明显的性能提升,同理亦适用于 GPU 上 2D 卷积运算。下面的 CUDA 代码实现了 **M** 矩阵常量存储优化的扩展法 2D 卷积(本书不提供共享存储优化 **M** 矩阵的代码,读者可自行实践对比)。

```
__constant__ float d_array_m1[m*m];
__global__ void convolution_2d_gpu2(float *array_n1,float *array_p,int n,int n1)
{
    int tidx=threadIdx.x;
    int tidy=threadIdx.y;
    int bidx=blockIdx.x;
    int bidy=blockIdx.y;
    int i,j,k,l;
    j=bidx*blockDim.x+tidx;
    i=bidy*blockDim.y+tidy;
    float temp=0.0;
    int kk,ll;
    for(k=0;k<m;k++)
    {
        for(l=0;l<m;l++)
        {
          kk=i+k;
            ll=j+l;
            temp+=d_array_m1[k*m+l] * array_n1[kk*n1+ll];
        }
    }
    array_p[i*n+j]=temp;
}
    cudaMemcpyToSymbol(d_array_m1,array_m,sizeof(float)*m*m);
    convolution_2d_gpu2<<<griddimnum,blockdimnum>>>(d_array_n1,d_array_p1,n,n1);
```

10.7.3 M 常量 N 共享 2D 卷积

在将掩码 **M** 矩阵存储在常量存储的基础上,根据 1D 卷积经验,采用共享存储优化 **N** 矩阵的子矩阵。此时读入共享存储中的矩阵元素在两个维度上都需要扩展,处理过程也变得比较复杂。图 10.3 展示了共享存储加载 **N** 矩阵的子矩阵的流程。

下面给出了用常量存储优化掩码矩阵 **M** 和用共享存储优化矩阵 **N** 的扩展法 2D 卷积 CUDA C 代码,代码中需要特别注意线程同步之前的部分,即共享存储数据加载过程。

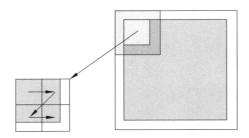

图 10.3 2D 卷积共享存储数据加载示意图

```
__global__ void convolution_2d_gpu5(float *array_n1,float *array_p,int n,int n1)
{
    __shared__ float array_ns[mm*mm];
    const int tidx=threadIdx.x;
    const int tidy=threadIdx.y;
    const int bidx=blockIdx.x;
    const int bidy=blockIdx.y;
    int i,j,k,l;
    j=bidx*blockDim.x;
    i=bidy*blockDim.y;

    int tidk,tidl;
    array_ns[tidy*mm+tidx]=array_n1[(i+tidy)*n1+j+tidx];
    if(tidy<(m-1))
    {
        tidk=tidy+block_width;
        array_ns[tidk*mm+tidx]=array_n1[(i+tidk)*n1+j+tidx];
    }
    else if(tidy<(m+m-2))
    {
        tidk=tidx;
        tidl=tidy-(m-1)+block_width;
        array_ns[tidk*mm+tidl]=array_n1[(i+tidk)*n1+j+tidl];
    }
    else if((tidy<(3*m-3))&&(tidx<(m-1)))
    {
        tidk=tidy-2*(m-1)+block_width;
        tidl=tidx+block_width;
        array_ns[tidk*mm+tidl]=array_n1[(i+tidk)*n1+j+tidl];
    }
    __syncthreads();
    float temp=0.0;
    int kk,ll;
```

```
        for(k=0;k<m;k++)
        {
            for(l=0;l<m;l++)
            {
                kk=tidy+k;
                ll=tidx+l;
                temp+=d_array_m1[k*m+l] * array_ns[kk*mm+ll];
            }
        }
        array_p[(i+tidy)*n+j+tidx]=temp;
}
convolution_2d_gpu5<<<griddimnum,blockdimnum>>>(d_array_n1,d_array_p1,
n,n1);
```

10.7.4 2D 卷积实验结果分析

编译执行各版本 2D 卷积代码,统计执行时间如表 10.5 所示。表中数据显示 GPU 上 2D 卷积的性能特性与 1D 卷积基本一致,读者可参考 10.6 节的分析,本节不再赘述。

表 10.5　2D 卷积执行时间　　　　　　　　　　　　　单位:ms

n	m	判断法	扩展法	gpu_1	gpu_2	gpu_5	计算＋通信
512	5	12.28	5.20	1.23	0.14	0.08	0.57
1024	5	57.97	19.32	1.63	0.46	0.26	1.64
2048	5	231.96	79.36	5.98	1.78	0.97	6.60
4096	5	771.66	318.40	23.32	7.08	3.82	25.23
8192	5	3088.35	1264.38	92.76	28.76	15.26	99.21
16 384	5	12 359.40	5059.40	370.50	113.82	61.57	396.23

10.8　知识点总结

卷积的边界处理:判断法和扩展法。1D 卷积和 2D 卷积的判断法和扩展法实现方式。

判断的移除:在卷积运算中,扩展法可移除判断法中的判断处理,从而提高运算性能。

1D 卷积特别是 2D 卷积的索引处理方法详见 10.1 节。

常量存储的使用:常量存储利用__contant__限定符声明,常量存储是只读的,普通方法无法修改。常量存储位于显存中,但拥有片上常量缓存。常量存储的访问必须是广播的,并且常量存储不支持并行访存。

常量存储数据传输:cudaMemcpyToSymbol()函数提供从主机存储到常量存储的复

制功能,详细介绍参见 5.4 节,实例见 10.3 节。

```
extern __host__ cudaError_t CUDARTAPI cudaMemcpyToSymbol(const char *symbol,
const void *src, size_t count, size_t offset __dv(0), enum cudaMemcpyKind kind __
_dv(cudaMemcpyHostToDevice));
```

```
cudaMemcpyToSymbol(d_array_m1,array_m,sizeof(float)*m);
```

共享存储使用:共享存储使用__shared__限定符声明,共享存储位于片上,其作用范围是 block 内所有 thread,生存周期是 kernel 函数。共享存储的访存延迟比全局存储的访存延迟小近 1 个量级。

共享存储读和写之间必须同步(__syncthreads())。

对于小规模只读数据的广播访问(掩码数组/掩码矩阵 M),常量存储优化比共享存储优化的性能更好。

常量存储或共享存储替代全局存储访问可大幅提升 GPU 程序性能。

无论 CPU 还是 GPU,扩展法比判断法的卷积运算速度更快、性能更好。

合理的线程维度可以提高 kernel 函数的执行性能,但在卷积运算中影响不大。

计算 GPU 程序的加速比时,不能仅考虑 kernel 函数的计算时间,还需要考虑通信等异构计算引入的开销。

1D 卷积共享存储数据加载:分两次加载数组到共享存储,该数组长度大于线程数量且小于两倍线程数量。

```
__shared__ float array_ns[mm];
array_ns[tidx]=array_n1[j+tidx];
if(tidx<(m-1))
{
    tidk=tidx+block_width;
    array_ns[tidk]=array_n1[tidk+j];
}
```

2D 卷积共享存储分块数据加载:分 4 块加载矩阵数据到共享存储中,其中该矩阵的长和宽均大于 block 中的 thread 维度。

```
__shared__ float array_ns[mm*mm];
...
    int tidk,tidl;
    array_ns[tidy*mm+tidx]=array_n1[(i+tidy)*n1+j+tidx];
    if(tidy<(m-1))
    {
        tidk=tidy+block_width;
        array_ns[tidk*mm+tidx]=array_n1[(i+tidk)*n1+j+tidx];
    }
```

```
    else if(tidy<(m+m-2))
    {
        tidk=tidx;
        tidl=tidy-(m-1)+block_width;
        array_ns[tidk*mm+tidl]=array_n1[(i+tidk)*n1+j+tidl];
    }
    else if((tidy<(3*m-3))&&(tidx<(m-1)))
    {
        tidk=tidy-2*(m-1)+block_width;
        tidl=tidx+block_width;
        array_ns[tidk*mm+tidl]=array_n1[(i+tidk)*n1+j+tidl];
    }
    ...
```

10.9 扩展练习

设计并实现 3D 卷积的 GPU 移植及优化。

第11章 曼德博罗特集

11.1 曼德博罗特集及其串行实现

曼德博罗特集是一种经典的图像绘制算法,利用一个复平面上的简单迭代公式的迭代次数绘图,获得一个奇异且绚丽的图案。曼德博罗特集的计算属于易并行问题,每个像素点的颜色值计算互不相关但计算量差异巨大。

曼德博罗特集的迭代公式为

$$z_{k+1} = (z_k)^2 + c$$

下面展示了串行曼德博罗特集 C 代码,取最大迭代次数为 255,即 unsigned char 类型上限。像素点的颜色规则为:若迭代次数未达到最大迭代次数,而计算的 z_n 的模已超过最大阈值 2,则停止迭代,以该迭代值为该点颜色值;若迭代次数达到最大迭代次数而 z_n 的模依然未超过阈值,则以 255 作为其颜色值。

```
#define MAX_ITE 255
typedef struct _LI_RGB
{
    unsigned char b,g,r;
}LI_RGB;
void ComputeColor(double x, double y, LI_RGB *color)
{
    double zx, zy;
    double tempx, tempy;
    int count;

    count=0;
    zx=0;
    zy=0;

    while((count<MAX_ITE) && ((zx*zx+zy*zy)<4.0))
    {
```

```
        tempx=zx*zx-zy*zy+x;
        tempy=2.0*zx*zy+y;
        zx=tempx;
        zy=tempy;
        count++;
    }

    (*color).b=(count<=8?(count*16-1) : (unsigned char)((count-8)/120*128+
127));
    (*color).g=(count<=32?(count*8-1) : 255);
    (*color).r=(count<=16?(count*16-1) : (unsigned char)((count-16)/239*
128+127));
}
for(j=0; j<n; j++)
{
    y=y1-j*dy;
    for(i=0; i<n; i++)
    {
        x=x0+i*dx;
        ComputeColor(x, y, &pRGB[j*n+i]);
    }
}
```

编译并执行上述串行曼德博罗特集代码,输出结果图像如图 11.1 所示,测试并统计运行时间如表 11.1 所示。当然,要画出结果图像,仅凭上述代码是不够的,还需要将结果图像矩阵写到图像文件中,完整代码可参见附录 B.1。

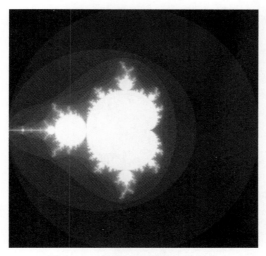

图 11.1 曼德博罗特集结果图像

表 11.1　曼德博罗特集串行执行时间　　　　　　　　　　单位：ms

n	1024	2048	4096	8192	16 384	32 768
serial	146.39	583.45	2330.93	9318.60	37 256.13	148 963.92

11.2　曼德博罗特集的 GPU 映射

由 11.1 节描述的曼德博罗特集理论和串行代码可知，结果图像的每个像素的计算都是独立的。因此，移植曼德博罗特集到 GPU 时，首先考虑将结果图像每个像素的计算任务分配给 GPU 的一个线程，通过大量线程同时计算来实现 GPU 并行计算。

下面展示了根据上述思想编写的曼德博罗特集 CUDA C 代码(标记为 cuda)，代码中还融入循环处理的思想，以便处理任意规模的曼德博罗特集计算任务。

```
typedef struct _LI_RGB
{
    unsigned char b,g,r;
}LI_RGB;
__device__ inline
void ComputeColor(double x, double y, LI_RGB *color)
{
    double zx, zy;
    double tempx, tempy;
    int count;
    count=0;
    zx=0;
    zy=0;
    while((count<MAX_ITE) && ((zx*zx+zy*zy)<4.0))
    {
        tempx=zx*zx-zy*zy+x;
        tempy=2.0*zx*zy+y;
        zx=tempx;
        zy=tempy;
        count++;
    }
    (*color).b=(count<=8?(count*16-1) : (unsigned char)((count-8)/120*128+
        127));
    (*color).g=(count<=32?(count*8-1) : 255);
    (*color).r=(count<=16?(count*16-1) : (unsigned char)((count-16)/239*128
        +127));
}
__global__
```

```
void compute_color_gpu(LI_RGB *pRGB,double x0,double dx,double y1,double dy,
int n)
{
    unsigned int tid=threadIdx.x;
    unsigned int bid=blockIdx.x;
    double y,x;
    while(bid<n)
    {
        tid=threadIdx.x;
        y=y1-bid*dy;
        while(tid<n)
        {
            x=x0+tid*dx;
            ComputeColor(x, y, &pRGB[bid*n+tid]);
            tid+=blockDim.x;
        }
        bid+=gridDim.x;
    }
}
```

```
    LI_RGB *pRGB;
    pRGB=(LI_RGB *) malloc(n*n*sizeof(LI_RGB));
    cudaHostRegister(pRGB,sizeof(LI_RGB)*n*n,0);
    LI_RGB *d_pRGB;
    cudaMalloc((void**) &d_pRGB, sizeof(LI_RGB)*n*n);
    ⋮
    compute_color_gpu<<<threadnum,threadnum>>>(d_pRGB,x0,dx,y1,dy,n);
    cudaMemcpy(pRGB, d_pRGB, sizeof(LI_RGB)*n*n, cudaMemcpyDeviceToHost);
```

编译并执行上述 CUDA C 代码,输出结果与串行输出结果一致。测试并统计执行时间,如表 11.2 所示。

表 11.2　cuda 版曼德博罗特集执行时间　　　　　　　　　单位:ms

n	kernel	all	n	kernel	all
1024	0.86	1.38	8192	37.61	67.68
2048	2.93	4.83	16 384	147.35	267.57
4096	9.75	17.29	32 768	581.78	1062.54

对比表 11.1 和表 11.2,GPU 上曼德博罗特集运算时间(包括通信时间等额外开销)比 CPU 串行运算时间快了 100 多倍。说明曼德博罗特集非常适合在 GPU 上计算。

观察表 11.2,可以发现一个很有意思的现象,即 kernel 函数的执行时间恰好接近全部时间的一半,即计算时间和通信时间(主机端已采用页锁定存储)相当,这也是 11.4 节研究计算通信重叠优化的缘由。

11.3 一些优化尝试及效果

本节将基于 11.2 节 GPU 上曼德博罗特集的 CUDA C 代码,展开一系列性能优化研究,并测评其优化效果。本节采用的优化策略均是常见的 GPU 优化策略,但值得关注的是,多数优化策略并没有获得正效果,甚至有些优化策略起还产生了负效果。由此可以得出这样的结论:各种 GPU 优化策略是有其适用范围的,合理运用并反复实践和验证才是获得最大性能收益的关键。

注意:本节涉及的代码版本标记可能比较乱,原因是笔者做了大量的优化尝试,本节代码只是笔者从这些版本中挑选的比较有代表性的代码,为了不搞混这些代码,故没有重新修订代码版本标记。

11.3.1 访存连续

在定义的结构体 LI_RGB 中,有 3 个非负字符型数据,其存储长度为 3B。而在 11.2 节代码中,对结构体 LI_RGB 中数据的访问是依次改写,此时两个相邻线程的访存不相邻,间隔 3B 数据。显然此时访存并不连续,本节试图解决该访存不连续问题。

针对 GPU 上曼德博罗特集运算中的访存不连续的问题,一种修改方法是将结构体拆开,使用 3 个 uchar 数组替换原来的结构体数组,声明如下:

```
unsigned char *color_b; unsigned char *color_g; unsigned char *color_r;
typedef struct _LI_RGB
{
    unsigned char b,g,r;
}LI_RGB;
LI_RGB *color;
```

经数组替换结构体后,相邻线程访问的地址空间是连续的。修改后的曼德博罗特集 CUDA C 代码记为 cuda_1_0,具体代码参见附录 B.2。

编译执行修改后的 cuda_1_0 代码,统计不同输入参数规模情况下的执行时间,并与原始版本对比,如表 11.3 所示。表中数据显示,访存连续优化前后执行时间变化极小,既没有正效果也没有负效果。没有正效果说明访存连续优化在曼德博罗特集计算中不适用,没有负效果说明访存连续优化策略本身是正确的。

表 11.3 访存连续优化效果   单位:ms

n	cuda	cuda_1_0	n	cuda	cuda_1_0
1024	0.86	0.85	8192	37.61	37.57
2048	2.93	2.95	16 384	147.35	147.40
4096	9.75	9.83	32 768	581.78	582.46

11.3.2 uchar4 访存合并

char 类型访问全局存储只能达到全部访存带宽的 1/4 左右,而通过合并访问(char4)可获得接近 GPU 访存带宽的访存性能(相关的实验论证见 17.6 节)。

在曼德博罗特集计算问题中,访存的数据是单个非负字符型,是否也只能获得 1/4 的访存带宽?那么将数据访问组织成 uchar4 类型合并访存能否获得更好的性能呢?

带着上述问题,笔者修改了相应的 CUDA C 代码,将修改后的代码标记为 cuda_0_2(详细代码见附录 B.3)。

CUDA 提供的 uchar4 数据类型结构如下:

```
struct __device_builtin__ __align__(4) uchar4
{
    unsigned char x, y, z, w;
};
```

执行 uchar4 访存合并优化后的曼德博罗特集程序,测试并统计不同规模计算时间,如表 11.4 所示。表中数据显示,uchar4 访存合并优化前后 kernel 函数执行时间几乎相等,说明该优化方法既没有获得正效果,也没有获得负效果。没有正效果说明 uchar4 访存合并优化在曼德博罗特集问题上不适用,没有负效果说明访存合并优化思想本身是正确的。

表 11.4 访存合并的优化效果   单位:ms

n	cuda	cuda_0_2	n	cuda	cuda_0_2
1024	0.86	0.85	8192	37.61	37.53
2048	2.93	2.96	16 384	147.35	147.30
4096	9.75	9.82	32 768	581.78	581.64

另外,如果统计传输时间,使用 uchar4 访存合并优化的 cuda_0_2 版本将增加 1/3 的通信开销。

11.3.3 4 种零拷贝

纵观整个曼德博罗特集计算过程,计算得到的像素点的值就是曼德博罗特集的最终结果,那么能否直接将结果写到主机端内存,而不通过 global memory 转存呢?此时需要零拷贝(zerocopy)技术。该项技术能让 GPU 或 CPU 获得对方的存储指针,直接读写远程数据。本节给出了 4 种不同的零拷贝实现方法。

1. cuda_zerocopy

下面是对 cuda 版本的曼德博罗特集进行 zerocopy 改造的关键代码,并将修改后的曼德博罗特集 CUDA C 代码标记为 cuda_zerocopy(详细代码见附录 B.4)。

```
cudaHostAlloc((void **) &pRGB, sizeof(LI_RGB) * n * n,cudaHostAllocMapped);
cudaHostGetDevicePointer((void **)&d_pRGB, (void *)pRGB, 0);
compute_color_gpu<<<blocknum,threadnum>>>(d_pRGB,x0,dx,y1,dy,n);
```

上述 cuda_zerocopy 版本的代码执行性能极差(后文有具体实验数据),为什么明明可行的优化思想反而导致性能变差呢?如何才能正确使用 zerocopy 技术来优化曼德博罗特集的计算过程呢?带着这几个问题笔者进行了以下几个实践探索。

2. cuda_1_0_zerocopy

对 cuda_1_0 版本进行了零拷贝改造,修改后的曼德博罗特集计算 CUDA C 代码标记为 cuda_1_0_zerocopy。下面展示了修改的核心部分代码,详细代码见附录 B.5。

```
cudaHostAlloc((void **) &pB, sizeof(char) * n * n,cudaHostAllocMapped);
cudaHostAlloc((void **) &pG, sizeof(char) * n * n,cudaHostAllocMapped);
cudaHostAlloc((void **) &pR, sizeof(char) * n * n,cudaHostAllocMapped);
unsigned char *d_pR, *d_pG, *d_pB;
cudaHostGetDevicePointer((void **)&d_pB, (void *)pB, 0);
cudaHostGetDevicePointer((void **)&d_pG, (void *)pG, 0);
cudaHostGetDevicePointer((void **)&d_pR, (void *)pR, 0);

compute_color_gpu<<<threadnum,threadnum>>>(d_pB,d_pG,d_pR,x0,dx,y1,dy,
n);
```

3. cuda_0_0_zerocopy

对 cuda 版本进行微调(标记为 cuda_0_0,读者可从附录 B.6 代码逆推得到),把结构体内 3 个变量的访存结合起来,而不是单独访存,这种优化思想兼顾 11.3.1 节的访存连续和 11.3.2 节的访存合并。以下是这种微调的核心代码。

```
__device__ inline
void ComputeColor(double x, double y, LI_RGB *color)
{
    ⋮
    LI_RGB rgb;
    rgb.b=(count<=8?(count * 16-1) : (unsigned char)((count-8)/120 * 128+
127));
    rgb.g=(count<=32?(count * 8-1) : 255);
    rgb.r=(count<=16?(count * 16-1) : (unsigned char)((count-16)/239 * 128+
127));
    *color=rgb;
}
```

在上述微调的基础上,进行 zerocopy 改造,修改后的曼德博罗特集计算 CUDA C 代码标记为 cuda_0_0_zerocopy,详细代码见附录 B.6。

4. cuda_0_2_zerocopy

对 uchar4 类型合并访存优化后的 cuda_0_2 版本进行了零拷贝修改,下面给出了 zerocopy 修改的关键代码,零拷贝改造后的相应曼德博罗特集计算代码标记为 cuda_0_2_zerocopy,详见附录 B.7。

```
cudaHostAlloc((void **) &pRGB, sizeof(uchar4) * n * n,cudaHostAllocMapped);
cudaHostGetDevicePointer((void **)&d_pRGB, (void *)pRGB, 0);
compute_color_gpu<<<blocknum,threadnum>>>(d_pRGB,x0,dx,y1,dy,n);
```

分别执行上述零拷贝版本,统计时间并与原始 cuda 版本执行时间对比,如表 11.5 所示。

表 11.5 零拷贝的使用与效果　　　　　　　　　　　　　单位:ms

n	1024	2048	4096	8192	16 384	32 768
serial	146.39	583.45	2330.93	9318.60	37 256.13	148 963.92
cuda	1.38	4.83	17.29	67.68	267.57	1062.54
cuda_zerocopy	73.32	290.73	1162.41	4653.66	18 637.97	74 596.48
cuda_1_0_zerocopy	1.00	3.75	15.07	64.40	285.40	1170.30
cuda_0_0_zerocopy	72.65	290.69	1162.41	4653.75	18 638.93	74 593.46
cuda_0_2_zerocopy	0.95	3.07	12.04	52.87	204.64	846.09

分析表 11.5 中的数据可知:

(1) 对比 cuda 和 cuda_zerocopy 两个版本,相同的访存模式,但执行时间却差异巨大。说明 GPU 对 CPU 端存储的零拷贝访问和 GPU 上全局存储的访问差异较大。

(2) cuda_zerocopy 和 cuda_0_0_zerocopy 两个版本都很慢。说明长为 3B 的数据不适合使用零拷贝技术。

(3) cuda_1_0_zerocopy 性能较好,能与 cuda 版本性能匹配。说明**零拷贝的访问必须是对齐的、连续的**。

(4) 对比 cuda_1_0_zerocopy 和 cuda_1_0(同 cuda),发现数据量小时采用了 zerocopy 技术后执行时间减少了,而当数据量大时却变慢了。这种情况也无法验证 zerocopy 优化的正优化效果。

(5) cuda_0_2_zerocopy 版本获得了最佳性能,执行速度比原来的 cuda 版本还快。说明**零拷贝的最佳数据长度之一是 4B**。

(6) GPU 上全局存储访问相对自由,访存模式兼容性强;而 zerocopy 访存比较严格,访存模式兼容性差,访存模式一旦不兼容将导致巨大的性能瓶颈。

11.3.4 总结分析

前文针对曼德博罗特集的 CUDA 代码采取了 3 种优化策略,其中零拷贝优化就有 4

个不同的实现版本。事实上笔者编写的优化版本远不止这些,有些只是笔者的一个简单想法,实现后没啥效果就没在书中提及。谈到这里,主要目的是告诉读者,GPU 代码优化中,优化方法是灵活多变的,优化策略与具体应用问题也有一定的匹配性,只有在真正实践后获得正效果才能验证优化方法的适用性。

回顾上文采取的 3 种优化策略,访存连续、访存合并、零拷贝都是针对访存进行的优化,事实上访存优化一般也是 GPU 优化的重中之重。然而,在曼德博罗特集计算问题中,访存是否是问题的核心呢?

下面设计了一组实验来验证访存在曼德博罗特集计算问题中所占的比重,该实验删除原代码中的关键计算代码,而仅剩下访存操作,关键代码修改如下:

```
__device__ inline
void ComputeColor(double x, double y, LI_RGB *color)
{
    LI_RGB rgb={255,0,0};
    *color=rgb;
}
```

统计执行时间,并与原来的曼德博罗特集计算时间对比(见表 11.6),表中数据显示,访存时间占 kernel 函数总执行时间的比重非常小,说明曼德博罗特集计算问题是一个计算受限问题,仅使用访存优化价值并不大。

表 11.6　访存时间比重分析　　　　　　　　　　单位:ms

n	writeonly	cuda	n	writeonly	cuda
1024	0.15	0.86	8192	2.30	37.61
2048	0.49	2.93	16 384	9.55	147.35
4096	0.61	9.75	32 768	37.19	581.78

分析结果指示了曼德博罗特集的优化重点是计算部分,针对计算部分的优化方法,其实应该算是关于曼德博罗特集计算本身的优化,而不能算是 GPU 的优化内容,因此本书不展开讨论,有兴趣的读者可以研究曼德博罗特集计算优化方法。

11.4　计算通信重叠优化

11.2 节实验分析得出结论,曼德博罗特集的 GPU 并行计算中,kernel 函数耗时与通信时间相当,因此可作为一个典型的计算通信重叠优化案例。本节探讨用计算通信重叠优化曼德博罗特集计算。

CPU/GPU 异构系统中的计算通信重叠采用多流并发技术实现,处于不同流的计算和通信能够同时进行。下面是基于曼德博罗特集开发计算通信重叠优化的关键代码。代码中展示了 CUDA 流的创建、使用和销毁,另外,值得一提的是计算通信重叠必须采用异步通信函数,且参与通信的 CPU 端存储必须是页锁定存储。详细的计算通信重叠优化后的曼德博罗特集计算 CUDA C 代码见附录 B.8(标记为 cuda_2)。

```
    cudaStream_t stream0,stream1;
    cudaStreamCreate(&stream0);
    cudaStreamCreate(&stream1);

    for(i=0;i<n;i+=blocknum*2)
    {
compute_color_gpu<<<blocknum,threadnum,0,stream0>>>(d_pRGB,x0,dx,y1,dy,n,i);
compute_color_gpu<<<blocknum,threadnum,0,stream1>>>(d_pRGB,x0,dx,y1,dy,n,i
+blocknum);
cudaMemcpyAsync(&pRGB[i*n], &d_pRGB[i*n], sizeof(LI_RGB) * n * blocknum,
cudaMemcpyDeviceToHost,stream0);
cudaMemcpyAsync(&pRGB[(i+blocknum)*n], &d_pRGB[(i+blocknum)*n], sizeof(LI_
RGB) * n * blocknum, cudaMemcpyDeviceToHost,stream1);
    }

    cudaStreamDestroy(stream0);
    cudaStreamDestroy(stream1);
```

上述代码中还需要注意:在计算通信重叠(多流并发)代码开发时,尽量将无依赖关系的不同流的计算和通信分开放在一起,这种函数布置时一定要注意其中的逻辑依赖关系。读者可以在编程实践时通过调整函数位置、观察实验结果,增加理解和体验。

统计执行时间并与前文两个最好的版本对比(见表 11.7)。表中数据显示,计算通信重叠优化能够取得明显的加速效果,但并没有得到理想的加速比,主要原因可能有:①多流并发引入了额外的开销(流的创建和销毁、多流引入的竞争开销等);②对计算的拆分导致了 kernel 函数执行性能的下降。

表 11.7 计算通信重叠效果　　　　　　　　　　　　　　　　　　单位:ms

n	1024	2048	4096	8192	16 384	32 768
cuda	1.38	4.83	17.29	67.68	267.57	1062.54
cuda_0_2_zerocopy	0.95	3.07	12.04	52.87	204.64	846.09
cuda_2	1.07	3.78	13.06	52.56	208.44	827.15

事实上,仅根据上面的结果并不能确切说明计算通信已经重叠,只能说明时间减少了。那么如何才能认定计算通信重叠了呢? 可以利用 NVIDIA visual profiler 工具来验证计算和通信是否已经确实重叠。如果是 Windows 平台,执行后通过性能分析工具提供的时间轴可以看出计算和通信是否重叠(工具的详细使用方法可参阅 17.8 节)。若是 Linux 平台,图形界面下 NVIDIA visual profiler 的使用方法与 Windows 平台一致;另外 Linux 平台下还提供了命令行的性能分析工具 nvprof。可以通过 nvprof --help 查看命令详细参数(参见附录 D)。下面是用 nvprof 命令开启选项 --print-gpu-trace 后执行得到的结果。

```
[fangmq@cn18%yhstar mandelbrot]$nvprof --print-gpu-trace ./cuda_2
n     kernel     kernel+communication
==11817==NVPROF is profiling process 11817, command: ./cuda_2
2048   5.661964
==11817==Profiling application: ./cuda_2
==11817==Profiling result:
Start      Duration    Grid Size     Block Size    Regs*    SSMem*    DSMem*
Size       Throughput  Device        Context       Stream Name
309.81ms   115.58μs    (512 1 1)     (512 1 1)     16       0B        0B
-          -           Tesla K20c (0) 1            8
 compute_color_gpu(_LI_RGB*, double, double, double, double, int, int) [172]
310.19ms   1.3754ms    (512 1 1)     (512 1 1)     16       0B        0B
-          -           Tesla K20c (0) 1            9
 compute_color_gpu(_LI_RGB*, double, double, double, double, int, int) [181]
310.24ms   475.04μs    -             -             -        -         -
3.1457MB   6.6220GB/s  Tesla K20c (0) 1            8
                                    [CUDA memcpy DtoH]
311.25ms   1.3521ms    (512 1 1)     (512 1 1)     16       0B        0B
-          -           Tesla K20c (0) 1            8
 compute_color_gpu(_LI_RGB*, double, double, double, double, int, int) [192]
311.58ms   474.09μs    -             -             -        -         -
3.1457MB   6.6353GB/s  Tesla K20c (0) 1            9
                                    [CUDA memcpy DtoH]
312.55ms   85.474μs    (512 1 1)     (512 1 1)     16       0B        0B
-          -           Tesla K20c (0) 1            9
 compute_color_gpu(_LI_RGB*, double, double, double, double, int, int) [201]
312.61ms   472.06μs    -             -             -        -         -
3.1457MB   6.6638GB/s  Tesla K20c (0) 1            8
                                    [CUDA memcpy DtoH]
313.08ms   472.49μs    -             -             -        -         -
3.1457MB   6.6578GB/s  Tesla K20c (0) 1            9
                                    [CUDA memcpy DtoH]
Regs: Number of registers used per CUDA thread.
SSMem: Static shared memory allocated per CUDA block.
DSMem: Dynamic shared memory allocated per CUDA block.
```

分析上述输出结果的加粗部分,其中计算开始于 310.19ms,持续 1.3754ms,而传输开始于 310.24ms,持续了 475.04μs,中间部分时间是重叠的。类似的重叠结果在上述输出中还能找到。

nvprof 分析结果显示了一个值得关注的现象:kernel 函数第一次执行时间是 115.78μs,接下来两次均为 1.35ms 左右,最后一次却变成了 88.29μs。通过仔细观察曼德博罗特集结果图像(见图 11.1),可以发现该图中的图案主要存在于图的中间两份(假设将图从上至下分成四份),此时由于相邻线程不同的分支处理和计算导致 kernel 函数耗时增长,而上下两份的计算在线程间的分支相同,执行速度自然较快。

11.5 突破 kernel 执行时间限制

如果实验平台配置的 GPU 性能较低，比如 GT610 或 GT705，在这类显卡上运行 CUDA 程序，一般 kernel 函数都会因为超时而强制退出。本节将给出两种突破 kernel 执行时间限制的方法，以解决在低端 GPU 上正常运行 CUDA 程序的问题。

首先，介绍下两种突破 kernel 函数执行时间限制的方法。

（1）修改环境变量。修改环境变量可以从根本上解除 kernel 函数执行时间限制，具体修改方法为：单击"开始"→"运行"命令，弹出"运行"对话框，在"打开"文本框中输入 regedit，单击"确定"按钮，打开注册表管理器，查找 HKLM\System\CurrentControlSet\Control\GraphicsDrivers 键下的 TdrLevel 项（若无则新建 DWORD 类型值），将该项设置为 0 并重启。

（2）拆分 kernel 函数。拆分 kernel 函数基于分而治之的思想，单个 kernel 函数执行超时，那么将其拆分为多个 kernel 函数，令每个子 kernel 函数都不超时。通过分析具体的 kernel 函数，将 kernel 函数中的任务进行人为拆分，将子任务分别写成 kernel 函数。

笔者在初写本节代码时，手头仅有一张 GT 705，仔细阅读过 2.4 节的读者肯定知道这显卡弱极了；当时无法上网，且忘记了需要修改的注册表项，就选择了拆分 kernel 函数的方法。当然，本书中的代码和实验结果是后来在 Tesla K20c GPU 环境下重新整理和测试得来的。下面介绍曼德博罗特集计算 kernel 函数的拆分办法。

从前文对曼德博罗特集的分析可知，每个像素元素值的计算都是独立的，而上文代码共启动一个 kernel 函数，计算整个图像的所有像素点元素值。在拆分时，可将其拆分为每个 kernel 函数计算图像的一行像素点值。下面是曼德博罗特集计算 kernel 函数拆分的核心代码，完整代码参见附录 B.9（标记为 cuda_1_2）。

```
__global__
void compute_color_gpu(unsigned char *color_b,unsigned char *color_g,unsigned char *color_r,double x0,double dx,double y1,double dy,int n,int j)
{
    unsigned int id=blockIdx.x * blockDim.x+threadIdx.x;
    double y,x;
    double zx, zy;
    double tempx, tempy;
    int count;
//  while(bid<n)
    y=y1-j * dy;
    if(id<n)
    {
        x=x0+id * dx;
        {
            ...
```

```
            }
        }
    }
        int blocknum= (n+threadnum-1)/threadnum;
        for(j=0; j<n; j++)
        {
            compute_color_gpu<<<blocknum,threadnum>>>(d_pB,d_pG,d_pR,x0,dx,y1,
            dy,n,j);
        }
```

kernel 函数拆分后性能一般会下降,表 11.8 展示了 kernel 函数拆分前后曼德博罗特集计算的执行时间。性能下降的原因主要有两方面：①拆分后单个 kernel 函数任务量变得很小,无法充满整个 GPU(无法发挥完整的 GPU 性能)；②kernel 函数调度开销远大于 GPU 内部 block 切换调度开销和 kernel 内部循环开销(15.8 节将展开详细讨论)。

表 11.8 kernel 拆分的影响   单位：ms

n	cuda	cuda_1_2	n	cuda	cuda_1_2
1024	0.86	28.18	8192	37.61	226.45
2048	2.93	56.58	16 384	147.35	470.55
4096	9.75	112.03	32 768	581.78	1195.16

最后,值得一提的是,kernel 函数的运行时间限制将在 Pascal 架构的 GPU 中得到解决,Pascal 架构中的计算抢占可以打破 kernel 函数的运行时间限制,GPU 程序开发时将不必专门考虑 kernel 函数运行是否会超时。

11.6 知识点总结

cudaHostRegister()函数：页锁定存储注册函数,可将 malloc()函数分配的可分页存储注册为页锁定存储。详细参数说明参考 5.4.2 节。

```
pRGB=(LI_RGB *)malloc(n*n*sizeof(LI_RGB));
cudaHostRegister(pRGB,sizeof(LI_RGB)*n*n,0);
```

cudaHostAlloc()函数：页锁定存储分配函数。详细参数说明参考 5.4.2 节。

```
cudaHostAlloc((void**) &pRGB, sizeof(LI_RGB)*n*n,
cudaHostAllocDefault);
```

访存连续：相邻的线程访问相邻的数据。

uchar4 合并访存：将 4 个 unsigned char 类型数据组合成结构体,整个结构体同时访存。

```
struct __device_builtin__ __align__(4) uchar4
{
    unsigned char x, y, z, w;
};
```

zerocopy（零拷贝）：GPU 直接访问位于 CPU 端的数据，或 CPU 直接访问位于 GPU 端的数据。

zerocopy（零拷贝）使用方法：将需要零拷贝访存的数据声明为页锁定存储，并声明存储分配为 cudaHostAllocMapped，即分配页锁定存储映射到 GPU。

```
cudaHostAlloc((void **) &pRGB, sizeof(LI_RGB) * n * n,cudaHostAllocMapped);
cudaHostGetDevicePointer((void **)&d_pRGB, (void *)pRGB, 0);
compute_color_gpu<<<blocknum,threadnum>>>(d_pRGB,x0,dx,y1,dy,n);
```

zerocopy 使用注意事项如下。

（1）zerocopy 存储访问与全局存储访问存在很大区别，zerocopy 存储访问更加严格。

（2）长为 3B 的数据不适合使用 zerocopy 访存。

（3）zerocopy 必须是连续访存，否则导致性能极差。

（4）长为 4B 的数据适合使用 zerocopy 访存。

zerocopy 操作隐含计算通信重叠。

GPU 优化前应先定位计算与访存的耗时比例，避免不必要的优化尝试。

流的声明、创建、使用和销毁：cudaStream_t、cudaStreamCreate()、cudaStreamDestroy()，详细的参数说明见 5.4.5 节，实例见 11.4 节。

计算通信重叠：CPU/GPU 异构系统中的计算通信重叠采用多流并发技术实现，处于不同流的计算和通信能够同时进行。计算通信重叠必须采用异步通信函数，且参与通信的主机端存储必须是页锁定存储。相关使用实例见 11.4 节。

nvprof：Linux 下的命令行 GPU 性能分析工具，基本使用方法如下所示。11.4 节有分析实例。

```
$nvprof --print-gpu-trace  ./cuda_2
```

NVIDIA visual profiler：图形化 GPU 性能分析工具，详见 17.8 节。

突破 kernel 函数执行时间限制办法如下。

（1）修改环境变量。修改环境变量可以从根本上解除 kernel 函数执行时间限制，具体修改方法为：单击"开始"→"运行"命令，弹出"运行"对话框，在"打开"文本框中输入 regedit，单击"确定"按钮，打开注册表管理器，查找 HKLM\System\CurrentControlSet\Control\GraphicsDrivers 键下的 TdrLevel 项（若无则新建 DWORD 类型值），将该项设置为 0 并重启。

（2）拆分 kernel 函数。基于分而治之的思想，单个 kernel 函数执行超时，那么将其拆分为多个 kernel 函数，令每个子 kernel 函数都不超时。通过分析具体的 kernel 函数，将 kernel 函数中的任务进行人为拆分，将子任务分别写成不同的 kernel 函数。

kernel 函数调度开销远大于 GPU 内部 block 切换调度开销和 kernel 内部循环开销，详见 15.8 节。

11.7 扩展练习

（1）利用本章提到的性能分析工具分析其他 GPU 程序的性能。

（2）针对某一超过 kernel 函数执行时间限制的 GPU 程序，分别采用本章提及的两种方法解决问题。

（3）寻找一个 GPU 代码实例，采用计算通信重叠方法进行优化。

第12章 扫描：前缀求和

12.1 前缀求和及其串行代码

前缀求和本质上是扫描算法，输入一个数组，返回一个同等长度的数组，新数组的每个元素是其对应的输入数组中索引小于等于该索引的元素和。例如，输入数组为$[x_0,x_1,\cdots,x_{n-1}]$，前缀求和运算后的输出数组为$[x_0,(x_0+x_1),\cdots,(x_0+x_1+\cdots+x_{n-1})]$。

扫描分为闭扫描和开扫描两种，前面介绍的前缀和为闭扫描，开扫描在求和时忽略对应元素本身，即输入数组为$[x_0,x_1,\cdots,x_{n-1}]$，前缀求和运算后的输出数组为$[0,x_0,(x_0+x_1),\cdots,(x_0+x_1+\cdots+x_{n-2})]$。

本章讨论的是闭扫描前缀和的 GPU 移植和优化，开扫描只需在闭扫描的基础上减去对应元素即可。

根据前缀求和计算公式，很容易得到相应的 C 程序实现源代码（标记为 serial_1）。

```
#define DATATYPE float
void scan_serial_1(DATATYPE *array_a,DATATYPE *array_b,int n)
{
    array_b[0]=array_a[0];
    for(int i=1;i<n;i++)
    {
        array_b[i]=array_a[i]+array_b[i-1];
    }
}
```

然而，上述代码存在漏洞（数据类型为 float 时）：当数据量 n 较小时，结果是正确的，而当数据量 n 逐渐增大后，结果越来越不准确。测试数据显示，当 $n=1024\times1024$ 时，结果已经偏离比较多了；而当 $n=1024\times1024\times128$ 时，结果则完全错误。详细的验证过程不在此赘述，有兴趣的读者可自行实践验证。

上述代码漏洞的原因是数据累加到 array_b[i] 上，每次累加都存在精度损失，当数据量达到一定程度后精度损失累积过多导致结果错误。针对这种原因,须对代码进行修改,修改后代码(标记为 serial)如下。修改思想是采用高精度中间变量,避免或减少精度损失的累积。

```
void scan_serial(DATATYPE *array_a,DATATYPE *array_b,int n)
{
    double sum=0;
    for(int i=0;i<n;i++)
    {
        sum+=array_a[i];
        array_b[i]=sum;
    }
}
```

再次验证修改后代码的计算结果,结果在精度范围内。说明上述分析和解决方案都是正确可行的。

扫描(前缀求和)计算不是一个简单的并行问题,无法通过简单分割来实现并行计算。针对前缀求和问题,本章利用 Kogge-Stone 和 Brent-Kung 两种方法来实现前缀求和的 GPU 移植。

12.2　Kogge-Stone 并行前缀和

Kogge-Stone 是一种最小深度的前缀和求解方法,图 12.1 展示了 Kogge-Stone 并行前缀和求解过程,其中,参与运算的两个数间隔随深度增加而指数增加(2 的幂次),仅索引大于等于运算间隔数的数据位置进行运算,最后一次运算的间隔为所有数据的一半。

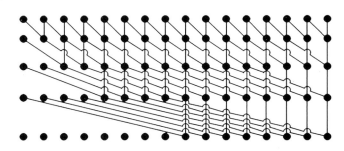

图 12.1　Kogge-Stone 并行前缀和

12.2.1　直接 Kogge-Stone 分段前缀和

下面给出 Kogge-Stone 并行前缀和的 CUDA C 代码。kernel 函数的功能是对每个 block 内的数据求前缀和,不同 block 内数据相互独立,即以 block 为单位分段求取前缀和。

```
#include<cuda_runtime.h>
#define shared_size 512
__global__ void scan_cuda_koggestone_1(const DATATYPE *array_a,DATATYPE
*array_b,long n)
{
    __shared__ DATATYPE temp[shared_size];
    const int tidx=threadIdx.x;
    int idx=blockIdx.x * blockDim.x+threadIdx.x;
    for(;idx<n;idx+=blockDim.x * gridDim.x)      //多block循环
//  if(idx<n)        //无block循环,仅限计算相应线程数量规模
    {
        temp[tidx]=array_a[idx];
        for(int stride=1;stride<=tidx;stride*=2)
        {
            __syncthreads();
            temp[tidx]+=temp[tidx-stride];
        }
        __syncthreads();
        array_b[idx]=temp[tidx];
    }
}
    scan_cuda_koggestone_1<<<shared_size,shared_size>>>(d_array_a,d_array_
b,n);
//  scan_cuda_koggestone_1<<<1,shared_size>>>(d_array_a,d_array_b,n);
```

从逻辑上看,上述代码并无漏洞,其中每次对共享存储的访问均包含同步操作,而运算过程也与 Kogge-Stone 并行前缀和计算过程相符合。但事实上实验结果显示程序存在问题,当数据量较小时,比如用单个 block 运算,结果是正确的;当数据量较大时(比如 512×512),前面十几个 block 计算结果是正确的,但后面开始出现莫名其妙的错误,而且出错位置是随机的(注:错误是对每个 block 的结果分别鉴别得出的结果)。

这个错误现象非常令人费解,若判定代码存在问题,但前部分求解是正确的;若代码正确,而后部分求解又出错,而且错误是随机的。虽然该程序很难定位错误,但基本可以判定这种随机出错现象是由于共享存储的使用不当引发的。通常情况下共享存储使用不当导致的结果出错一般是由于下次读取的数据不是本次修改的数据,一般采用 __syncthreads() 同步操作即可避免,然而在上述代码中,即使使用了同步操作也没有完全保证结果的正确性。该怎么解决这种问题呢?请阅读 12.2.2 节。

12.2.2 交错 Kogge-Stone 分段前缀和

针对 12.2.1 节描述的同步操作无法完全保证共享存储正确访存的情况,一个解决方案是再次申请相同尺寸的共享存储,交错地将新的数据和未修改的数据写到不同的共享存储空间。

下面是修改后的交错 Kogge-Stone 前缀求和 kernel 函数（标记为 koggestone）。其中，申请了两倍的共享存储空间，通过 pin 和 pout 来交错定位读取和存储的共享存储空间位置。运行结果显示，无论数据量多大，该版本的执行结果均正确。

```
__global__ void scan_cuda_koggestone_2(const DATATYPE *array_a,DATATYPE
*array_b,long n)
{
    __shared__ DATATYPE temp[shared_size * 2];
    int tidx=threadIdx.x;
    int idx=blockIdx.x * blockDim.x+threadIdx.x;
    for(;idx<n;idx+=blockDim.x * gridDim.x)
    {
        temp[tidx]=array_a[idx];
        int pout=0;
        int pin=1;
        for(int stride=1;stride<blockDim.x;stride*=2)
        {
            pout=1-pout;
            pin=1-pout;
            __syncthreads();
            temp[pout * shared_size+tidx]=temp[pin * shared_size+tidx];

            if(tidx>=stride)
            {
                temp[pout * shared_size+tidx]+=temp[pin * shared_size+tidx-
                    stride];
            }
        }
        __syncthreads();
        array_b[idx]=temp[pout * shared_size+tidx];
    }
}
    scan_cuda_koggestone_2<<<shared_size,shared_size>>>(d_array_a,d_array_
b,n);
```

这种方法虽然保证了结果的正确性，但数据访存次数明显增加了，必然会在一定程度上影响程序性能。

12.2.3 完整 Kogge-Stone 前缀和

当数据量大于 GPU 一个 block 内 thread 数量上限时，单个 kernel 函数无法完成完整的前缀求和运算，此时需要在前文所述分段求前缀和的基础上增加一系列步骤来完成完整的前缀和计算。图 12.2 展示了完整的 Kogge-Stone 前缀求和流程图。首先利用 Kogge-Stone 方法求分段前缀和，接着取每个分段的最后一个前缀和组成新分段，然后求新分段前缀和（若新分段仍大于 block 内 thread 数量，还需要再次分段求解），最后将新分段的前缀和扇出（Step4 所示的操作形似一把扇子）到相应分段并累加到该分段的每个元素。

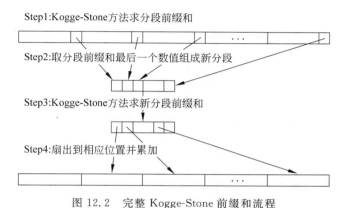

图 12.2　完整 Kogge-Stone 前缀和流程

下面是完整的 Kogge-Stone 前缀求和 CUDA C 代码。kernel 函数 scan_cuda_between_11 从分段前缀和取最后一个结果组成新分段；kernel 函数 scan_cuda_end_1 完成了扇出到相应分段位置并累加的计算任务。

```
__global__ void scan_cuda_between_11(const DATATYPE *array_b,DATATYPE *array_tmp,int mn)
{
    int idx=blockIdx.x * blockDim.x+threadIdx.x;
    if(idx<mn)
    {
        array_tmp[idx]=array_b[(idx+1) * shared_size-1];
    }
}

__global__ void scan_cuda_end_1(const DATATYPE *array_tmp,DATATYPE *array_b,long n)
{
    __shared__ DATATYPE temp[1];
    int tidx=threadIdx.x;
    int idx=blockIdx.x * blockDim.x+threadIdx.x;
    for(;idx<n;idx+=blockDim.x * gridDim.x)
    {
        if(tidx==0)
        {
            temp[0]=array_tmp[idx/shared_size-1];
```

```
        }
        __syncthreads();
        if(idx>=shared_size)
        {
            array_b[idx]=array_b[idx]+temp[0];
        }
    }
}
long n=1024*1024*4;
int mn=n/shared_size;
int mmn=mn/shared_size;
DATATYPE *d_array_a,*d_array_b,*d_array_tmp,*d_array_tmp1,*d_array_tmp_
tmp,*d_array_tmp_tmp1;
cudaMalloc((void **) &d_array_a, sizeof(DATATYPE) * n);
cudaMalloc((void **) &d_array_b, sizeof(DATATYPE) * n);
cudaMalloc((void **) &d_array_tmp, sizeof(DATATYPE) * mn);
cudaMalloc((void **) &d_array_tmp1, sizeof(DATATYPE) * mn);
cudaMalloc((void **) &d_array_tmp_tmp, sizeof(DATATYPE) * mmn);
cudaMalloc((void **) &d_array_tmp_tmp1, sizeof(DATATYPE) * mmn);
 ⋮
scan_cuda_koggestone_2<<<shared_size,shared_size>>>(d_array_a,d_array_b,n);
scan_cuda_between_11<<<mmn,shared_size>>>(d_array_b,d_array_tmp,mn);
scan_cuda_koggestone_2<<<mmn,shared_size>>>(d_array_tmp,d_array_tmp1,mn);
scan_cuda_between_11<<<1,mmn>>>(d_array_tmp1,d_array_tmp_tmp,mmn);
scan_cuda_koggestone_2<<<1,mmn>>>(d_array_tmp_tmp,d_array_tmp_tmp1,mmn);
scan_cuda_end_1<<<mmn,shared_size>>>(d_array_tmp_tmp1,d_array_tmp1,mn);
scan_cuda_end_1<<<shared_size,shared_size>>>(d_array_tmp1,d_array_b,n);
 ⋮
```

上述代码包含 3 次分段求前缀和的过程，能够处理的前缀求和计算的最大数据量为 $1024 \times 1024 \times 1024$，根据数据类型不同可能需要的存储空间为 4GB 或 8GB，基本上已相当于当前高端计算 GPU 的显存空间。再大的数据将超出 GPU 显存的容量，不在本章讨论范围之内。

12.3 Brent-Kung 并行前缀和

Brent-Kung 方法是另一种经典的并行前缀求和方法，如图 12.3 所示。该方法包含两个过程，分别是自底向上和自顶向下。自底向上过程中，跨度值从 1 开始以 2 的幂次增长，直到达到数据量的一半。自顶向下过程，跨度值从最大值的 1/4 开始以 2 的幂次减少，直到 1。

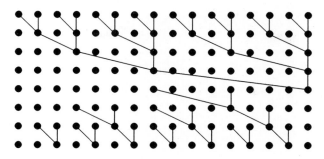

图 12.3 Brent-Kung 并行前缀求和

12.3.1 Brent-Kung 分段前缀和

下面的 kernel 函数实现了 Brent-Kung 分段前缀和的求解（标记为 brentkung）。代码中最外层 for 循环用于数据量巨大情况下的循环数据处理；内层第一个 for 循环实现 Brent-Kung 前缀求和的自底向上过程；内层第二个 for 循环实现 Brent-Kung 前缀求和的自顶向下的过程。

```
__global__ void scan_cuda_brentkung_1(const DATATYPE *array_a, DATATYPE *array
_b, long n)
{
    __shared__ DATATYPE temp[shared_size];
    int tidx=threadIdx.x;
    int idx=blockIdx.x * blockDim.x+threadIdx.x;
    for(;idx<n;idx+=blockDim.x * gridDim.x)
    {
        temp[tidx]=array_a[idx];
        for(int stride=1;stride<=blockDim.x;stride*=2)
        {
            __syncthreads();
            int index=(tidx+1) * 2 * stride-1;
            if(index<blockDim.x)
            {
                temp[index]+=temp[index-stride];
            }
        }
        for(int stride=blockDim.x/4; stride>0; stride/=2)
        {
            __syncthreads();
            int index=(tidx+1) * 2 * stride-1;
            if((index+stride)<blockDim.x)
            {
                temp[index+stride]+=temp[index];
```

```
            }
        }

        __syncthreads();
        array_b[idx]=temp[tidx];
    }
}
```

上述代码中需要关注索引的计算,即"int index=(tidx+1)*2*stride-1;"。下面的程序中利用一段代码打印数组长为 16 时的索引变化情况,可以直观地从中找到规律。

```
#define shared_size 16
    ⋮
    for(int stride=1;stride<=shared_size;stride*=2)
    {
        printf("stride:%d\t",stride);
        for(int tidx=0;tidx<shared_size;tidx++)
        {
            int index=(tidx+1) * 2 * stride-1;
            if(index<shared_size)
            {
                printf("%d\t",index);
            }
        }
        printf("\n");
    }
[fangmq@cn18%yhstar 11_scan]$./test
stride: 1 1 3 5 7 9 11 13 15
stride: 2 3 7 11 15
stride: 4 7 15
stride: 8 15
stride: 16
```

12.3.2 两倍数据的 Brent-Kung 分段前缀和

从图 12.3 和 12.3.1 节打印的 index 数量可知,16 个数据的数组其实仅需 8 个线程即能完成计算。因此,线程可以处理线程数量两倍数据的 Brent-Kung 分段前缀和。本节讨论通过增加一倍的数据来提升 Brent-Kung 分段前缀和 kernel 函数的线程利用率。

下面是修改后的两倍数据的 Brent-Kung 分段前缀和 CUDA C 代码(标记为 brentkung_2)。

```
__global__ void scan_cuda_brentkung_2(const DATATYPE *array_a,DATATYPE *array
_b,long n)
{
    __shared__ DATATYPE temp[shared_size*2];
    int tidx=threadIdx.x;
    int idx=(blockIdx.x*2)*blockDim.x+threadIdx.x;
    for(;idx<n;idx+=(blockDim.x*2)*gridDim.x)
    {
        temp[tidx]=array_a[idx];
        temp[shared_size+tidx]=array_a[idx+shared_size];

        for(int stride=1;stride<=blockDim.x;stride*=2)
        {
            __syncthreads();
            int index=(tidx+1)*2*stride-1;
            if(index<(blockDim.x*2))
            {
                temp[index]+=temp[index-stride];
            }
        }
        for(int stride=(blockDim.x*2)/4; stride>0; stride/=2)
        {
            __syncthreads();
            int index=(tidx+1)*2*stride-1;
            if((index+stride)<(blockDim.x*2))
            {
                temp[index+stride]+=temp[index];
            }
        }

        __syncthreads();
        array_b[idx]=temp[tidx];
        array_b[idx+shared_size]=temp[tidx+shared_size];
    }
}
```

12.3.3 避免 bank conflict 的两倍数据 Brent-Kung 分段前缀和

无论是 1 倍还是 2 倍数据的 Brent-Kung 分段前缀和计算，运算和访存均位于索引 index 位置及其与跨度的差值位置，当 block 内 thread 数量大于 32 时，会产生 bank conflict。避免 bank conflict 的方法之一是在每个 warp（32 个数据）后填充 1 个无效数据。

基于两倍数据 Brent-Kung 分段前缀和代码进行避免 bank conflict 的修改（实验结果显示 2 倍数据版本比 1 倍数据性能更好，因此选择该版本为基准进行改进），修改后 kernel 函数代码（标记为 brentkung_3）如下：

```
__global__ void scan_cuda_brentkung_3(const DATATYPE *array_a,DATATYPE *array_b,long n)
{
    __shared__ DATATYPE temp[shared_size * 2+(shared_size * 2)/32];
    int tidx=threadIdx.x;
    int idx=(blockIdx.x * 2) * blockDim.x+threadIdx.x;
    for(;idx<n;idx+=(blockDim.x * 2) * gridDim.x)
    {
        temp[tidx+tidx/32]=array_a[idx];
        temp[shared_size+tidx+(shared_size+tidx)/32]=array_a[idx+shared_size];

        for(int stride=1;stride<=blockDim.x;stride*=2)
        {
            __syncthreads();
            int index=(tidx+1) * 2 * stride-1;
            if(index<(blockDim.x * 2))
            {
                temp[index+index/32]+=temp[index-stride+(index-stride)/32];
            }
        }
        for(int stride=(blockDim.x * 2)/4; stride>0; stride/=2)
        {
            __syncthreads();
            int index=(tidx+1) * 2 * stride-1;
            if((index+stride)<(blockDim.x * 2))
            {
                temp[index+stride+(index+stride)/32]+=temp[index+index/32];
            }
        }

        __syncthreads();
        array_b[idx]=temp[tidx+tidx/32];
        array_b[idx+shared_size]=temp[tidx+shared_size+(tidx+shared_size)/32];
    }
}
```

12.3.4 完整 Brent-Kung 前缀和

与 Kogge-Stone 前缀求和类似，在 Brent-Kung 前缀求和中，当数据量大于 1024（单个 block 内线程上限）时，单个 kernel 函数无法一次性完成 Brent-Kung 前缀求和计算，需要一系列额外的辅助计算（见图 12.4）。首先利用 Brent-Kung 方法求分段前缀和，再取各分段前缀和的最后一个结果组建新分段，继续求新分段前缀和，后将新分段的前缀和扇出到相应分段并与该分段的所有元素相加。

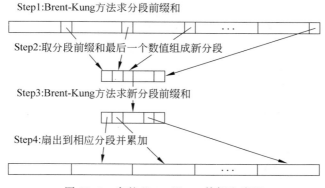

图 12.4　完整 Brent-Kung 前缀和流程

下面是基于 12.3.1 节 1 倍数据 Brent-Kung 分段前缀和开发的完整版 Brent-Kung 前缀求和 CUDA C 代码。scan_cuda_between_11()函数从分段前缀和中取最后一个结果组成新分段；scan_cuda_end_1()函数扇出新分段前缀和元素到相应分段位置并与该分段所有元素相加。

```
    ⋮
scan_cuda_brentkung_1<<<shared_size,shared_size>>>(d_array_a,d_array_b,n);
scan_cuda_between_11<<<mmn,shared_size>>>(d_array_b,d_array_tmp,mn);
scan_cuda_brentkung_1<<<mmn,shared_size>>>(d_array_tmp,d_array_tmp1,mn);
scan_cuda_between_11<<<1,mmn>>>(d_array_tmp1,d_array_tmp_tmp,mmn);
scan_cuda_brentkung_1<<<1,mmn>>>(d_array_tmp_tmp,d_array_tmp_tmp1,mmn);
scan_cuda_end_1<<<mmn,shared_size>>>(d_array_tmp_tmp1,d_array_tmp1,mn);
scan_cuda_end_1<<<shared_size,shared_size>>>(d_array_tmp1,d_array_b,n);
    ⋮
```

在 2 倍数据参与的 Brent-Kung 前缀求和运算中（12.3.2 节或 12.3.3 节），上述 scan_cuda_between_11()函数和 scan_cuda_end_1()函数不再适用，修改后的适用于 2 倍数据运算的完整版 Brent-Kung 前缀求和 CUDA C 代码如下：

```
__global__ void scan_cuda_between_22(const DATATYPE *array_b,DATATYPE *array_tmp,int mn)
{
```

```
        int idx=blockIdx.x*blockDim.x+threadIdx.x;
        if(idx<mn)
        {
            array_tmp[idx]=array_b[(idx+1)*(shared_size*2)-1];
        }
    }

    __global__ void scan_cuda_end_2(const DATATYPE *array_tmp,DATATYPE *array_b,
    long n)
    {
        __shared__ DATATYPE temp[1];
        int tidx=threadIdx.x;
        int idx=blockIdx.x*blockDim.x+threadIdx.x;
        for(;idx<n;idx+=blockDim.x*gridDim.x)
        {
            if(tidx==0)
            {
                temp[0]=array_tmp[idx/blockDim.x-1];
            }
            __syncthreads();
            if(idx>=blockDim.x)
            {
                array_b[idx]=array_b[idx]+temp[0];
            }
        }
    }
```

```
        ⋮
    scan_cuda_brentkung_3<<<shared_size,shared_size>>>(d_array_a,d_array_b,n);
    scan_cuda_between_22<<<mmn,(shared_size*2)>>>(d_array_b,d_array_tmp,mn);
    scan_cuda_brentkung_3<<<mmn,shared_size>>>(d_array_tmp,d_array_tmp1,mn);
    scan_cuda_between_22<<<1,mmn>>>(d_array_tmp1,d_array_tmp_tmp,mmn);
    scan_cuda_brentkung_3<<<1,mmn>>>(d_array_tmp_tmp,d_array_tmp_tmp1,mmn);
    scan_cuda_end_2<<<mmn,(shared_size*2)>>>(d_array_tmp_tmp1,d_array_
    tmp1,mn);
    scan_cuda_end_2<<<shared_size,(shared_size*2)>>>(d_array_tmp1,d_array_
    b,n);
        ⋮
```

12.4　warp 分段的 Kogge-Stone 前缀求和

12.2 节和 12.3 节的前缀求和都是按 block 分段的，若缩短分段性能又会如何呢？比如按 warp 分段，从理论上分析，由于 warp 内所有 thread 执行相同的指令，可以避免使用同步操作,进而提升性能。下面展示了按 warp(32 个数据)分段时,Kogge-Stone 前缀求和的 CUDA C 代码(标记为 koggestone_32),该代码见参考文献[5]。

```
__device__ DATATYPE scanwarp(volatile DATATYPE *spartials)
{
    const int tid=threadIdx.x;
    const int lane=tid&31;
    if(lane>=1)
    {   spartials[0]+=spartials[-1];}
    if(lane>=2)
    {   spartials[0]+=spartials[-2];}
    if(lane>=4)
    {   spartials[0]+=spartials[-4];}
    if(lane>=8)
    {   spartials[0]+=spartials[-8];}
    if(lane>=16)
    {   spartials[0]+=spartials[-16];}
    return spartials[0];
}
__device__ DATATYPE scanblock(volatile DATATYPE *spartials)
{
    extern __shared__ DATATYPE warppartials[];
    const int tid=threadIdx.x;
    const int lane=tid&31;
    const int warpid=tid>>5;
    DATATYPE sum=scanwarp(spartials);
    __syncthreads();
    if(lane==31)
    {   warppartials[16+warpid]=sum;}
    __syncthreads();
    if(warpid==0)
    {   scanwarp(16+warppartials+tid);}
    __syncthreads();
    if(warpid>0)
    {   sum+=warppartials[16+warpid-1];}
    __syncthreads();
    *spartials=sum;
    __syncthreads();
    return sum;
}
__global__ void scanandwritepartials(DATATYPE *out,DATATYPE *gpartials,const DATATYPE *in,size_t n,size_t numblocks,bool bwritespine)
{
    extern volatile __shared__ DATATYPE spartials[];
```

```
        const int tid=threadIdx.x;
        volatile DATATYPE *myshared=spartials+tid;
        for(size_t iblock=blockIdx.x;iblock<numblocks;iblock+=gridDim.x)
        {
            size_t index=iblock*blockDim.x+tid;
            *myshared=(index<n)?in[index]:0;
            __syncthreads();
            DATATYPE sum=scanblock(myshared);
            __syncthreads();
            if(index<n)
            {
                out[index]=*myshared;
            }
            if(bwritespine&&(threadIdx.x==(blockDim.x-1)))
            {
                gpartials[iblock]=sum;
            }
        }
    }
    __global__ void scanaddbasesums(DATATYPE *out,DATATYPE *gbasesums,size_t n,
size_t numblocks)
    {
        const int tid=threadIdx.x;
        DATATYPE fan_value=0;
        for(size_t iblock=blockIdx.x;iblock<numblocks;iblock+=gridDim.x)
        {
            size_t index=iblock*blockDim.x+tid;
            if(iblock>0)
            {
                fan_value=gbasesums[iblock-1];
            }
            out[index]+=fan_value;
        }
    }
```
```
    void scanfan(DATATYPE *out,const DATATYPE *in,size_t n,int b)
    {
        if(n<=b)
        {
            scanandwritepartials<<<1,b,b*sizeof(DATATYPE)>>>(out,0,in,n,1,
            false);
            return;
        }
        DATATYPE *gpartials=0;
```

```
    size_t numpartials=n/b;
    const unsigned int maxblocks=150;
    unsigned int numblocks=(numpartials<maxblocks)?numpartials : maxblocks;
    cudaMalloc(&gpartials,numpartials * sizeof(DATATYPE));
    scanandwritepartials<<<numblocks,b,b*sizeof(DATATYPE)>>>(out,
    gpartials,in,n,numpartials,true);
    scanfan(gpartials,gpartials,numpartials,b);
    scanaddbasesums<<<numblocks,b>>>(out,gpartials,n,numpartials);
}
```

```
long n=1024 * 1024 * 4;
DATATYPE *array_a,*array_b;
array_a=(DATATYPE *)malloc(sizeof(DATATYPE) * n);
array_b=(DATATYPE *)malloc(sizeof(DATATYPE) * n);
  ⋮
DATATYPE *d_array_a,*d_array_b;
cudaMalloc((void **) &d_array_a, sizeof(DATATYPE) * n);
cudaMalloc((void **) &d_array_b, sizeof(DATATYPE) * n);
cudaMemcpy(d_array_a,array_a,sizeof(DATATYPE ) * n,cudaMemcpyHostToDevice);
scanfan(d_array_b,d_array_a,n,512);
cudaMemcpy(array_b,d_array_b,sizeof(DATATYPE ) * n,cudaMemcpyDeviceToHost);
cudaFree(d_array_a);
cudaFree(d_array_b);
free(array_a);
free(array_b);
```

除了上述代码，参考文献[5]还提供了 warp 为分段单位的其他几种实现方法，可实践对比不同实现方法间的性能优劣，本书不再赘述。

12.5 实验结果分析与结论

配置数据类型为 int，输入数据规模为 $1024 \times 1024 \times 4$，统计各函数执行时间（包括 CPU 与 GPU 通信时间）在表 12.1 中。其中，当输入数据类型为 int 时，serial_1 的计算结果是正确的。计算 1～计算 7 分别指代前文代码中的 7 个 kernel 函数。CPU 端存储为可分页存储，若声明为页锁定存储，H2D 和 D2H 的传输时间可缩减近半。

表 12.1 前缀求和执行时间　　　　　　　　　　　　　　　　单位：ms

step	serial	serial_1	koggestone	brentkung	brentkung_2	brentkung_3	koggestone_32
计算1	6.86	3.52	0.99	1.94	1.26	0.96	0.73
计算2			9.31E−03	9.18E−03	7.04E−03	6.91E−03	6.05E−03
计算3			7.17E−03	1.32E−02	1.53E−02	1.17E−02	4.74E−03
计算4			2.78E−03	2.78E−03	2.46E−03	2.50E−03	3.65E−03

续表

step	serial	serial_1	koggestone	brentkung	brentkung_2	brentkung_3	koggestone_32
计算 5			3.78E−03	4.67E−03	4.03E−03	4.74E−03	
计算 6			3.42E−03	3.39E−03	4.03E−03	3.94E−03	
计算 7			0.24	0.25	0.30	0.30	0.21
H2D			5.10	5.06	5.14	5.17	5.05
D2H			7.73	7.63	7.72	8.67	8.67
总计算	6.86	3.52	1.26	2.22	1.59	1.29	0.95
总时间	6.86	3.52	14.09	14.92	14.45	15.12	14.67

表 12.1 中数据显示：

(1) 对比 serial 和 serial_1 两列数据，serial 版本比 serial_1 版本慢。性能差异是由于高精度（double）类型数据参与运算引入的，若改回原来的数据类型则两者相差无几。当高精度类型数据参与运算时，真正参与运算的指令就变成了双精度指令。

(2) 比较 brentkung 和 brentkung_2 两列数据，显然 brentkung_2 计算性能更好，说明利用两倍数据来充分利用线程是划算的。

(3) 比较 brentkung_2 和 brentkung_3 两列数据，brentkung_3 版本的计算耗时更短，说明在 brentkung_2 确实存在 bank conflict，而在 brentkung_3 版本成功避免了 bank conflict，并获得了性能收益。

(4) 对比 koggestone 和 brentkung_3 两列数据，发现 kernel 函数计算总时间相当接近，其中分段求前缀和过程 brentkung_3 略占优势，而扇出累加过程则是 koggestone 更好。说明两种方法的性能基本一致。

一些 GPU 书籍[5]认为 Kogge-Stone 是低效率方法，而 Brent-Kung 是高效率方法，但实验结果并非如此。笔者认为在理论上 Kogge-Stone 性能应该更好，原因是 12.2.2 节的代码中多了数据复制并多占用了 1 倍的共享存储，而 12.2.1 节的代码虽然执行出错但逻辑是正确的，若能很好地解决该问题，Kogge-Stone 的性能优势应更为突出。

(5) 对比 koggestone 和 koggestone_32 两列，koggestone_32 版本的前缀求和性能显然更高。说明以 warp 为单位的分段前缀和比更大分段（512）的前缀求和性能更好。原因可能包括：①warp 为单位时用 volatile 声明共享变量为敏感变量可避免同步操作；②避免了 12.2.2 节两倍共享存储使用和多余的数据复制过程。

(6) 对比总计算时间，发现尽管 GPU 并行版本比 CPU 串行版本快，但是性能提升非常有限，最高仅 3.7 倍。究其原因应该是前缀求和问题本身的计算量与数据量关系是 $n:n$，GPU 上的并行并不是简单并行，而是需要复杂的运算，还包括不少逻辑判断。

(7) 从总时间看，CPU 与 GPU 间的通信时间占比较大，且仅通信时间（即便是采用页锁定存储）就超过了 CPU 端串行计算时间。与向量加法和向量内积类似，单纯地将前缀求和过程移植到 GPU 是不划算的。

若参与运算的数据类型为 float，为保证计算结果精度，还需要对本章的 kernel 函数进行改进，采用 Kahan's Summation Formula 方法，该方法的用法效果及其在 GPU 上的性能已在第 8 章详细论述。

12.6 知识点总结

前缀求和闭扫描：输入一个数组，返回一个同等长度的数组，新数组的每个元素是对应的输入数组中索引小于等于该索引的元素和。例如，输入数组为 $[x_0, x_1, \cdots, x_{n-1}]$，前缀求和运算后的输出数组为 $[x_0, (x_0+x_1), \cdots, (x_0+x_1+\cdots+x_{n-1})]$。

前缀求和开扫描：开扫描求和时，在闭扫描基础上忽略对应元素本身，即输入数组为 $[x_0, x_1, \cdots, x_{n-1}]$，前缀求和运算后的输出数组为 $[0, x_0, (x_0+x_1), \cdots, (x_0+x_1+\cdots+x_{n-2})]$。

Float 类型数据运算时会累积误差，编程时需要注意，解决方法有两种：高精度中间变量和 Kahan's Summation Formula 方法。

前缀求和的两种并行化方法：Kogge-Stone 和 Brent-Kung。

Kogge-Stone 前缀求和：详见 12.2 节。

Brent-Kung 前缀求和：详见 12.3 节。

共享存储运算结果出错而逻辑正确时，一种解决方案是检查同步操作，若还出错则可以利用交错复制的方法，每次更新在额外存储中，交错进行（详见 12.2.2 节）。

GPU 上完整前缀求和：①求分段前缀和；②取分段前缀和最后一个元素组成新分段；③再次求新分段前缀和（若新分段较大还需再继续分段）；④扇出到相应分段并与该分段所有元素相加（详见 12.2.3 节和 12.3.4 节）。

Brent-Kung 前缀和索引运算：index＝(tidx＋1)＊2＊stride－1(详见 12.3.1 节)。

两倍数据的 Brent-Kung 分段前缀和的修改（见 12.3.2 节）。

两倍数据的 Brent-Kung 分段前缀和如何避免 bank conflict（见 12.3.3 节）。

两倍数据的 Brent-Kung 分段前缀和的配套取值函数和扇出累加函数（见 12.3.4 节）。

以 warp 为分段单位的 Kogge-Stone 前缀求和方法的 GPU 实现（见 12.4 节）。

CPU 运算中引入 double 类型可能导致增加 1 倍的耗时。

两倍数据的 Brent-Kung 前缀求和比普通 1 倍数据版本性能要好。

Brent-Kung 前缀求和存在 bank conflict，通过在共享存储每 32 个数据后填充 1 个无效数据方法可以避免 bank conflict，避免 bank conflict 后性能获得了明显的提升。

Kogge-Stone 和 Brent-Kung 两种方法在 GPU 上性能相当。

在前缀求和问题中，GPU 计算比 CPU 计算性能提升有限，主要原因是前缀求和在 CPU 上串行计算本身运算量较少，而在 GPU 中并非简单并行，需要复杂的运算甚至包含不少逻辑判断操作。

串行前缀求和时间少于通信时间。若在数据位于 CPU 端，将前缀求和过程移植到 GPU 不可取；若数据驻留在 GPU 端，前缀求和作为复杂应用的一个步骤，则采用 GPU 加速前缀求和可以获得性能收益。

12.7 扩展练习

（1）实现其他方法的 warp 分段前缀求和。
（2）完成避免 bank conflict 的 1 倍数据 Brent-Kung 分段前缀和。
（3）将 kernel 函数数据类型改为 float 型，测试并分析实验结果。
（4）实现前缀求和开扫描的 GPU 移植和优化。
（5）选取其他扫描算法（非加法运算）进行 GPU 移植和优化研究。
（6）扩展 12.2 节的方法，实现任意规模数据的前缀求和。

第13章 排 序

13.1 串行排序及其性能

排序是计算机中常见的经典问题,排序方法有很多,比如选择排序、冒泡排序、快速排序、双调排序网络、归并排序和基数排序等。大量数据结构书籍都对各种排序方法进行了详细的分析论述,本书不再赘述。本节试图通过实验分析测评这些排序算法的实际性能。

13.1.1 选择排序

选择排序是最简单的排序方法之一,其核心思想是每次从待排序数列中选择最小(或最大)元素交换到已排序数列的末尾。下面给出选择排序的C语言代码。

```c
void select_sort(DATATYPE *a, unsigned int n)
{
    DATATYPE temp;
    unsigned int id;
    for(unsigned int i=0;i<n;i++)
    {
        temp=(DATATYPE)RAND_MAX;
        for(unsigned int j=i;j<n;j++)
        {
            if(temp>a[j])
            {
                temp=a[j];
                id=j;
            }
        }
        a[id]=a[i];
        a[i]=temp;
    }
}
select_sort(array_b1,n);
```

13.1.2 冒泡排序

冒泡排序的思想是逐个对比相邻元素，将较大（或较小）的元素往后移，一趟结束后最大（或最小）元素就像气泡一样冒到了最后，如此往复即可获得有序序列。下面给出冒泡排序的 C 语言实现代码。

```c
void bubbling_sort(DATATYPE *arr,unsigned int n)
{
    DATATYPE temp;
    unsigned int temp1;
    for(unsigned int i=n-1;i>0;i--)
    {
        for(unsigned int j=0;j<i;j++)
        {
            if(arr[j]>arr[j+1])
            {
                temp=arr[j];
                arr[j]=arr[j+1];
                arr[j+1]=temp;
            }
        }
    }
}
    bubbling_sort(array_b2,n);
```

13.1.3 快速排序

快速排序的思想是选择第一个元素为基数，通过头尾两个指针的比较、交换和移动，使其定位在正确的位置上，此时其左边的数列均小于（或大于）该基数，右边的数列均大于（或小于）该基数。下面给出快速排序 C 语言实现代码。

```c
void quick_sort(DATATYPE *arr,unsigned int startPos, unsigned int endPos)
{
    unsigned int i,j;
    DATATYPE key;
    key=arr[startPos];
    i=startPos;
    j=endPos;
    while(i<j)
    {
        while(arr[j]>=key && i<j)--j;
        arr[i]=arr[j];
```

```
            while(arr[i]<=key && i<j)++i;
            arr[j]=arr[i];
        }
        arr[i]=key;
        if(i-1>startPos) quick_sort(arr,startPos,i-1);
        if(endPos>i+1) quick_sort(arr,i+1,endPos);
    }
    quick_sort(array_b3,0,n-1);
```

13.1.4 基数排序

基数排序的思想是对数列各元素的二进制位进行对比，从最低位到最高位，逐位比较，将该值为 0 的元素排在值为 1 的元素前，完成所有位的比较后即可得到从小到大的数列（反之为从大到小的数列）。下面给出基数排序的 C 语言实现代码。

```
#define base_int unsigned int
void radix_sort(base_int *arr,base_int n)
{
    base_int *tmp_0,*tmp_1;
    tmp_0=(base_int *)malloc(sizeof(base_int)*n);
    tmp_1=(base_int *)malloc(sizeof(base_int)*n);

    for(base_int bit=0;bit<32;bit++)
    {
        base_int base_0=0;
        base_int base_1=0;
        for(base_int i=0;i<n;i++)
        {
            base_int x=arr[i];
            base_int bit_mask=(1<<bit);

            if((x & bit_mask)>0)
            {
                tmp_1[base_1]=x;
                base_1++;
            }
            else
            {
                tmp_0[base_0]=x;
                base_0++;
            }
        }
```

```
            for(base_int i=0;i<base_0;i++)
            {
                arr[i]=tmp_0[i];
            }
            for(base_int i=0;i<base_1;i++)
            {
                arr[i+base_0]=tmp_1[i];
            }
        }
        free(tmp_0);
        free(tmp_1);
    }
    radix_sort(array_b,n);
```

在上述代码中存在冗余数据移动,即每位值为 0 的数据先移动到 tmp_0 数组,后复制回 arr 数组,事实上 arr 数组前面空闲(已移动)的空间足够存放 tmp_0 数组,故可移除 tmp_0 数组,既节省存储空间又减少数据移动。下面给出优化后的基数排序 C 语言实现代码。

```
void radix_sort1(base_int *arr,base_int n)
{
    base_int *tmp_1;
    tmp_1=(base_int * )malloc(sizeof(base_int) * n);

    for(base_int bit=0;bit<32;bit++)
    {
        base_int base_0=0;
        base_int base_1=0;
        for(base_int i=0;i<n;i++)
        {
            base_int x=arr[i];
            base_int bit_mask=(1<<bit);

            if((x & bit_mask)>0)
            {
                tmp_1[base_1]=x;
                base_1++;
            }
            else
            {
                arr[base_0]=x;
                base_0++;
            }
        }
```

```
        for(base_int i=0;i<base_1;i++)
        {
            arr[i+base_0]=tmp_1[i];
        }
    }
    free(tmp_1);
}
radix_sort1(array_b11,n);
```

13.1.5 双调排序网络

排序网络是对任何输入进行相同处理,输出一个有序的序列,属于排序算法。双调排序网络是一种较快的排序网络,特别适合于并行处理。双调序列包含两个单调的子序列:一个非递减,另一个非递增,中间值为其最大值或最小值,可抽象看成∧或∨。13.3 节将详细阐述双调排序网络的思想。下面给出双调排序网络的 C 语言实现代码。

```
void bitonic_sort(DATATYPE *arr,int n)
{
    DATATYPE temp;
    for(int k=2;k<=n;k*=2)
    {
        for(int j=k/2;j>0;j/=2)
        {
            for(int i=0;i<n;i++)
            {
                int iXj=i^j;
                if(iXj>i)
                {
                    if((i&k)==0 && arr[i]>arr[iXj])
                    {
                        temp=arr[i];
                        arr[i]=arr[iXj];
                        arr[iXj]=temp;
                    }
                    if((i&k)!=0 && arr[i]<arr[iXj])
                    {
                        temp=arr[i];
                        arr[i]=arr[iXj];
                        arr[iXj]=temp;
                    }
                }
            }
        }
    }
}
bitonic_sort(array_b4,n);
```

上述代码摘自《GPU高性能运算之CUDA》[2]一书,串行执行结果正确。但该书中的GPU代码中,将i层循环做成GPU并行,即用线程索引替换i,此时却得不到正确的执行结果(相关实验见13.3.2节)。笔者认为原因可能是在i层循环存在依赖,当条件符合时i和i×j(i和j的异或)交换,循环后面当$i^{(n)}=i×j$时就有了依赖。

13.1.6 合并排序

合并排序是将两个有序的序列合并成一个有序的序列。在排序时首先需要将原始序列不断拆分,直到仅剩1个元素(有序序列),然后再进行合并。下面给出合并排序的C语言实现代码。

```c
void merge(DATATYPE *arr,int low,int m,int high)
{
    int i=low,j=m+1,p=0;
    DATATYPE *brr;
    brr=(DATATYPE *)malloc((high-low+1) * sizeof(DATATYPE));
    if(!brr)
    {
        return;
    }
    while(i<=m&&j<=high)
    {
        brr[p++]=(arr[i]<=arr[j])? arr[i++]:arr[j++];
    }

    while(i<=m)
    {
        brr[p++]=arr[i++];
    }
    while(j<=high)
    {
        brr[p++]=arr[j++];
    }

    for(p=0,i=low;i<=high;p++,i++)
    {
        arr[i]=brr[p];
    }
    free(brr);
}
void merge_sort(DATATYPE *arr,int low,int high)
{
    int mid;
```

```
    if(low<high)
    {
        mid=(low+high)/2;
        merge_sort(arr,low,mid);
        merge_sort(arr,mid+1,high);
        merge(arr,low,mid,high);
    }
}
merge_sort(array_b5,0,n-1);
```

13.1.7 串行排序性能对比

在对比实验结果前,先定义输入数据,本章输入的待排序数据是利用rand()函数生成的随机数据,可能无法覆盖所有的排序情况,但也有一定的代表性。下面是输入数据生成代码。

```
#define DATATYPE unsigned int
void vector_init(DATATYPE *a,unsigned int n)
{
    for(unsigned int i=0;i<n;i++)
    {
        a[i]=(DATATYPE)rand();
    }
}
```

排序后结果的验证主要通过以下代码完成。代码检验的是从小到大的排序序列。

```
void sort_err(DATATYPE *a,unsigned int n,char sortname[])
{
    for(unsigned int i=0;i<(n-1);i++)
    {
        if(a[i]>a[i+1])
        {
            printf("Sorry,%s is error!!!\n",sortname);
            return;
        }
    }
    printf("Congratulates, %s is right!\n",sortname);
}
sort_err(array_b,n,"radix sort");
```

下面给出程序的执行结果,其中待排序数列拥有1 048 576个uint元素。

```
[fangmq@cn18%yhstar book]$./sort
1048576      radix sort          200.409174(ms)
1048576      radix1 sort         187.321901(ms)
1048576      select sort         548425.249100(ms)
1048576      bubbling sort       1956573.269844(ms)
1048576      quick sort          107.421875(ms)
1048576      bitonic sort        430.958033(ms)
1048576      merge sort          181.921005(ms)
Congratulates, radix sort is right!
Congratulates, radix1 sort is right!
Congratulates, select sort is right!
Congratulates, bubbling sort is right!
Congratulates, quick sort is right!
Congratulates, bitonic sort is right!
Congratulates, merge sort is right!
```

上述结果数据显示：

（1）串行排序方法的性能相差极大，基数排序、快速排序、双调排序网络和合并排序的性能远强于选择排序和冒泡排序。后文基于 GPU 的并行排序也主要基于这些快速的排序方法展开研究。

（2）从算法上分析，选择排序和冒泡排序性能相当，但实验结果显示选择排序明显占优，主要原因除了输入数据的因素外，还因为选择排序相对冒泡排序减少了大量数据移动，减少了存储访问，而访存往往是影响计算机程序执行性能的关键因素。

（3）排序算法的串行性能从高到低分别为快速排序、合并排序、基数排序、双调排序网络、选择排序、冒泡排序。

（4）基数排序中，优化后的版本（radix1）比原始版本（radix）耗时稍短，说明 13.1.4 节阐述的优化是有效的。

13.2 基数排序

13.2.1 基数排序概述

基数排序是对待排序数列各元素的二进制位进行对比，从最低位到最高位，逐位比较，每次都将二进制位值为 0 的元素排在值为 1 的元素前，完成所有位的比较后即可得到升序序列（反之为降序序列）。图 13.1 给出了一个基数排序的实例，通过 8 个步骤可完成 8 位二进制元素的序列排序。

并行排序时，可将序列等分成若干分段，每个子序列可以并行进行基数排序，排序后可得到一些有序的子序列，此时需要进行合并，将单独的有序子序列合并为一个完整的有序序列。图 13.2 展示了基数排序的合并过程，首先比较每个子序列的最小值（首个元素），取其中最小值移到整个数列的首个元素；接着再次比较所有子序列的最小值，再次取

	step1	step2	step3	step4	step5	step6	step7	step8	
124	01111100	01111100	01111100	11110000	11110000	01000100	01000100	00000101	00000101	5
20	00010100	00010100	00010100	00010010	00010010	00000101	00000101	00010010	00010010	18
5	00000101	01010110	11110000	01111100	00010100	11110000	00010010	00010100	00010100	20
86	01010110	11110000	01000100	00010100	01000100	00010010	00010100	10110111	01000100	68
240	11110000	01000100	00000101	01000100	00000101	00010100	01010110	01000100	01010110	86
183	10110111	00010010	01010110	00000101	01010110	01010110	11110000	01010110	01111100	124
68	01000100	00000101	00010010	01010110	10110111	10110111	10110111	11110000	10110111	183
18	00010010	10110111	10110111	10110111	01111100	01111100	01111100	01111100	11110000	240

图 13.1 基数排序实例

最小值移到数列列尾；以此类推，移完所有的元素即可最终得到有序的完整序列。

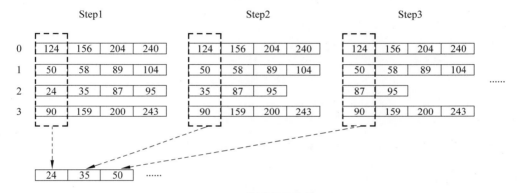

图 13.2 基数排序的合并过程

13.2.2 单 block 基数排序

基数排序要在 GPU 上并行执行，首先进行任务分割，将任务分配给大量线程，每个线程进行基数排序，然后将基数排序的结果合并，得到最终结果。

下面代码实现了 GPU 的基数排序过程。由于 GPU 全局存储的访问特性，分段时并不连续分段，而是按线程索引号进行分段（即数据访问为 arr[i＋tid]）。

```
#define base_int unsigned int
__device__ void radix_sort(base_int *arr,base_int n,base_int *tmp_1,base_int tid,base_int tdim)
{
    for(base_int bit=0;bit<32;bit++)
    {
        base_int base_0=0;
        base_int base_1=0;
        base_int bit_mask= (1<<bit);
        for(base_int i=0;i<n;i+=tdim)
        {
```

```
            base_int x=arr[i+tid];
            if((x & bit_mask)>0)
            {
                tmp_1[base_1+tid]=x;
                base_1+=tdim;
            }
            else
            {
                arr[base_0+tid]=x;
                base_0+=tdim;
            }
        }
        for(base_int i=0;i<base_1;i+=tdim)
        {
            arr[i+base_0+tid]=tmp_1[i+tid];
        }
    }
}
```

基数排序后的结果合并有多种实现方案。下面分别介绍基数排序后序列的单线程合并（merge_array）、原子操作多线程合并（merge_array1）、归约多线程合并（merge_array2）和分段原子操作多线程合并（merge_array3）。

单线程合并（**merge_array**）：即图 13.2 所示的合并过程，该过程中仅使用单个线程。下面给出相应的 CUDA C 实现代码。

```
__device__ void merge_array(base_int *arr,base_int *brr,base_int tid,base_int tdim,base_int n)
{
    __shared__ base_int list_index[max_num_lists];
    list_index[tid]=0;
    __syncthreads();
    if(tid==0)
    {
        base_int nl=n/tdim;
        for(base_int i=0;i<n;i++)
        {
            base_int min_val=0xFFFFFFFF;
            base_int min_idx=0;
            for(base_int list=0;list<tdim;list++)
            {
                if(list_index[list]<nl)
                {
```

```
                    base_int idx=list+(list_index[list]*tdim);
                    base_int x=arr[idx];
                    if(x<=min_val)
                    {
                        min_val=x;
                        min_idx=list;
                    }
                }
            }
            list_index[min_idx]++;
            brr[i]=min_val;
        }
    }
}
```

原子操作多线程合并（merge_array1）：图 13.2 的合并过程可分解为数据读取、求最小值。在原子操作多线程合并中，数据读取过程是并行的，而求最小值则利用了原子操作函数 atomicMin()，该函数本质上是一个串行访问数据的过程。

```
__device__ void merge_array1(base_int *arr,base_int *brr,base_int tid,base_
int tdim,base_int n)
{
    base_int nl=n/tdim;
    __shared__ base_int list_index[max_num_lists];
    list_index[tid]=0;
    __syncthreads();
    for(base_int i=0;i<n;i++)
    {
        __shared__ base_int min_val;
        __shared__ base_int min_idx;
        base_int x;
        if(list_index[tid]<nl)
        {
            base_int idx=tid+(list_index[tid]*tdim);
            x=arr[idx];
        }
        else
        {
            x=0xFFFFFFFF;
        }
        if(tid==0)
        {
```

```
            min_val=0xFFFFFFFF;
            min_idx=0xFFFFFFFF;
        }
        __syncthreads();
        atomicMin(&min_val,x);
        __syncthreads();
        if(min_val==x)
        {
            atomicMin(&min_idx,tid);
        }
        __syncthreads();
        if(tid==min_idx)
        {
            list_index[tid]++;
            brr[i]=x;
        }
    }
}
```

归约多线程合并(merge_array2)：该版本中,数据读取过程等同 merge_array1 版本,也是并行的。求最小值的过程采用了归约运算,即代码中的 while 循环,归约过程是并行的,因此在理论上该版本的性能应该优于 merge_array1 版本。

```
__device__ void merge_array2(base_int *arr,base_int *brr,base_int tid,base_int tdim,base_int n)
{
    base_int nl=n/tdim;
    __shared__ base_int list_index[max_num_lists];
    __shared__ base_int red_val[max_num_lists];
    __shared__ base_int red_idx[max_num_lists];
    list_index[tid]=0;
    red_val[tid]=0;
    red_idx[tid]=0;
    __syncthreads();
    for(base_int i=0;i<n;i++)
    {
        base_int tid_max=tdim>>1;
        base_int x;
        if(list_index[tid]<nl)
        {
            base_int idx=tid+(list_index[tid] * tdim);
            x=arr[idx];
        }
```

```
        else
        {
            x=0xFFFFFFFF;
        }
        red_val[tid]=x;
        red_idx[tid]=tid;
        __syncthreads();
        while(tid_max!=0)
        {
            if(tid<tid_max)
            {
                base_int idx1=tid+tid_max;
                base_int val1=red_val[idx1];
                if(red_val[tid]>val1)
                {
                    red_val[tid]=val1;
                    red_idx[tid]=red_idx[idx1];
                }
            }
            tid_max>>=1;
            __syncthreads();
        }
        if(tid==0)
        {
            list_index[red_idx[0]]++;
            brr[i]=red_val[0];
        }
        __syncthreads();
    }
}
```

分段原子操作多线程合并（merge_array3）：首先，该版本的数据读取过程是并行的；其次，最小值的求解时，采用了分段思想进行原子操作求最值。分段原子操作的思想如图 13.3 所示，在图中将 32 个线程分为 4 份，每份 8 个线程，第一阶段，利用原子操作函数将所有线程上的值归到第 1 份上，即得到 8 个最小值；接着第二阶段是对这 8 个线程中的数据仅需原子操作，最终在 0 号线程上求得最小值。

图 13.3　分段原子操作

下面给出相应的 CUDA C 代码。

```
#define reduction_size 8
#define reduction_bit 3
#define max_red_num ((max_num_lists)/reduction_size)
__device__ void merge_array3(base_int *arr,base_int *brr,base_int tid,base_int tdim,base_int n)
{
    base_int x=arr[tid];
    base_int s_idx=tid>>reduction_bit;
    base_int num_red=tdim>>reduction_bit;
    base_int nl=n/tdim;
    __shared__ base_int list_index[max_num_lists];
    list_index[tid]=0;
    __syncthreads();

    for(base_int i=0;i<n;i++)
    {
        __shared__ base_int min_val[max_red_num];
        __shared__ base_int min_tid;
        if(tid<tdim)
        {
            min_val[s_idx]=0xFFFFFFFF;
        }
        if(tid==0)
        {
            min_tid=0xFFFFFFFF;
        }
        __syncthreads();
        atomicMin(&min_val[s_idx],x);

        if(num_red>0)
        {
            __syncthreads();
            if(tid<num_red)
            {
                atomicMin(&min_val[0],min_val[tid]);
            }
            __syncthreads();
        }
        if(min_val[0]==x)
        {
            atomicMin(&min_tid,tid);
        }
```

```
        __syncthreads();
        if(tid==min_tid)
        {
            list_index[tid]++;
            brr[i]=x;
            if(list_index[tid]<nl)
            {
                x=arr[tid+(list_index[tid] * tdim)];
            }
            else
            {
                x=0xFFFFFFFF;
            }
        }
    }
}
```

接着通过一个 kernel 函数将上述基数排序和归约过程结合起来,详细代码如下。前面其他几个版本的函数及调用过程类似,只需要修改下面代码中的函数调用即可。

```
__global__ void merge_sort_gpu(base_int *arr,base_int tdim,base_int n)
{
    base_int tid=blockDim.x * blockIdx.x+threadIdx.x;
    __shared__ base_int tmp[num_elem];
    __shared__ base_int tmp1[num_elem];

    for(base_int i=0;i<n;i+=tdim)
    {
        tmp[i+tid]=arr[i+tid];
    }
    __syncthreads();
    radix_sort(tmp,n,tmp1,tid,tdim);
    merge_array(tmp,arr,tid,tdim,n);
}
    base_int tdim=32;
    merge_sort_gpu<<<1,tdim>>>(d_a,tdim,n);
```

编译(若包含原子操作,编译时需要指定计算能力)并执行上述代码,执行结果(其中耗时包含 kernel 函数执行时间和数据传输时间)如下:

```
[fangmq@cn18%yhstar book]$./radix
there are 2 GPUs!
GPU 0: Tesla K20c
```

```
1024 radix sort on GPU    9.890079 (ms)
1024 radix1 sort on GPU   8.460045 (ms)
1024 radix2 sort on GPU   3.077030 (ms)
1024 radix3 sort on GPU   4.068851 (ms)
Congratulates, radix/merge sort is right!
Congratulates, radix/merge1 sort is right!
Congratulates, radix/merge2 sort is right!
Congratulates, radix/merge3 sort is right!
```

结果显示，首先排序结果是正确的，但性能较差，排序 1024 个数据最少需要 3.07ms（13.1.4 节串行基数排序处理 1024 个数据仅需 0.21ms）。另外，上述代码仅使用了 1 个 block，能够处理的数据量有限。结果显示，单线程合并（merge_array）、原子操作多线程合并（merge_array1）、归约多线程合并（merge_array2）和分段原子操作多线程合并（merge_array3）4 种合并方法中，归约多线程合并的方法（merge_array2）性能最佳。merge_array3 比 merge_array1 快了近 1 倍，说明将这种分段的思想融入原子操作中，可以获得一定的性能提升。

那么能否利用多个 block 做基数排序呢？答案是肯定的，当然代码会复杂很多。首先在任务分解时需要考虑不同 block 内 thread 的子任务及其数据加载过程，单独线程的基数排序过程基本不变，重点是最后的合并过程，由于 block 间无法同步，如有必要可能需要二次合并。

13.2.3 基于 thrust 库的基数排序

Thrust 函数库是 CUDA Toolkit 提供的函数库之一，其提供了转置、归约、前缀和、排序、再排序等函数。本节主要利用其中的排序函数实现 GPU 并行基数排序。

下面是基于 thrust 函数库的基数排序源代码。

```
#include<thrust/host_vector.h>
#include<thrust/device_vector.h>
#include<thrust/sort.h>
#include<thrust/copy.h>
#include<thrust/sequence.h>
#include<thrust/random.h>
#include<thrust/generate.h>
#include<thrust/detail/type_traits.h>
    unsigned int n=1024 * 1024;
    DATATYPE *array_a,*array_b;
    array_a=(DATATYPE * )malloc(sizeof(DATATYPE) * n);
    array_b=(DATATYPE * )malloc(sizeof(DATATYPE) * n);
    vector_init(array_a,n);

    thrust::host_vector<base_int>h_keys(n);
```

```
    thrust::host_vector<base_int>h_keysSorted(n);
    thrust::host_vector<base_int>h_keysSorted1(n);
    for(base_int i=0;i<n;i++)
    {
        h_keys[i]=array_a[i];
    }
    thrust::device_vector<base_int>d_keys;

    d_keys=h_keys;
    thrust::sort(d_keys.begin(), d_keys.end());
    thrust::copy(d_keys.begin(), d_keys.end(), h_keysSorted.begin());

    for(base_int i=0;i<n;i++)
    {
        array_b[i]=h_keysSorted[i];
        array_b1[i]=h_keysSorted1[i];
    }
    sort_err(array_b,n,"thrust_radix sort");
⋮
```

编译执行上述代码，输出结果如下。为了更好地分析结果，测试时令排序过程执行了两遍，结果显示，第一遍排序耗时巨大（比串行还长），而第二遍的排序时间为正常排序时间。主要原因是调用 thrust 库函数亦如 CUBLAS 库函数一般，存在首次库函数调用开销。

```
[fangmq@cn18%yhstar book]$./thrust_radix
there are 2 GPUs!
GPU 0: Tesla K20c
1048576    thrust_radix sort on GPU   1554.267168   (ms)
1048576    thrust_radix sort on GPU   6.505966      (ms)
Congratulates, thrust_radix sort is right!
Congratulates, thrust_radix sort is right!
```

13.3 双调排序网络

13.3.1 双调排序网络概述

本节将重点介绍双调排序网络的 3 个重要概念。理解了这 3 个概念也就理解了双调排序网络。

1. 双调序列

双调序列是指，对于一个偶数个元素的序列，前一半是升序（降序），后一半是降序（升序),或者能够通过循环移位实现前升后降或前降后升。图 13.4 展示了几种双调序列的情况。图 13.4(a)为先升后降双调序列;图 13.4(b)为先降后升双调序列;图 13.4(c)中,将左侧虚线左边的降序循环移到右侧虚线的右边,就构成了先升后降的双调序列。

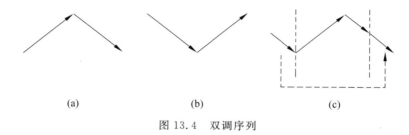

图 13.4 双调序列

注意:双调序列仅关注分段的升降序情况,并没有规定升序的最大值一定比降序的最小值大。

双调序列是双调归并网络工作的基础,当然并非任意序列都是双调序列,而任意两个元素组成的序列必然是双调序列,这是双调排序网络工作的原理。

2. 双调归并网络

双调归并网络是对双调序列中上升段和下降段元素逐一比较并排序,得到的两个序列也分别是双调序列,继续对这两个序列的上升段和下降段元素逐一比较并排序,以此类推直至比较的元素是相邻元素,最终获得有序的序列。图 13.5 展示了一个双调归并网络的实例,该实例中的比较器均为升序比较器,最终获得一个升序序列。

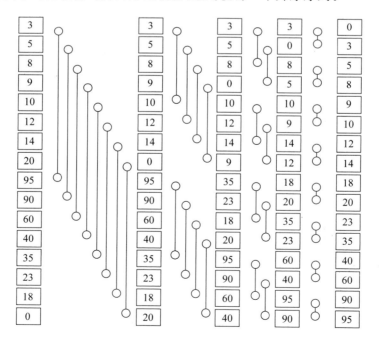

图 13.5 双调归并网络

3. 双调排序网络

双调排序网络的输入序列可以是任意序列,而不必是双调序列,双调排序网络以两个元素的序列必然为双调序列为基础,不断进行双调归并排序并获得双调序列,逐步迭代直至整个序列是一个双调序列,最后对这个双调序列采用双调归并网络,从而获得有序序列。图 13.6 简单地展示了双调排序网络的流程,其中 BM 表述双调归并网络,↑指代升序,↓指代降序。

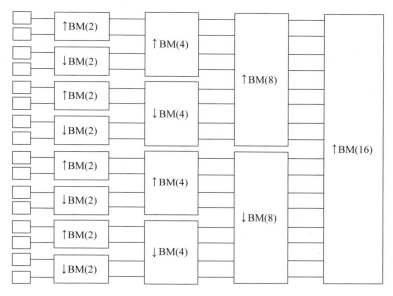

图 13.6 双调排序网络

图 13.7 展示了一个双调排序网络的实例,对于任一非双调数列,任意相邻两个元素可组成双调序列,通过双调归并网络(BM(2))可得到有序的序列,与相邻的有序(逆序)序列共同构建双调序列,再进行更大规模的双调归并网络(BM(4)),以此类推,最终构建得到一个覆盖所有元素的双调序列,最后对整个双调序列进行双调归并网络,即图中的 BM(16),也就是图 13.5 的过程,最终获得有序序列。

再进一步扩展下,最开始输入到图 13.6 中的每个序列元素可认为是一个有序序列,此时整个双调排序网络就变得可扩展、可拆分,这也是 13.3.3 节的多 block 双调排序网络的理论基础。

13.3.2　单 block 双调排序网络

下面是单 block 内双调排序网络的 CUDA C 源代码(代码参考自 *CUDA Toolkit Samples*),其中,sortDir 的值表示双调排序网络的升序(1)和降序(0)。Comparator() 函数是升(降)序比较器,升序和降序由 arrowDir 变量值控制。Bitonicsortblock() 函数是双调排序网络,函数中双层循环里的 size 循环即图 13.6 中双调排序网络的循环,而 stride 循环则是图 13.5 中的双调归并网络;双层循环结束后得到的整个数列是双调数列,最后

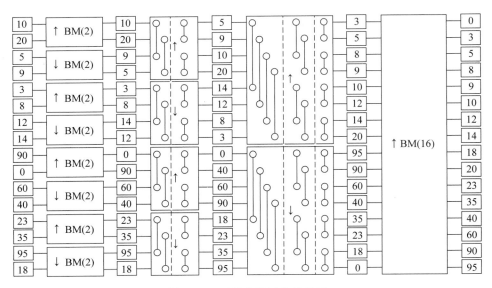

图 13.7 双调排序网络的实例

利用 stride 的一层循环进行双调归并网络(类似图 13.7 中的 BM(16)),计算得到最终的有序序列;其中,dir 参数用来控制升降序。

```
#define threadnum 1024
inline __device__ void comparator(DATATYPE &keyA, DATATYPE &keyB, uint arrowDir)
{
    DATATYPE t;
    if((keyA>keyB)==arrowDir)
    {
        t=keyA;
        keyA=keyB;
        keyB=t;
    }
}
__global__ void bitonicsortblock (DATATYPE * arr, uint arrayLength, uint sortDir)
{
    __shared__ uint tmp[threadnum];
    tmp[threadIdx.x    ]=arr[blockIdx.x*blockDim.x+threadIdx.x    ];
    tmp[threadIdx.x+(threadnum/2)]=arr[blockIdx.x*blockDim.x+threadIdx.x+(threadnum/2)];
    __syncthreads();
    for(uint size=2; size<arrayLength; size<<=1)
    {
```

```
        uint dir=(threadIdx.x & (size/2)) !=0;
        for(uint stride=size/2; stride>0; stride>>=1)
        {
            __syncthreads();
            uint pos=2 * threadIdx.x- (threadIdx.x & (stride-1));
            comparator(tmp[pos],tmp[pos+stride],dir);
        }
    }
    {
        for(uint stride=arrayLength/2; stride>0; stride>>=1)
        {
            __syncthreads();
            uint pos=2 * threadIdx.x- (threadIdx.x & (stride-1));
            comparator(tmp[pos],tmp[pos+stride],sortDir);
        }
    }
    __syncthreads();
    arr[blockIdx.x * blockDim.x+threadIdx.x]=tmp[threadIdx.x];
    arr[blockIdx.x * blockDim.x+threadIdx.x+ (threadnum/2)]=tmp[threadIdx.x+
    (threadnum/2)];
}
```

```
    DATATYPE *d_b;
    cudaMalloc((void**) &d_b, sizeof(DATATYPE) * n);
    cudaMemcpy(d_b, array_a, sizeof(DATATYPE) * n, cudaMemcpyHostToDevice);
    bitonic_sort_block<<<1,threadnum>>>(d_b,n);
    cudaMemcpy(array_b, d_b, sizeof(DATATYPE) * n, cudaMemcpyDeviceToHost);
    cudaMemcpy(d_b, array_a, sizeof(DATATYPE) * n, cudaMemcpyHostToDevice);
    bitonicsortblock<<<1,threadnum/2>>>(d_b,n,1);
    cudaMemcpy(array_b1, d_b, sizeof(DATATYPE) * n, cudaMemcpyDeviceToHost);
    cudaMemcpy(d_b, array_a, sizeof(DATATYPE) * n, cudaMemcpyHostToDevice);
    bitonicsortblock<<<1,threadnum/2>>>(d_b,n,0);
    cudaMemcpy(array_b2, d_b, sizeof(DATATYPE) * n, cudaMemcpyDeviceToHost);
    sort_err(array_b,n,"bitonic_block sort");
    sort_err(array_b1,n,"bitonic block sort");
    sort_err_down(array_b2,n,"bitonic block sort");
```

bitonic_sort_block()：摘自《GPU 高性能运算之 CUDA》[2]一书中的 kernel 函数，详细代码见附录 F.1。

调用 bitonicsortblock()函数时，通过控制 sortDir 参数的值，令其分别进行升序和降序排序。

下面展示了上述代码的执行结果。结果显示 bitonic_sort_block 函数的执行结果是错误的,前面 13.1.5 节分析了错误原因,这里的执行结果验证了笔者的判断。

```
[fangmq@cn18%yhstar book]$./bitonic_block
there are 2 GPUs!
GPU 0: Tesla K20c
1024 bitonic_block sort on GPU 0.194073 (ms)
1024 bitonic_block sort on GPU 0.079870 (ms)
1024 bitonic_block sort on GPU 0.054121 (ms)
Sorry,bitonic_block sort is error!!! i=5
Congratulates, bitonic block sort is right!
Congratulates, bitonic block sort is right(down)!
```

13.3.3 多 block 双调排序网络

13.3.2 节代码仅能处理单个 block 的双调排序网络,无法充分发挥 GPU 的性能,亦无法处理大量数据。本节进行扩展,使用多 block 来处理大规模数列的双调排序网络。

下面是扩展后的多 block 双调排序网络的 CUDA C 代码(代码参考自 *CUDA Toolkit Samples*)。代码根据数据量大小分情况处理:当数据量小于单 block 处理极限时,采用单 block 双调排序 kernel 函数;当数据量大于单 block 处理极限时,采用多 block 双调排序网络。多 block 双调排序网络中,先进行单 block 双调排序,然后将各 block 的排序结果作为单独元素重新输入到双调排序网络中(参见图 13.6)。类似于单 block 双调排序网络,其中 dir 值控制排序后数列的升降序。

```
typedef unsigned int uint;
#define SHARED_SIZE_LIMIT 1024
__device__ inline void Comparator(DATATYPE& keyA,DATATYPE& keyB,uint dir)
{
    DATATYPE t;
    if((keyA>keyB)==dir)
    {
        t=keyA; keyA=keyB; keyB=t;
    }
}
__global__ void bitonicSortShared(DATATYPE *d_DstKey,DATATYPE *d_SrcKey,uint arrayLength,uint dir)
{
    __shared__ DATATYPE s_key[SHARED_SIZE_LIMIT];
    d_SrcKey+=blockIdx.x * SHARED_SIZE_LIMIT+threadIdx.x;
    d_DstKey+=blockIdx.x * SHARED_SIZE_LIMIT+threadIdx.x;
    s_key[threadIdx.x+0]=d_SrcKey[0];
    s_key[threadIdx.x+(SHARED_SIZE_LIMIT/2)]=d_SrcKey[(SHARED_SIZE_LIMIT/2)];
    for(uint size=2; size<arrayLength; size<<=1)
    {
```

```
            uint ddd=dir ^ ((threadIdx.x & (size/2)) !=0);
            for(uint stride=size/2; stride>0; stride>>=1)
            {
                __syncthreads();
                uint pos=2 * threadIdx.x-(threadIdx.x & (stride-1));
                Comparator(s_key[pos+0], s_key[pos+stride], ddd);
            }
        }
        for(uint stride=arrayLength/2; stride>0; stride>>=1)
        {
            __syncthreads();
            uint pos=2 * threadIdx.x-(threadIdx.x & (stride-1));
            Comparator(s_key[pos+0], s_key[pos+stride], dir);
        }
        __syncthreads();
        d_DstKey[0]=s_key[threadIdx.x+0];
        d_DstKey[(SHARED_SIZE_LIMIT/2)]=s_key[threadIdx.x+(SHARED_SIZE_LIMIT/2)];
}
__global__ void bitonicSortShared1(DATATYPE *d_DstKey,DATATYPE *d_SrcKey)
{
    __shared__ DATATYPE s_key[SHARED_SIZE_LIMIT];
    d_SrcKey+=blockIdx.x * SHARED_SIZE_LIMIT+threadIdx.x;
    d_DstKey+=blockIdx.x * SHARED_SIZE_LIMIT+threadIdx.x;
    s_key[threadIdx.x+0]=d_SrcKey[0];
    s_key[threadIdx.x+(SHARED_SIZE_LIMIT/2)]=d_SrcKey[(SHARED_SIZE_LIMIT/2)];
    for(uint size=2; size<SHARED_SIZE_LIMIT; size<<=1)
    {
        uint ddd=(threadIdx.x & (size/2)) !=0;
        for(uint stride=size/2; stride>0; stride>>=1)
        {
            __syncthreads();
            uint pos=2 * threadIdx.x-(threadIdx.x & (stride-1));
            Comparator(s_key[pos+0], s_key[pos+stride],ddd);
        }
    }
    uint ddd=blockIdx.x & 1;
    {
        for(uint stride=SHARED_SIZE_LIMIT/2; stride>0; stride>>=1)
        {
            __syncthreads();
            uint pos=2 * threadIdx.x-(threadIdx.x & (stride-1));
            Comparator(s_key[pos+0], s_key[pos+stride], ddd);
        }
```

```
        }
        __syncthreads();
        d_DstKey[0]=s_key[threadIdx.x+0];
        d_DstKey[(SHARED_SIZE_LIMIT/2)]=s_key[threadIdx.x+(SHARED_SIZE_LIMIT/2)];
}
__global__ void bitonicMergeGlobal(DATATYPE *d_DstKey, DATATYPE *d_SrcKey,
uint arrayLength,uint size,uint stride,uint dir)
{
    uint global_comparatorI=blockIdx.x*blockDim.x+threadIdx.x;
    uint comparatorI=global_comparatorI & (arrayLength/2-1);
    uint ddd=dir ^ ((comparatorI & (size/2)) !=0);
    uint pos=2*global_comparatorI-(global_comparatorI & (stride-1));
    DATATYPE keyA=d_SrcKey[pos+0];
    DATATYPE keyB=d_SrcKey[pos+stride];
    Comparator(keyA, keyB, ddd);
    d_DstKey[pos+0]=keyA;
    d_DstKey[pos+stride]=keyB;
}
__global__ void bitonicMergeShared(DATATYPE *d_DstKey, DATATYPE *d_SrcKey,
uint arrayLength,uint size,uint dir)
{
    __shared__ DATATYPE s_key[SHARED_SIZE_LIMIT];
    d_SrcKey+=blockIdx.x*SHARED_SIZE_LIMIT+threadIdx.x;
    d_DstKey+=blockIdx.x*SHARED_SIZE_LIMIT+threadIdx.x;
    s_key[threadIdx.x+0]=d_SrcKey[0];
    s_key[threadIdx.x+(SHARED_SIZE_LIMIT/2)]=d_SrcKey[(SHARED_SIZE_LIMIT/2)];
    uint comparatorI=(blockIdx.x*blockDim.x+threadIdx.x) & ((arrayLength/2)-1);
    uint ddd=dir ^ ((comparatorI & (size/2)) !=0);
    for(uint stride=SHARED_SIZE_LIMIT/2; stride>0; stride>>=1)
    {
        __syncthreads();
        uint pos=2*threadIdx.x-(threadIdx.x & (stride-1));
        Comparator(s_key[pos+0], s_key[pos+stride], ddd);
    }
    __syncthreads();
    d_DstKey[0]=s_key[threadIdx.x+0];
    d_DstKey[(SHARED_SIZE_LIMIT/2)]=s_key[threadIdx.x+(SHARED_SIZE_LIMIT/2)];
}
    ⋮
    DATATYPE *d_b1,*d_b2;
    cudaMalloc((void**) &d_b1, sizeof(DATATYPE) * n);
    cudaMalloc((void**) &d_b2, sizeof(DATATYPE) * n);
    cudaMemcpy(d_b1, array_a, sizeof(DATATYPE) * n, cudaMemcpyHostToDevice);
```

```
    unsigned int dir=1;              //控制升降,1为递增,0为递减
    unsigned int batchSize=1;
    unsigned int arrayLength=n;
    DATATYPE * d_DstKey,* d_SrcKey;
    d_DstKey=d_b2;
    d_SrcKey=d_b1;
    uint  blockCount=batchSize * arrayLength/SHARED_SIZE_LIMIT;
    uint threadCount=SHARED_SIZE_LIMIT/2;
    if(arrayLength<=SHARED_SIZE_LIMIT)
    {
        bitonicSortShared<<<blockCount, threadCount>>>(d_DstKey, d_SrcKey,
        arrayLength, dir);
    }else
    {
        bitonicSortShared1<<<blockCount, threadCount>>>(d_
        DstKey, d_SrcKey);
        for(uint size=2 * SHARED_SIZE_LIMIT; size<=arrayLength; size<<=1)
        {
            for(unsigned stride=size/2; stride>0; stride>>=1)
            {
                if(stride>=SHARED_SIZE_LIMIT)
                {
                    bitonicMergeGlobal<<<(batchSize * arrayLength)/512, 256>>>
                    (d_DstKey, d_DstKey, arrayLength, size, stride, dir);
                }else{
                    bitonicMergeShared<<<blockCount, threadCount>>>(d_
                    DstKey, d_DstKey, arrayLength, size, dir);
                    break;
                }
            }
        }
    }
    cudaMemcpy(array_b, d_b2, sizeof(DATATYPE) * n, cudaMemcpyDeviceToHost);
    sort_err(array_b,n," bitonic sort");
⋮
```

bitonicSortShared()函数：完整的双调排序网络，但只适用于单 block 情况，等同 13.3.2 节的 bitonicsortblock()函数。

bitonicSortShared1()函数：每个 block 完成单独的双调排序网络，函数结束后每 SHARED_SIZE_LIMIT 个元素的序列是有序序列。

主程序中的 size 和 stride 两层循环继续进行双调排序网络，只是这里双调排序网络的每个输入不再仅仅是一个元素，而是一系列双调序列(或者说是长为 SHARED_SIZE_

LIMIT 的有序序列)。同样,size 层循环是图 13.6 中的双调排序网络的循环,而 stride 层循环则是图 13.5 中的双调归并网络的循环。

bitonicMergeGlobal()函数:该函数可视为一个超级比较器,由于这里的单个输入是有序的序列而非单独的元素,因此,原来的升降序比较器就不再适用,而是利用该函数实现。条件 stride>=SHARED_SIZE_LIMIT 用于避免当 stride 过小时频繁的 kernel 函数启动。

bitonicMergeShared()函数:完成 stride<SHARED_SIZE_LIMIT 时的双调归并网络。

编译并执行上述代码,执行结果如下:

```
[fangmq@cn18%yhstar book]$./bitonic
there are 2 GPUs!
GPU 0: Tesla K20c
1048576   bitonic sort on GPU    8.265972   (ms)
Congratulates, bitonic sort is right!
```

13.4 快速排序

基于 CUDA Toolkit Samples 中的快速排序代码进行简单封装,可得到快速排序的 CUDA 实现,且令调用部分更加简明直观。完整代码见附录 F.2,下面展示了其中关键部分的调用代码。

```
void run_quicksort_cdp(DATATYPE *d_a, DATATYPE *d_buff, unsigned int n)
{
    unsigned int stacksize=QSORT_STACK_ELEMS;
    qsortAtomicData *gpustack;
    cudaMalloc((void **)&gpustack, stacksize * sizeof(qsortAtomicData));
    cudaMemset(gpustack, 0, sizeof(qsortAtomicData));
    qsortRingbuf buf;
    qsortRingbuf *ringbuf;
    cudaMalloc((void **)&ringbuf, sizeof(qsortRingbuf));
    buf.head=1;
    buf.tail=0;
    buf.count=0;
    buf.max=0;
    buf.stacksize=stacksize;
    buf.stackbase=gpustack;
    cudaMemcpy(ringbuf, &buf, sizeof(buf), cudaMemcpyHostToDevice);
    unsigned int numblocks=(unsigned int)(n+(512-1))/512;
    qsort_warp<<<numblocks, 512>>>(d_a, d_buff, 0U, n, gpustack, ringbuf,
        true, 0);
```

```
    cudaFree(ringbuf);
    cudaFree(gpustack);
}
    ⋮
    DATATYPE *d_a,*d_buff;
    cudaMalloc((void **)&d_a, n * sizeof(DATATYPE));
    cudaMalloc((void **)&d_buff, n * sizeof(DATATYPE));
    cudaMemcpy(d_a, array_a, n * sizeof(DATATYPE), cudaMemcpyHostToDevice);
    run_quicksort_cdp(d_a,d_buff,n);
    cudaMemcpy(array_b, d_a, n * sizeof(DATATYPE), cudaMemcpyDeviceToHost);
    sort_err(array_b,n," quick sort");
    ⋮
```

由于上述代码中的一些特殊调用(__global__ 声明的 kernel 函数直接调用 __global__ 声明的 kernel 函数),普通编译无法成功,编译时需要分步进行,首先编译为中间文件,然后将中间文件链接为可执行文件。下面给出了相关的编译命令。

```
$nvcc quick.cu -c -dc -gencode arch=compute_35,code=\"sm_35,compute_35\"
$nvcc quick.o -o quick -gencode arch=compute_35,code=\"sm_35,compute_35\"
-lcudadevrt
```

执行编译成功的可执行程序,执行结果如下:

```
[fangmq@cn18%yhstar book]$./quick
there are 2 GPUs!
GPU 0: Tesla K20c
1048576    quick sort on GPU    16.690016(ms)
Congratulates, quick sort is right!
```

13.5 合并排序

本节的 GPU 并行合并排序 CUDA C 代码同样参考自 *CUDA Toolkit Samples*,下面是简化后的 GPU 合并排序调用代码,完整的 GPU 合并排序代码参见附录 F.3。

```
    ⋮
    DATATYPE *d_a,*d_b,*d_buff;
    cudaMalloc((void **)&d_a, n * sizeof(DATATYPE));
    cudaMalloc((void **)&d_b, n * sizeof(DATATYPE));
    cudaMalloc((void **)&d_buff, n * sizeof(DATATYPE));
    cudaMemcpy(d_a, array_a, n * sizeof(DATATYPE), cudaMemcpyHostToDevice);
    mergeSort(d_b,d_buff,d_a,n,1);
```

```
            cudaMemcpy(array_b, d_b, n * sizeof(DATATYPE), cudaMemcpyDeviceToHost);
            sort_err(array_b,n,"merge sort");
        ⋮
```

编译并执行上述代码,执行结果如下:

```
[fangmq@cn18%yhstar book]$./merge
there are 2 GPUs!
GPU 0: Tesla K20c
1048576    merge sort on GPU    8.430004    (ms)
Congratulates, merge sort is right!
```

13.6 实验结果分析与结论

设置 n 为不同的数值,分别测试几种快速排序方法的串行和 GPU 并行版本耗时信息及排序结果,其中,快速排序在数据量达到 2 097 152 后出现段错误,GPU 版本的双调排序网络在数据量达到 33 554 432 后排序结果出现随机错误,GPU 版本的合并排序在数据量达到 8 388 608 后无法执行。相应的执行时间和加速比信息如表 13.1~表 13.3 所示。

表 13.1 串行排序时间统计表　　　　　　　　　　　　　单位:ms

n	bitonic	merge	quick	radix1
1 048 576	418.23	191.76	112.53	194.90
2 097 152	903.06	394.17		391.12
4 194 304	1946.60	817.59		794.37
8 388 608	4432.63	1700.96		1659.18
16 777 216	9539.18	3546.33		3375.92
33 554 432	20 445.30	7373.67		6789.49
67 108 864	43 712.58	15 318.92		13 663.34
134 217 728	93 255.98	31 742.43		27 298.47

表 13.2 GPU 排序时间统计表　　　　　　　　　　　　　单位:ms

n	bitonic	merge	quick	thrust_radix
1 048 576	8.67	8.67	18.58	6.26
2 097 152	18.45	16.97	33.94	11.39
4 194 304	37.81	34.18	71.40	21.54
8 388 608	81.45		137.11	41.77
16 777 216	175.27		268.91	81.35
33 554 432			715.15	162.59
67 108 864			1192.58	322.84
134 217 728			2665.65	642.74

表 13.3　排序加速比

n	bitonic	merge	quick	thrust_radix
1 048 576	48.3	22.1	6.1	31.1
2 097 152	49.0	23.2		34.3
4 194 304	51.5	23.9		36.9
8 388 608	54.4			39.7
16 777 216	54.4			41.5
33 554 432				41.8
67 108 864				42.3
134 217 728				42.5

表 13.1～表 13.3 中的数据显示：

（1）串行排序中，数据量小时快速排序耗时少，但数据量增大后无法处理，有兴趣的读者可以尝试查找原因和解决方案。

（2）若在串行排序中选择，基数排序是最佳选择。

（3）GPU 并行排序中双调排序网络和合并排序均存在漏洞，数据量达到一定程度将导致结果错误或无法执行。有兴趣的读者可尝试查找并解决该问题。

（4）Thrust 库函数提供的基数排序在上述 4 种 GPU 排序方法中性能最佳。

（5）从加速比上看，双调排序网络的加速比最大，但却不是最佳 GPU 并行排序方法。基数排序加速比可达到 31～42 倍，是本章最佳的 GPU 排序方案。

13.7　知识点总结

选择排序、冒泡排序、快速排序、基数排序、双调排序网络、合并排序（略）。

排序算法的串行性能从高到低分别为快速排序、合并排序、基数排序、双调排序网络、选择排序、冒泡排序。

基数排序的 GPU 并行方案（见 13.2.2 节）。

基数排序的 4 种合并方法：单线程合并（merge_array）、原子操作多线程合并（merge_array1）、归约多线程合并（merge_array2）和分段原子操作多线程合并（merge_array3），其中归约多线程合并（merge_array2）性能最佳（详见 13.2.2 节）。

Thrust 库函数的调用方法（实例见 13.2.3 节）。

Thrust 库函数存在首次调用开销，约为 1548 ms。

双调排序网络的 block 内排序和归并方法（见 13.3.2 节）。

基于 13.1.5 节中双调排序网络串行代码的 i 层并行是不可行的，依赖分析见 13.1.5 节，实验结果见 13.3.2 节。

双调排序网络的多 block 并行方案区别于单 block 并行的关键是将合并子序列的移到了 kernel 函数外进行（详见 13.3.3 节）。

结构体的对齐声明实例如下，目的是辅助 GPU 访存。

```
typedef struct __align__(128) qsortAtomicData_t
{
    ⋮
} qsortAtomicData;
```

PTX 指令代码在 CUDA C 的 kernel 函数中引用方法如下：

```
asm volatile("bfind.u32 %0, %1;": "=r"(ret): "r"(word));
```

适用于多种数据类型的模板函数定义方法如下，其中，T 可在调用时确定数据类型。

```
template<typename T>
static __device__ void ringbufFree(qsortRingbuf *ringbuf, T *data)
{
    ⋮
}
```

一些常用的位运算符回顾：按位与（&）、按位或（|）、按位异或（^）、按位取反（~）、向左移位（<<）、向右移位（>>）。

__global__ 声明的 kernel 函数直接调用 __global__ 声明的 kernel 函数时，直接编译无法成功，需要分步编译，方法如下：

```
$nvcc quick.cu -c -dc -gencode arch=compute_35,code=\"sm_35,compute_35\"
$nvcc quick.o -o quick -gencode arch=compute_35,code=\"sm_35,compute_35\"
-lcudadevrt
```

13.8 扩展练习

（1）文中部分排序仅支持 unsigned int 类型的数列，修改并使其适合不同类型数据的排序，比如 float。

（2）文中排序代码执行后仅获得排序后的结果序列，而不知道各元素在原始数列中的位置，通过添加代码，完善此功能。

（3）实现基数排序的多 block 并行版本。

第14章 几种简单图像处理

本章将讨论几种简单图像处理算法的 GPU 移植和优化,包括图像直方图统计、中值滤波和均值滤波。

14.1 图像直方图统计

图像直方图是对图像灰度矩阵的统计,每个像素的灰度值在 0~255 之间,直方图统计所有像素灰度值的分布情况。图像直方图统计时,利用一个 256 长度的数组,逐一分析图像中的各像素,根据灰度值大小令数组相应的元素加 1。

本节探讨的图像直方图统计使用的是灰度图像,而非彩色图像,若用彩色图像则需要转化为灰度图像,转化公式如下:

$$gray = 0.299 \times red + 0.587 \times green + 0.114 \times blue$$

14.1.1 串行直方图统计

根据图像直方图统计流程,笔者实现了串行直方图统计代码,完整的串行直方图统计代码见附录 C.1。下面的串行直方图统计代码中,img[]存储了图像灰度矩阵,hist[]是统计的直方图数组。

```
for(int i=0;i<imagesize;i++)
{
    hist[img[i]]++;
}
```

14.1.2 并行直方图统计

下面是直方图统计的 CUDA C 代码,首先在 block 内利用原子加法操作对图像灰度值进行统计,统计结果存储于共享存储,然后再次采用原子加法操作将各 block 内共享存储子结果加到最终结果 hist 中。其中,关于原子加法操作的论述见 7.6 节。

```
__global__ void hist_compute(unsigned char *img,int *hist,int n)
{
    __shared__ int temp[256];
    int tidx=threadIdx.x;
    int idx=blockIdx.x * blockDim.x+threadIdx.x;
    temp[tidx]=0;
    for(;idx<n;idx+=blockDim.x * gridDim.x)
    {
        atomicAdd(&temp[img[idx]],1);
    }
    __syncthreads();
    atomicAdd(&hist[tidx],temp[tidx]);
}
```

```
    ⋮
unsigned char *d_img;
cudaMalloc((void**)&d_img,sizeof(char) * imagesize);
int *d_hist;
cudaMalloc((void**)&d_hist,sizeof(int) * 256);
cudaMemset(d_hist,0,sizeof(int) * 256);
cudaMemcpy(d_img,img,sizeof(char) * imagesize,cudaMemcpyHostToDevice);
hist_compute<<<256,256>>>(d_img,d_hist,imagesize);
cudaMemcpy(hist,d_hist,sizeof(int) * 256,cudaMemcpyDeviceToHost);
    ⋮
```

笔者在做结果验证时发现一个问题：新编译的可执行程序执行结果与串行结果有些出入，而遗留的可执行文件的执行结果是完全吻合的。经研究发现，问题出在 hist_compute 函数的线程（块）配置上，依据代码，thread 数量 256 是不能变的，而 block 数量是可以变化的，比如可取 64、128、256、512、1024 等，且实验结果显示该问题中 block 数量越多性能越好。当 block 取值为 64、128、256 时，执行结果是完全正确的，而 block 取到 512 和 1024 时，执行结果就出现错误。

为什么 block 数量大于等于 512 时执行结果出错呢？笔者猜测可能原因是 GPU 中的全局存储只能支持 256 个（或者小于 512 个）原子操作，超过极限将导致错误。具体的实验探索分析就不在此展开了，有兴趣的读者可以开展深入研究。

14.1.3　实验结果与分析

编译串行和 GPU 并行直方图统计代码，输入图 14.1(a)图像数据，输出结果画图可得到图 14.1(b)直方图，串行和 GPU 并行版本输出结果一致（设定线程（块）维度为 <<<256,256>>>）。统计串行和 GPU 并行直方图统计时间信息，见表 14.1。其中 serial 和 cuda 的 all_1 两列是通过 gettimeofday()函数测得，kernel、H2D 和 D2H 是利用 nvprof 命令行性能分析工具获得，all_0 是 kernel、H2D 和 D2H 的累加和。表 14.2 则计算了直方图统计中的计算加速比(kernel)和总加速比。

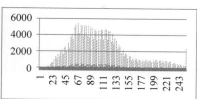

(a)　　　　　　　　　(b)

图 14.1　输入图像及其直方图统计

表 14.1　直方图统计执行时间　　　　　　　　　　单位：ms

n	serial	cuda				
		kernel	H2D	D2H	all_0	all_1
512	0.236	0.036	0.044	0.003	0.083	0.147
1024	0.960	0.103	0.168	0.003	0.274	0.336
2048	3.747	0.440	0.671	0.003	1.114	1.172
4096	15.624	2.038	2.651	0.003	4.692	4.766

表 14.2　直方图统计加速比

n	kernel	all	n	kernel	all
512	6.60	1.60	2048	8.52	3.20
1024	9.35	2.86	4096	7.67	3.28

分析表 14.1 和表 14.2 中的数据可知：

(1) 对比表 14.1 的 serial 和 kernel 的两列数据以及表 14.2 的 kernel 行，可知 GPU 计算图像直方图统计比 CPU 串行版本快，加速比为 6.6～9.35。

(2) 对比表 14.1 的 serial 和 all_1 两列数据以及表 14.2 的 all 行，可知将图像直方图统计移植到 GPU 上可以获得正性能收益，GPU 执行时间缩减能弥补通信等开销，加速比约为 1.6～3.28。

(3) 对比 all_0 和 all_1 两列数据，发现用 gettimeofday() 测得的时间比 kernel 函数和通信函数的执行时间总和多了 0.06～0.08ms，这是 kernel 函数或通信函数的启动开销。该现象在 14.2 节中值滤波和 14.3 节均值滤波均中存在。

14.2　中值滤波

中值滤波是图像滤波的常用方法之一，其思想是取图像元素周围的 9 个点（包括其自身），然后对这 9 个点排序得到最中间的值，即为中值滤波结果。

14.2.1　串行中值滤波

根据中值滤波思想，实现串行中值滤波代码如下。完整的串行中值滤波（包括图像读写）代码见附录 C.3。在下面代码中，排序仅排一半，获得中间值后即停止，避免多余的运算。

```
    ⋮
memset(filter, 0, sizeof(char) * imagesize);
unsigned char p[9],temp;
int l,m;
float tmp_f;
for(j=1;j<imageheight-1;j++)
{
    for(k=1;k<imagewidth-1;k++)
    {
        p[0]=img[(j-1) * imagewidth+ (k-1)];
        p[1]=img[(j-1) * imagewidth+ (k)];
        p[2]=img[(j-1) * imagewidth+ (k+1)];
        p[3]=img[(j) * imagewidth+ (k-1)];
        p[4]=img[(j) * imagewidth+ (k)];
        p[5]=img[(j) * imagewidth+ (k+1)];
        p[6]=img[(j+1) * imagewidth+ (k-1)];
        p[7]=img[(j+1) * imagewidth+ (k)];
        p[8]=img[(j+1) * imagewidth+ (k+1)];
        for(l=0;l<5;l++)
        {
            for(m=0;m<8-l;m++)
            {
                if(p[m]>p[m+1])
                {
                    temp=p[m];
                    p[m]=p[m+1];
                    p[m+1]=temp;
                }
            }
        }
        filter[j * imagewidth+k]=p[4];
    }
}
    ⋮
```

14.2.2 1D 并行中值滤波

在 1D 线程维度配置的情况下,令 block 内的所有 thread 处理图像滤波的一行,令 grid 内的 block 对应图像滤波的行。为了令 kernel 函数能够处理尽可能大的图像数据,采用循环处理的方式。下面给出相应的 1D 并行中值滤波 CUDA C 代码(标记为 cuda_1d)。

```
__global__ void median_filter_0(unsigned char *img,unsigned char *filter,int imagewidth,int imageheight)
{
    int bid=blockIdx.x;
    int tid=threadIdx.x;
    unsigned char p[9],temp;
    float tmp_f;
    int m,l;
    for(;bid<imageheight;bid+=gridDim.x)
    {
        if((bid>0)&&(bid<(imageheight-1)))
        {
            for(tid=threadIdx.x;tid<imagewidth;tid+=blockDim.x)
            {
                if((tid>0)&&(tid<(imagewidth-1)))
                {
                    p[0]=img[(bid-1) * imagewidth+ (tid-1)];
                    p[1]=img[(bid-1) * imagewidth+ (tid)];
                    p[2]=img[(bid-1) * imagewidth+ (tid+1)];
                    p[3]=img[(bid) * imagewidth+ (tid-1)];
                    p[4]=img[(bid) * imagewidth+ (tid)];
                    p[5]=img[(bid) * imagewidth+ (tid+1)];
                    p[6]=img[(bid+1) * imagewidth+ (tid-1)];
                    p[7]=img[(bid+1) * imagewidth+ (tid)];
                    p[8]=img[(bid+1) * imagewidth+ (tid+1)];
                    for(l=0;l<5;l++)
                    {
                        for(m=0;m<8-l;m++)
                        {
                            if(p[m]>p[m+1])
                            {
                                temp=p[m];
                                p[m]=p[m+1];
                                p[m+1]=temp;
                            }
                        }
```

```
            }
            filter[bid * imagewidth+tid]=p[4];
        }
      }
    }
  }
}
    ⋮
  cudaMalloc((void**)&d_img,sizeof(char) * imagesize);
  cudaMalloc((void**)&d_filter,sizeof(char) * imagesize);
  cudaMemset(d_filter,0,sizeof(char) * imagesize);
  cudaMemcpy(d_img,img,sizeof(char) * imagesize,cudaMemcpyHostToDevice);
  median_filter_0<<<512,512>>>(d_img,d_filter,imagewidth,imageheight);
  cudaMemcpy(filter,d_filter,sizeof(char) * imagesize,cudaMemcpyDeviceToHost);
    ⋮
```

14.2.3 共享 1D 中值滤波

配置 1D 线程维度求解图像滤波时，存在明显的数据复用现象，如图 14.2 所示。计算中间 1 行图像元素的滤波时，需要用到上下各一行数据，此时三行数据中的任一元素（图中深灰色的数据）都会被 3 个滤波计算（浅灰色的 3 个滤波计算）使用，即存在数据复用。

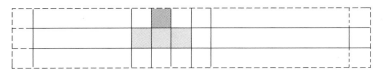

图 14.2　1D 数据复用示意图

上述 1D 数据复用现象具体到 kernel 函数代码中，即图像数据 img[] 的访问。笔者声明 3 个共享存储数组存储三行数据。由于滤波计算需要的数据多于 block 内 thread 数量，因此在数据加载时的边界处理就尤为重要。如图 14.3 所示，数据加载时先加载 thread 对应部分数据（实线部分），然后分别加载前端和后端。前端加载需要判断是否超过图像左边界，后端加载需要判断是否超出图像右边界（图像上下边界判断 blockIdx.x 值）。

图 14.3　1D 数据加载示意图

下面给出相应的数据加载代码。

```
srow0[threadIdx.x+1]=img[(bid-1)*imagewidth+tid];
srow1[threadIdx.x+1]=img[(bid)*imagewidth+tid];
srow2[threadIdx.x+1]=img[(bid+1)*imagewidth+tid];
if((tid==0)&&(threadIdx.x==0))         //前端处理
{
    srow0[threadIdx.x]=0;
    srow1[threadIdx.x]=0;
    srow2[threadIdx.x]=0;
}
else if((tid>0)&&(threadIdx.x==0))
{
    srow0[threadIdx.x]=img[(bid-1)*imagewidth+tid-1];
    srow1[threadIdx.x]=img[(bid)*imagewidth+tid-1];
    srow2[threadIdx.x]=img[(bid+1)*imagewidth+tid-1];
}
if(tid==(imagewidth-1))                //后端处理
{
    srow0[threadIdx.x+2]=0;
    srow1[threadIdx.x+2]=0;
    srow2[threadIdx.x+2]=0;
}
else if((threadIdx.x+1)==blockDim.x)
{
    srow0[threadIdx.x+2]=img[(bid-1)*imagewidth+tid+1];
    srow1[threadIdx.x+2]=img[(bid)*imagewidth+tid+1];
    srow2[threadIdx.x+2]=img[(bid+1)*imagewidth+tid+1];
}
```

共享存储数据加载成功后,另一个关键是正确使用索引。下面是完整的共享 1D 中值滤波 CUDA C 代码(标记为 cuda_1d_1),其中需要注意共享存储数据的使用。

```
__global__ void median_filter_1(unsigned char *img,unsigned char *filter,int imagewidth,int imageheight)
{
    __shared__ unsigned char srow0[514];
    __shared__ unsigned char srow1[514];
    __shared__ unsigned char srow2[514];
    int bid=blockIdx.x;
    int tid=threadIdx.x;
    unsigned char p[9],temp;
    float tmp_f;
    int m,l;
    for(;bid<imageheight;bid+=gridDim.x)
    {
```

```
            if((bid>0)&&(bid<(imageheight-1)))
            {
                for(tid=threadIdx.x;tid<imagewidth;tid+=blockDim.x)
                {
                    __syncthreads();
                    … //数据加载代码
                    __syncthreads();
                    if((tid>0)&&(tid<(imagewidth-1)))
                    {
                        p[0]=srow0[threadIdx.x];
                        p[1]=srow0[threadIdx.x+1];
                        p[2]=srow0[threadIdx.x+2];
                        p[3]=srow1[threadIdx.x];
                        p[4]=srow1[threadIdx.x+1];
                        p[5]=srow1[threadIdx.x+2];
                        p[6]=srow2[threadIdx.x];
                        p[7]=srow2[threadIdx.x+1];
                        p[8]=srow2[threadIdx.x+2];
                        for(l=0;l<5;l++)
                        {
                            for(m=0;m<8-l;m++)
                            {
                                if(p[m]>p[m+1])
                                {
                                    temp=p[m];
                                    p[m]=p[m+1];
                                    p[m+1]=temp;
                                }
                            }
                        }
                        filter[bid*imagewidth+tid]=p[4];
                    }
                }
            }
        }
```

14.2.4 双重共享 1D 中值滤波

利用 16.3 节阐述的局部存储查看方法，14.2.3 节代码中的 p[9] 数组是声明在局部存储中的，第 16 章数据显示，局部存储的访存延迟与全局存储的访存延迟基本相当。而 16.3 节显示排序的数据只能声明在局部存储中。此时计算 block 内 p[] 所需空间为 $9\times 512B$，远小于共享存储容量，那么能否将排序过程放在共享存储中进行呢？

在 14.2.3 节代码的基础上进行修改,将排序数据空间 p[]存放在共享存储中,修改后的双重共享 1D 中值滤波 CUDA C 代码(标记为 cuda_1d_2)如下:

```
__global__ void median_filter_2(unsigned char *img,unsigned char *filter,int imagewidth,int imageheight)
{
    __shared__ unsigned char srow0[514];
    __shared__ unsigned char srow1[514];
    __shared__ unsigned char srow2[514];
    __shared__ unsigned char sp[512*9];
    int bid=blockIdx.x;
    int tid=threadIdx.x;
    unsigned char temp;//p[9],
    float tmp_f;
    int m,l;
    for(;bid<imageheight;bid+=gridDim.x)
    {
        if((bid>0)&&(bid<(imageheight-1)))
        {
            for(tid=threadIdx.x;tid<imagewidth;tid+=blockDim.x)
            {
                __syncthreads();
                srow0[threadIdx.x+1]=img[(bid-1)*imagewidth+tid];
                srow1[threadIdx.x+1]=img[(bid)*imagewidth+tid];
                srow2[threadIdx.x+1]=img[(bid+1)*imagewidth+tid];
                if((tid==0)&&(threadIdx.x==0))
                {
                    srow0[threadIdx.x]=0;
                    srow1[threadIdx.x]=0;
                    srow2[threadIdx.x]=0;
                }
                else if((tid>0)&&(threadIdx.x==0))
                {
                    srow0[threadIdx.x]=img[(bid-1)*imagewidth+tid-1];
                    srow1[threadIdx.x]=img[(bid)*imagewidth+tid-1];
                    srow2[threadIdx.x]=img[(bid+1)*imagewidth+tid-1];
                }
                if(tid==(imagewidth-1))
                {
                    srow0[threadIdx.x+2]=0;
                    srow1[threadIdx.x+2]=0;
                    srow2[threadIdx.x+2]=0;
                }
```

```
                    else if((threadIdx.x+1)==blockDim.x)
                    {
                        srow0[threadIdx.x+2]=img[(bid-1) * imagewidth+tid+1];
                        srow1[threadIdx.x+2]=img[(bid) * imagewidth+tid+1];
                        srow2[threadIdx.x+2]=img[(bid+1) * imagewidth+tid+1];
                    }
                    __syncthreads();
                    if((tid>0)&&(tid<(imagewidth-1)))
                    {
                        sp[0 * 512+threadIdx.x]=srow0[threadIdx.x];
                        sp[1 * 512+threadIdx.x]=srow0[threadIdx.x+1];
                        sp[2 * 512+threadIdx.x]=srow0[threadIdx.x+2];
                        sp[3 * 512+threadIdx.x]=srow1[threadIdx.x];
                        sp[4 * 512+threadIdx.x]=srow1[threadIdx.x+1];
                        sp[5 * 512+threadIdx.x]=srow1[threadIdx.x+2];
                        sp[6 * 512+threadIdx.x]=srow2[threadIdx.x];
                        sp[7 * 512+threadIdx.x]=srow2[threadIdx.x+1];
                        sp[8 * 512+threadIdx.x]=srow2[threadIdx.x+2];
                        for(l=0;l<5;l++)
                        {
                            for(m=0;m<8-l;m++)
                            {
                                if(sp[m * 512+threadIdx.x]>sp[(m+1) * 512+threadIdx.x])
                                {
                                    temp=sp[m * 512+threadIdx.x];
                                    sp[m * 512+threadIdx.x]=sp[(m+1) * 512+
                                    threadIdx.x];
                                    sp[(m+1) * 512+threadIdx.x]=temp;
                                }
                            }
                        }
                        filter[bid * imagewidth+tid]=sp[4 * 512+threadIdx.x];
                    }
                }
            }
        }
    }
}
```

值得一提的是,上述代码中 sp[] 的数据排布考虑了共享存储的连续访问以及 bank conflict 的避免,其索引为 sp[0 * 512+threadIdx.x]。而另一种数据排布的索引为 sp[0+threadIdx.x * 9],显然这种数据排列没有考虑连续访问以及 bank conflict 的避免。通过实验比较,这两种排列方式的性能相差很大。读者可自行实践验证,在此不再赘述。

14.2.5 2D 并行中值滤波

配置线程维度为 2D(2D 线程配置详见 5.6 节),即分块划分滤波图像。下面展示了 2D 并行中值滤波的 CUDA C 代码(标记为 cuda_2d)。

```
__global__ void median_filter_2d(unsigned char *img,unsigned char *filter,int imagewidth,int imageheight)
{
    int yy=blockIdx.y * blockDim.y+threadIdx.y;
    int xx=blockIdx.x * blockDim.x+threadIdx.x;
    int xz=blockDim.x * gridDim.x;
    int yz=blockDim.y * gridDim.y;
    unsigned char p[9],temp;//
    float tmp_f;
    int m,l;
    for(;yy<imageheight;yy+=yz)
    {
        if((yy>0)&&(yy<(imageheight-1)))
        {
            for(xx=blockIdx.x * blockDim.x+threadIdx.x;xx<imagewidth;xx+=xz)
            {
                if((xx>0)&&(xx<(imagewidth-1)))
                {
                    p[0]=img[(yy-1) * imagewidth+ (xx-1)];
                    p[1]=img[(yy-1) * imagewidth+ (xx)];
                    p[2]=img[(yy-1) * imagewidth+ (xx+1)];
                    p[3]=img[(yy) * imagewidth+ (xx-1)];
                    p[4]=img[(yy) * imagewidth+ (xx)];
                    p[5]=img[(yy) * imagewidth+ (xx+1)];
                    p[6]=img[(yy+1) * imagewidth+ (xx-1)];
                    p[7]=img[(yy+1) * imagewidth+ (xx)];
                    p[8]=img[(yy+1) * imagewidth+ (xx+1)];
                    for(l=0;l<5;l++)
                    {
                        for(m=0;m<8-l;m++)
                        {
                            if(p[m]>p[m+1])
                            {
                                temp=p[m];
                                p[m]=p[m+1];
                                p[m+1]=temp;
                            }
                        }
                    }
                    filter[yy * imagewidth+xx]=p[4];
                }
```

```
            }
          }
       }
    }
    dim3 threads(th_size,th_size,1);
    dim3 blocks(imagewidth/th_size,imageheight/th_size,1);
    median_filter_2d<<<blocks,threads>>>(d_img,d_filter,imagewidth,
imageheight);
```

上述 kernel 函数代码中，yy 和 xx 的循环可以省略，因为根据 2D 分块的线程配置方案，blocks(bx,by,bz)最大维度为(2147483647,65535,65535)，故无须循环即可处理大规模的输入数据(最大高为 32×65535)。

14.2.6 共享 2D 中值滤波

中值滤波的 2D 线程维度配置方案中，存在明显的数据复用现象，如图 14.4 所示。对于一般数据(深灰色网格)，会被其自身及周围线程(浅灰色网格)使用，共复用 9 次；对于边界数据，根据邻居的数量不同而复用次数有所不同(复用次数包括 1、2、3、4、6 等)，如图 14.4 所示。显然 2D 线程维度配置数据复用次数更多，因此考虑使用共享存储来优化该部分复用数据。

共享存储 2D 数据的加载过程比 1D 情况更加复杂，需要处理的边界情况也更多。2D 数据加载中，需要处理的边界情况包括以下两个方面：①需要读取的部分数据位于图像数据之外；②图像数据加载流程。

当需要读取的部分数据位于图像数据之外时，如图 14.5 所示，大的实线矩阵代表图像数据，小虚线矩阵表示某个 block 需要加载的数据。图 14.5 中显示了部分需要读取的数据位于对应的图像之外的情况，此时将这部分数据赋值为 0，相应的代码如下。

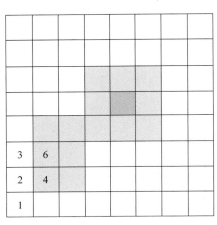

图 14.4　2D 数据复用示意图

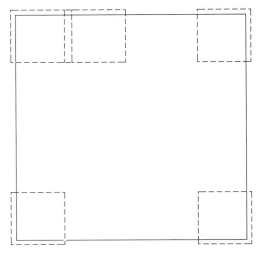

图 14.5　部分数据位于图像数据之外

```
    __shared__ unsigned char shb[th_size+2][th_size+2];
    int yy=blockIdx.y*blockDim.y+threadIdx.y;
    int xx=blockIdx.x*blockDim.x+threadIdx.x;
⋮
    if((yy==0)&&(threadIdx.y==0))
    {
        shb[threadIdx.y][threadIdx.x]=0;
        if(threadIdx.x<2)
        {
            shb[threadIdx.y][threadIdx.x+th_size]=0;
        }
    }
    if(yy==(imageheight-1)&&(threadIdx.y+1==th_size))
    {
        shb[threadIdx.y+2][threadIdx.x]=0;
        if(threadIdx.x<2)
        {
            shb[threadIdx.y+2][threadIdx.x+th_size]=0;
        }
    }
    if((xx==0)&&(threadIdx.x==0))
    {
        shb[threadIdx.y][threadIdx.x]=0;
        if(threadIdx.y<2)
        {
            shb[threadIdx.y+th_size][threadIdx.x]=0;
        }
    }
    if(xx==(imagewidth-1)&&(threadIdx.x+1==th_size))
    {
        shb[threadIdx.y][threadIdx.x+2]=0;
        if(threadIdx.y<2)
        {
            shb[threadIdx.y+th_size][threadIdx.x+2]=0;
        }
    }
    if(((yy-1)>=0)&&((xx-1)>=0))
    {
        shb[threadIdx.y][threadIdx.x]=img[(yy-1)*imagewidth+(xx-1)];
    }
```

在加载图像数据时,线程格为图14.6(a)中的实线矩阵,需要加载的数据为大的虚线矩阵。将实线矩阵向左和向上各平移一个单位,然后加载其覆盖的部分数据。观察

图 14.6(b),发现还有 3 个部分数据待加载,此时笔者在线程矩阵中划分出 3 个矩阵区域分别处理这 3 部分数据,并且由于所使用的线程区域没有重叠,这 3 部分区域的数据加载可以并行执行。但是在实践时发现,明明合理的逻辑,执行总是出错,实在解决不了笔者只能换成图 14.6(c)中的方案,该方案中有部分线程分块重叠,因此加载过程无法同步进行(这里的加载过程类似于 10.7.3 节阐述的 2D 卷积数据加载,有兴趣的读者可尝试实践解决该问题)。

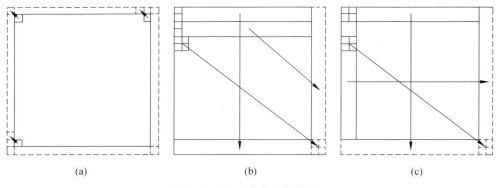

图 14.6 2D 图像数据加载流程

下面给出了 2D 图像数据加载代码,由于采用图 14.6(b)加载代码出错,笔者采用图 14.6(c)加载代码。当然,笔者也曾深入分析图 14.6(b)加载代码出错的位置,后定位到 yy 变量的替换,图 14.6(c)加载代码可以正确运行,但在注释处用 blockIdx.y * blockDim.y+threadIdx.y 替换 yy 变量,该 kernel 函数就无法执行了。尽管找到了出错位置,但笔者却没有找到解决方案,有兴趣的读者可以尝试解决该问题。

```
    __shared__ unsigned char shb[th_size+2][th_size+2];
    int yy=blockIdx.y * blockDim.y+threadIdx.y;
    int xx=blockIdx.x * blockDim.x+threadIdx.x;
    if(((yy-1)>=0)&&((xx-1)>=0))
    {
        shb[threadIdx.y][threadIdx.x]=img[(yy-1) * imagewidth+(xx-1)];
    }
```

```
//图 14.6(b)加载代码
    ⋮
    if(threadIdx.y<2){⋯}
    else if(threadIdx.y<4)
    {
        int tidy=threadIdx.x;
        int tidx=(threadIdx.y-2)+th_size;
        if(((blockDim.y * blockIdx.y+tidy-1)>=0)&&((blockDim.x * blockIdx.x+
        tidx-1)<imagewidth))
        {
```

```
shb[tidy][tidx]=img[(blockDim.y*blockIdx.y+tidy-1)*imagewidth+blockDim.x
*blockIdx.x+tidx-1];
        }
    }
    else if((threadIdx.x<2)&&(threadIdx.y<6)){…}
⋮
//图 14.6(c)加载代码
    if(threadIdx.y<2)
    {
        int tidy=threadIdx.y+th_size;
        if(((blockDim.y*blockIdx.y+tidy-1)<imageheight)&&((xx-1)>=0))
        {
shb[tidy][threadIdx.x]=img[(blockDim.y*blockIdx.y+tidy-1)*imagewidth+(xx
-1)];
        }
    }
    else if((threadIdx.x<2)&&(threadIdx.y<6)&&(threadIdx.y>=4))
    {
        int tidx=threadIdx.x+blockDim.x;
        int tidy=(threadIdx.y-4)+blockDim.x;
        if(((blockIdx.y*blockDim.y+tidy-1)<imageheight)&&((blockDim.x*
blockIdx.x+tidx-1)<imagewidth))
        {
shb[tidy][tidx]=img[(blockDim.y*blockIdx.y+tidy-1)*imagewidth+blockDim.x
*blockIdx.x+tidx-1];
        }
    }
    if(threadIdx.x<2)  //用 yy=blockIdx.y*blockDim.y+threadIdx.y 替换 yy 就出错
    {
        int tidx=threadIdx.x+th_size;
        int tidy=threadIdx.y;
        if(((yy-1)>=0)&&((blockDim.x*blockIdx.x+tidx-1)<imagewidth))
        {
            shb[tidy][tidx]=img[(yy-1)*imagewidth+blockDim.x*blockIdx.x+
            tidx-1];
        }
    }
```

为简化数据加载逻辑,这里删除了 kernel 函数中的循环处理。下面给出了共享 2D 中值滤波 CUDA C 代码(标记为 cuda_2d_1)。其中,在数据加载过程后有一句注释的语句,将共享存储数组对应元素赋值给滤波数组,该注释语句在调试时非常有用,将 kernel 函数分为前后两部分分别调试,能辅助程序员定位 kernel 函数出错位置和调试得到正确结果(笔者调试 CUDA C 代码未使用任何工具,而是采用直接修改编译执行的方式进行)。

```c
__global__ void median_filter_2d_1(unsigned char *img,unsigned char *filter,
int imagewidth,int imageheight)
{
    __shared__ unsigned char shb[th_size+2][th_size+2];
    int yy=blockIdx.y*blockDim.y+threadIdx.y;
    int xx=blockIdx.x*blockDim.x+threadIdx.x;
    int xz=blockDim.x*gridDim.x;
    int yz=blockDim.y*gridDim.y;
    unsigned char p[9],temp;//
    float tmp_f;
    int m,l;
    __syncthreads();
    …//数据加载过程
    __syncthreads();
//    filter[yy*imagewidth+xx]=shb[threadIdx.y+1][threadIdx.x+1];    //调试代码

    //for(;yy<imageheight;yy+=yz)
    {
        if((yy>0)&&(yy<(imageheight-1)))
        {
            //for(xx=blockIdx.x*blockDim.x+threadIdx.x;xx<imagewidth;xx+=xz)
            {
                if((xx>0)&&(xx<(imagewidth-1)))
                {
                    p[0]=shb[threadIdx.y][threadIdx.x];
                    p[1]=shb[threadIdx.y][threadIdx.x+1];
                    p[2]=shb[threadIdx.y][threadIdx.x+2];
                    p[3]=shb[threadIdx.y+1][threadIdx.x];
                    p[4]=shb[threadIdx.y+1][threadIdx.x+1];
                    p[5]=shb[threadIdx.y+1][threadIdx.x+2];
                    p[6]=shb[threadIdx.y+2][threadIdx.x];
                    p[7]=shb[threadIdx.y+2][threadIdx.x+1];
                    p[8]=shb[threadIdx.y+2][threadIdx.x+2];
                    for(l=0;l<5;l++)
                    {
                        for(m=0;m<8-l;m++)
                        {
                            if(p[m]>p[m+1])
                            {
                                temp=p[m];
                                p[m]=p[m+1];
                                p[m+1]=temp;
                            }
```

```
                    }
                }
                filter[yy*imagewidth+xx]=p[4];
            }
        }
    }
}
```

14.2.7 共享 2D 中值滤波的改进

本节基于 14.2.6 节中值滤波代码,并提出两种改进策略:分别是部分数据位于图像之外时的数据加载优化(见图 14.5)和图像数据加载优化(图 14.6(c)改为图 14.6(b)的策略)。

再次返回分析 14.2.6 节 kernel 函数代码,滤波计算前有两个判断语句,排除了图像边界的滤波计算,因此超出图像边界的 0 数据就无须再加载了,因此该部分代码可以删除。删除超出图像数据之外数据读取后的 kernel 函数代码标记为 cuda_2d_2。

另一方面,超出图像边界数据的读取包含很多判断语句,逻辑复杂且有可能导致性能损失。而且图像数据加载时图 14.6(b)的策略无法实现。此时可考虑另外一种实现方式——扩展法(类似于 2D 卷积),即在输入的图像数据外套一圈 0 数据,读取图像数据到共享存储时就无须考虑超出图像数据边界的情况,而图像数据的加载逻辑也变得相对简单。

下面是图像数据外部套 0 方案的共享 2D 中值滤波 CUDA C 代码(标记为 cuda_2d_9)。由代码可知,共享存储数据加载过程采用了图 14.6(b)的方案。而为了行数据对齐访问,套 0 时每行增加 32 个元素(已知输入数据是 2 的幂次情况,其他情况需要根据需要调整增加的元素个数)。

```
__global__ void median_filter_2d_9(unsigned char *img,unsigned char *filter,
int imagewidth,int imageheight)
{
    __shared__ unsigned char shb[th_size+2][th_size+2];
    int yy=blockIdx.y*blockDim.y+threadIdx.y;
    int xx=blockIdx.x*blockDim.x+threadIdx.x;
    int xz=blockDim.x*gridDim.x;
    int yz=blockDim.y*gridDim.y;
    unsigned char p[9],temp;
    float tmp_f;
    int m,l;
    int tidx,tidy;
    shb[threadIdx.y][threadIdx.x]=img[(yy)*(imagewidth+32)+(xx)];
```

```
    if(threadIdx.y<2)
    {
        tidy=threadIdx.y+blockDim.y;
        shb[tidy][threadIdx.x]=img[(blockIdx.y*blockDim.y+tidy)*
        (imagewidth+32)+xx];
    }
    else if(threadIdx.y<4)
    {
        tidy=threadIdx.x;
        tidx=(threadIdx.y-2)+blockDim.x;
 shb[tidy][tidx] = img[(blockIdx.y * blockDim.y + tidy) * (imagewidth + 32) +
blockIdx.x * blockDim.x+tidx];
    }
    else if((threadIdx.y<6)&&(threadIdx.x<2))
    {
        tidx=threadIdx.x+blockDim.x;
        tidy=(threadIdx.y-4)+blockDim.y;
 shb[tidy][tidx] = img[(blockIdx.y * blockDim.y + tidy) * (imagewidth + 32) +
blockIdx.x * blockDim.x+tidx];
    }
    __syncthreads();
//    filter[yy*imagewidth+xx]=shb[threadIdx.y+1][threadIdx.x+1];
    if((yy>0)&&(yy<(imageheight-1))&&(xx>0)&&(xx<(imagewidth-1)))
    {
        p[0]=shb[threadIdx.y][threadIdx.x];
        p[1]=shb[threadIdx.y][threadIdx.x+1];
        p[2]=shb[threadIdx.y][threadIdx.x+2];
        p[3]=shb[threadIdx.y+1][threadIdx.x];
        p[4]=shb[threadIdx.y+1][threadIdx.x+1];
        p[5]=shb[threadIdx.y+1][threadIdx.x+2];
        p[6]=shb[threadIdx.y+2][threadIdx.x];
        p[7]=shb[threadIdx.y+2][threadIdx.x+1];
        p[8]=shb[threadIdx.y+2][threadIdx.x+2];
        for(l=0;l<5;l++)
        {
            for(m=0;m<8-l;m++)
            {
                if(p[m]>p[m+1])
                {
                    temp=p[m];
                    p[m]=p[m+1];
                    p[m+1]=temp;
                }
```

```
            }
        }
        filter[yy*imagewidth+xx]=p[4];
    }
}
```

```
long imageheight,imagewidth;
    ⋮
long imagesize=imagewidth*imageheight;
long imagesize1=(imagewidth+32)*(imageheight+2);
unsigned char * img=(unsigned char *)malloc(sizeof(char) * imagesize1);
unsigned char * filter=(unsigned char *)malloc(sizeof(char) * imagesize);
cudaHostRegister(img,sizeof(char) * imagesize1,0);
cudaHostRegister(filter,sizeof(char) * imagesize,0);
memset(img,0,sizeof(char) * imagesize1);
for(int i=0;i<imageheight;i++)
{
    for(int j=0;j<imagewidth;j++)
    {
        img[(i+1)*(imagewidth+32)+(j+1)]=Himage->gray[i*imagewidth+j];
    }
}
unsigned char *d_img,*d_filter;
cudaMalloc((void**)&d_img,sizeof(char) * imagesize1);
cudaMalloc((void**)&d_filter,sizeof(char) * imagesize);
cudaMemset(d_filter,0,sizeof(char) * imagesize);
cudaMemcpy(d_img,img,sizeof(char) * imagesize1,cudaMemcpyHostToDevice);
dim3 threads(th_size,th_size,1);
dim3 blocks(imagewidth/th_size,imageheight/th_size,1);
median_filter_2d_9<<<blocks,threads>>>(d_img,d_filter,imagewidth,
    imageheight);
cudaMemcpy(filter,d_filter,sizeof(char) * imagesize,cudaMemcpyDeviceToHost);
    ⋮
cudaFree(d_img);
cudaFree(d_filter);
cudaHostUnregister(img);
cudaHostUnregister(filter);
free(img);
free(filter);
```

14.2.8 实验结果与分析

编译串行和 GPU 并行中值滤波代码，输入图 14.7(a)中的图像数据，输出滤波结果图像为图 14.7(b)，串行和 GPU 并行版本输出滤波图像一致。统计串行和 GPU 并行中值滤波时间,如表 14.3 所示。计算最佳方案的计算加速比和总时间加速比,如表 14.4 所示。

表 14.3 中值滤波计算时间 单位:ms

n	serial	cuda_1d	cuda_1d_1	cuda_1d_2	cuda_2d	cuda_2d_1	cuda_2d_2	cuda_2d_9
512	18.418	0.188	0.178	0.241	0.194	0.184	0.181	0.175
1024	57.948	0.744	0.737	0.986	0.762	0.712	0.705	0.682
2048	172.910	2.964	2.950	3.901	3.029	2.825	2.802	2.705
4096	563.718	11.795	11.695	15.639	12.092	11.279	11.185	10.794

表 14.4 最优中值滤波方案加速比

n	kernel	all	n	kernel	all
512	105.4	38.1	2048	63.9	25.5
1024	85.0	33.3	4096	52.2	20.9

(a)　　　　　　　　　　(b)

图 14.7 中值滤波结果

分析表 14.3 和表 14.4 中的数据可知:

(1) 对比 cuda_1d 和 cuda_1d_1 两行数据,采用共享存储存放图像数据优化的 cuda_1d_1 版本性能稍好,说明利用共享存储存放图像数据的优化方法是可行的。

(2) 对比 cuda_1d_1 和 cuda_1d_2 两行数据,cuda_1d_2 版本性能明显差了很多,说明利用共享存储优化排序数组 p[] 的方案不可行。可能原因是大量共享存储的使用影响 GPU 的活跃 warp 数量。

(3) 对比 cuda_1d 和 cuda_2d 两行数据,发现 cuda_1d 版本执行时间更短,说明在 2D 线程配置中若没有采用共享存储优化数据复用,仅访问全局存储,1D 线程配置的执行性能更好一些。原因应该是 2D 线程配置在访存时,每个 warp 访问的数据不连续存储,导致数据访问无法合并。

(4) 对比 cuda_2d 和 cuda_2d_1 两行数据,cuda_2d_1 版本相对性能更好,说明共享存储优化对 2D 中值滤波是有效果的。

(5) 对比 cuda_2d_1 和 cuda_2d_2 两行数据,cuda_2d_2 版本执行时间稍短,且差值

基本固定。这些时间差值是删除的边界数据处理过程耗时,且该部分处理过程是固定的。

(6) 对比 cuda_2d_2 和 cuda_2d_9 两行数据,cuda_2d_9 版本性能明显较好,说明在输入图像周围填 0 的方法在中值滤波中很有效。该方法既能优化边界数据加载过程,亦能优化一般图像数据加载过程。

(7) 版本 cuda_2d_9 的 kernel 函数执行时间最快,对比串行版本的加速比(见表 14.4),仅考虑计算部分 GPU 能加速 52～105 倍,总时间上 GPU 处理中值滤波能加速 20～38 倍。

(8) 表 14.4 中的加速比数据显示,输入图像越大加速比反而减少,这种现象有点反常。可能原因是 block 数量随数据增加而增加,申请的共享存储总量增加活跃 warp 数量减少,最终导致性能下降。解决方法是采用 kernel 函数内循环的方法来减少总共享存储的使用。

除了上述实验结果,2D 线程维度配置情况下还发现如下规律:共享存储使用、block 内 thread 的配置数量和 kernel 函数性能相关。表 14.5 统计了 block 内 thread 配置数量为(16,16)和(32,32)两种情况的各 kernel 函数处理 512×512 的图像滤波执行时间。表中数据显示,当采用了共享存储优化时(cuda_2d_1、cuda_2d_2 和 cuda_2d_9),block 内 thread 配置为(16,16)性能较好;而直接访问全局存储情况(cuda_2d),block 内 thread 配置为(32,32)执行更快(表 14.3 的结果已采取最佳 2D 线程配置)。

表 14.5 共享存储、thread 配置和性能的关系

512	cuda_2d	cuda_2d_1	cuda_2d_2	cuda_2d_9
(16,16)	203.78	183.55	181.18	175.04
(32,32)	192.10	193.31	188.77	183.30

14.3 均值滤波

均值滤波是另一种常见的图像滤波方法,其与中值滤波有点类似,都需要取像元及其周围元素共 9 点参与运算,区别在中间值排序获得中间值,而均值滤波则求 9 点平均值。计算公式可表示为

$$FX[i][j]=(X[i-1][j-1]+X[i-1][j]+X[i-1][j+1]\\+X[i][j-1]+X[i][j]+X[i][j+1]\\+X[i+1][j-1]+X[i+1][j]+X[i+1][j+1])/9$$

14.3.1 串行均值滤波

根据均值滤波计算思想,实现串行均值滤波代码。图像数据的输入和结果输出过程与中值滤波相似,不再赘述。

```
   ⋮
   float tmp_f;
   for(j=1;j<imageheight-1;j++)
   {
```

```
            for(k=1;k<imagewidth-1;k++)
            {
                tmp_f=0.0;
                tmp_f+=img[(j-1)*imagewidth+(k-1)];
                tmp_f+=img[(j-1)*imagewidth+(k)];
                tmp_f+=img[(j-1)*imagewidth+(k+1)];
                tmp_f+=img[(j)*imagewidth+(k-1)];
                tmp_f+=img[(j)*imagewidth+(k)];
                tmp_f+=img[(j)*imagewidth+(k+1)];
                tmp_f+=img[(j+1)*imagewidth+(k-1)];
                tmp_f+=img[(j+1)*imagewidth+(k)];
                tmp_f+=img[(j+1)*imagewidth+(k+1)];
                tmp_f/=9;
                filter[j*imagewidth+k]=(unsigned char) tmp_f;
            }
        }
    ⋮
```

14.3.2 并行均值滤波

均值滤波的 GPU 移植和优化方案与中值滤波类似,只需修改滤波过程即可。下面是基于 14.2.2 节代码修改得到的 1D 并行均值滤波 kernel 函数(标记为 cuda_1d)。

```
__global__ void mean_filter_0(unsigned char *img,unsigned char *filter,int imagewidth,int imageheight)
{
    int bid=blockIdx.x;
    int tid=threadIdx.x;
    float tmp_f;
    for(;bid<imageheight;bid+=gridDim.x)
    {
        if((bid>0)&&(bid<(imageheight-1)))
        {
            for(tid=threadIdx.x;tid<imagewidth;tid+=blockDim.x)
            {
                if((tid>0)&&(tid<(imagewidth-1)))
                {
                    tmp_f=0.0f;
                    tmp_f+=img[(bid-1)*imagewidth+(tid-1)];
                    tmp_f+=img[(bid-1)*imagewidth+(tid)];
                    tmp_f+=img[(bid-1)*imagewidth+(tid+1)];
                    tmp_f+=img[(bid)*imagewidth+(tid-1)];
```

```
                    tmp_f+=img[(bid) * imagewidth+(tid)];
                    tmp_f+=img[(bid) * imagewidth+(tid+1)];
                    tmp_f+=img[(bid+1) * imagewidth+(tid-1)];
                    tmp_f+=img[(bid+1) * imagewidth+(tid)];
                    tmp_f+=img[(bid+1) * imagewidth+(tid+1)];
                    tmp_f/=9;
                    filter[bid * imagewidth+tid]=(unsigned char)tmp_f;
                }
            }
        }
    }
}
```

利用类似的方法,可以获得 cuda_1d_1 版本、cuda_2d 版本、cuda_2d_2 版本和 cuda_2d_9 版本的均值滤波 kernel 函数。cuda_2d_1 版本代码在本节均值滤波不再讨论。各版本详细 kernel 函数代码参见附录 C.4。

14.3.3 实验结果与分析

编译串行和 GPU 并行均值滤波代码,输入图 14.8(a)的图像数据,输出滤波结果图像如图 14.8(b)所示,串行和 GPU 并行版本输出结果一致。统计串行和 GPU 并行均值滤波时间信息如表 14.6 所示。表 14.7 统计了性能较好的 cuda_1d 和 cuda_2d_9 两个版本的 kernel 函数加速比和总时间(利用 gettimeofday 测试通信和计算总时间)加速比。

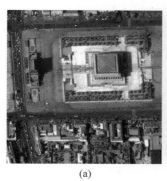 

(a)　　　　　　　(b)

图 14.8　均值滤波结果

表 14.6　均值滤波计算时间　　　　　　　　　　　　　　单位:μs

n	serial	cuda_1d	cuda_1d_1	cuda_2d	cuda_2d_2	cuda_2d_9	cuda_1d(all)	cuda_2d_9(all)
512	882.86	29.54	33.73	36.03	36.77	30.59	189.07	180.01
1024	3525.02	102.62	129.54	135.94	139.23	116.58	611.07	609.88
2048	14 151.10	398.27	524.99	541.28	551.74	457.25	2303.12	2274.04
4096	66 877.84	1531.40	2044.10	2152.70	2196.50	1815.80	9022.00	8951.90

表 14.7 均值滤波加速比

n	kernel		all	
	cuda_1d	cuda_2d_9	cuda_1d	cuda_2d_9
512	29.9	28.9	4.7	4.9
1024	34.4	30.2	5.8	5.8
2048	35.5	30.9	6.1	6.2
4096	43.7	36.8	7.4	7.5

表 14.6 和表 14.7 中的数据显示：

(1) 对比 cuda_1d 和 cuda_1d_1 两列数据，cuda_1d_1 版本执行较慢，说明均值滤波采取共享存储优化反而性能下降了。

(2) 对比 cuda_2d 和 cuda_2d_2 两列数据，cuda_2d 版本耗时较少；而对比 cuda_2d 和 cuda_2d_9 两列数据，cuda_2d_9 版本执行更快。说明在 2D 线程维度配置的均值滤波中，共享存储优化还是能起到加速效果的，但其共享存储数据加载过程就显得尤为重要了，若处理不好将导致性能下降。

(3) 对比 cuda_1d 和 cuda_2d_9 两列数据、cuda_1d(all) 和 cuda_2d_9(all) 两列数据，或观察表 14.7 中的加速比数据，可以发现，从 kernel 函数执行时间看，cuda_1d 版本执行较快，而利用 gettimeofday() 函数测试总时间时，反而 cuda_2d_9 版本总时间较少。

第四篇 核心篇

第 15 章　GPU 执行核心
第 16 章　GPU 存储体系
第 17 章　GPU 关键性能测评
第 18 章　CPUs 和 GPUs 协同

读完前三篇，相信读者对 GPU 已有一定理解，对于 GPU 的编程和优化也有一些认识了，但要真正透彻理解 GPU，还需要深入探究以下问题：kernel 函数在 GPU 中如何执行？真正影响 GPU 程序执行性能的关键因素是什么？如何评价 GPU 的好坏？CPU 与 GPU 如何协同计算？如何使用多块 GPU 同时计算？本篇将逐一回答这些疑问，主要内容如下。

第 15 章着重介绍 GPU 执行核心，从 GPU 对算术运算的支持和性能，到 GPU 的分支处理，GPU 上的同步与计时，再到 GPU 支持的数学函数，warp 和 block 原语，最后设计实验比较了 kernel 函数启动开销、线程切换开销和循环处理开销。

第 16 章详细介绍 GPU 复杂的存储体系，包括寄存器、局部存储、共享存储、常量存储、全局存储、纹理存储、主机端内存的使用方式以及零拷贝操作等。值得重点关注的内容有：①16.1 节测定了各存储单元的访存延迟；②16.3 节探索了影响局部变量数组在寄存器和局部存储间分配的因素；③主机端内存的使用方式（页锁定和可分页）的实测耗时分析和总结；④零拷贝操作的应用技巧。

第 17 章对 GPU 的关键性能进行了测评，具体包括 GPU 预热和启动开销、GPU 浮点运算能力、GPU 访存带宽、PCI－E 通信带宽等 GPU 关键性能参数。此外还介绍 GPU 参数获取、精准测时方法、性能分析工具 Visual Profiler 的使用和性能对比约定。

第 18 章介绍 CPU 和 GPU 协同并行的多种方法，以矩阵乘法为对象，分别进行了 CPU/GPU 协同计算、多 GPU 协同计算、多 CPU 和多 GPU 协同 3 大类协同并行运算研究，使用了 CUDA、MPI、OpenMP 的相互混合组合，各种组合的混合并行模式基本完全覆盖。

第15章 GPU 执行核心

由于 NVIDIA 未公开很多 GPU 技术细节，GPU 程序开发人员无法深入知悉 kernel 函数如何在 GPU 上执行，笔者当然也不得而知。本章主要从程序员角度阐述 GPU 执行核心的工作机理以及一些基于实验结果的合理推论。阅读本章前，请先回顾第 3 章内容。

15.1 概述

GPU 程序开发时，程序员关注的焦点包括功能的实现和性能的提升两个方面。

GPU 作为协处理器，与 CPU 是两个完全不同概念的产品，此外 GPU 真正支持可编程开发的历史毕竟较短，故 GPU 无法兼容 CPU 的全部功能，甚至应该说 GPU 支持的功能极其有限。因此了解 GPU 支持的运算功能是学习 GPU 编程的第一步。

另一方面，性能是 GPU 的第一生命线。当一个 kernel 函数被程序调用时，根据其调用时的线程维度配置<<<blocks,threads>>>，以 block 为单位映射到 SM(SMX、SMM)中，真正执行时又以 warp 为单位解析指令并分发到具体的运算单元(core/SP、DP、SFU)或访存单元(LD/ST)。

一般情况下，在 kernel 函数调用的线程维度配置中，block 数量和 thread 数量均远超硬件 SM 数量和 core 数量。此时每个 SM 需要运行很多个 block，但 SM 内资源又无法支持这么多 block 的同时执行，因此需要频繁的线程切换。

SM 中活动的 warp 数量占物理 warp 数量的比率为 occupancy(占用率)。

另外，warp 作为基本执行单元，遇到分支指令时，若 warp 内的线程处于不同的分支，则无法同时执行。warp 只能串行逐一执行各分支，而分支的执行顺序也会影响程序性能。

综上所述，kernel 函数执行性能的相关因素包括 core 或 DP 的算术运算能力、SFU 的特殊功能处理能力、分支处理、线程切换开销、存储访问、occupancy 等。

本章从 GPU 支持的算术运算、GPU 核心的运算能力、分支处理、常用命令的性能、并行引入开销的权衡(kernel 函数启动、线程切换和循环处理开销的权衡)等方面展开论述。存储访问的相关内容将在第 16 章详述。

15.2 算术运算支持

15.2.1 整数运算

Fermi 架构开始完整支持 32 位整数运算,包括加、减、乘、除、逻辑、条件、类型转换、位操作、窄整型 SIMD 操作等。这些 32 位整数运算基本上都支持标准 C 操作符调用(即普通符号运算)。另外还支持内置函数访问。

Tesla 架构不支持 32 位整数乘法,仅支持 24 位整数乘法,其中,32 位整数乘法需要 4 条指令才能实现。而从 Fermi 架构开始支持 32 位整数乘法。因此,在计算能力 1.X 的设备上最好用 24 位整数乘法,利用内置函数 __mul24 实现。

```
extern __device__ __device_builtin__ int __mul24(int x, int y);
```

GPU 中的位运算除了常见的与运算(&)和或运算(|)运算外,还提供了许多内置函数,例如 __brev()、__clz()、__ffs()、__popc()和__sad()等。这些内置函数在 Fermi(计算能力 2.X)之后的架构中分别对应一条指令,而在 Tesla 架构则被编译为多条指令。另外,对于 64 位整型数据(long long),其相应的内置函数改为 __brevll()、__clzll()、__ffsll()和 __popcll()。

```
extern __device__ __device_builtin__ unsigned int __brev(unsigned int x);
                                                            //位取反
extern __device__ __device_builtin__ unsigned int __byte_perm(unsigned int x,
unsigned int y, unsigned int s);      //根据参数 s,从 x、y 中选择数据构成 32 位结果
extern __device__ __device_builtin__ int __clz(int x);    //计算前导 0 的个数
extern __device__ __device_builtin__ int __ffs(int x);    //返回第一个有效位
extern __device__ __device_builtin__ int __popc(unsigned int x);   //返回 1 的个数
extern __device__ __device_builtin__ unsigned int __sad(int x, int y, unsigned
int z);//z=|x-y|
```

Kepler 架构的 GK110 增加了 64 位漏斗移位指令,用来连接两个 32 位整数形成 64 位整数,并左移或右移,返回高 32 位(左移)或低 32 位(右移)。漏斗移位指令最早是在计算能力 3.2 引入,但市场上并没有相应产品,而是应用在计算能力 3.5 的产品上。在头文件 sm_32_intrinsics.h 中可以找到相应的声明。

```
static __device__ inline unsigned int __funnelshift_l(unsigned int lo, unsigned
int hi, unsigned int shift);    //[hi:lo]连成 64 位整数,左移 shift&31 位,返回高 32 位
```

```
static _ _device_ _ inline unsigned int _ _funnelshift_lc(unsigned int lo,
unsigned int hi, unsigned int shift);
                    //[hi:lo]连成64位整数,左移min()shift,31()位,返回高32位
static _ _device_ _ inline unsigned int _ _funnelshift_r(unsigned int lo,
unsigned int hi, unsigned int shift);
                    //[hi:lo]连成64位整数,右移shift&31位,返回低32位
static _ _device_ _ inline unsigned int _ _funnelshift_rc(unsigned int lo,
unsigned int hi, unsigned int shift);
                    //[hi:lo]连成64位整数,右移min(shift,31)位,返回低32位
```

15.2.2 浮点运算

高性能浮点运算是 GPU 相对 CPU 的重大优势之一,特别是其支持的超越函数的精度和性能均强于 CPU。在 Tesla 架构中,SP 采用 IEE754－1985 标准,仅支持单精度运算,在计算能力 1.3 的设备中引入 DP 来支持双精度运算,但性能较差。自 Fermi 架构开始,core(SP)采用 IEE754－2008 标准,能同时支持单精度运算和双精度运算。

浮点数运算同样支持标准 C 操作符调用,也支持内置函数调用。表 15.1 罗列了一些浮点运算内置函数,其中[rn|rz|ru|rd]指代舍入模式,rn 表示舍入到最近偶数(即最近舍入),rz 表示向 0 舍入(亦称截断),ru 表示向下舍入(向负无穷大舍入),rd 表示向上舍入(向正无穷大舍入)。

表 15.1　浮点运算内置函数

内置函数	运算	内置函数	运算
_ _fadd_[rn|rz|ru|rd]	加	_ _dadd_[rn|rz|ru|rd]	加
_ _fmul_[rn|rz|ru|rd]	乘	_ _dmul_[rn|rz|ru|rd]	乘
_ _fmaf_[rn|rz|ru|rd]	乘加	_ _fma_[rn|rz|ru|rd]	乘加
_ _frcp_[rn|rz|ru|rd]	倒数	_ _drcp_[rn|rz|ru|rd]	倒数
_ _fdiv_[rn|rz|ru|rd]	除	_ _ddiv_[rn|rz|ru|rd]	除
_ _fsqrt_[rn|rz|ru|rd]	平方根	_ _dsqrt_[rn|rz|ru|rd]	平方根

GPU 除了提供整数和浮点类型标准 C 语言转换外,还提供了丰富的内置转换函数。例如,float 与 int 转换(_ _float2int_[rn|rz|ru|rd]、_ _int2float_[rn|rz|ru|rd])、float 和 unsigned int 转换(_ _float2uint_[rn|rz|ru|rd]、_ _uint2float_[rn|rz|ru|rd])、float 和 64 位 int 转换(_ _float2ll_[rn|rz|ru|rd]、_ _ll2float_[rn|rz|ru|rd])、double 到 float 转换(_ _double2float_[rn|rz|ru|rd])、double 与 int 转换(_ _double2int_[rn|rz|ru|rd]、_ _int2double_[rn|rz|ru|rd])、double 与 unsigned int 转换(_ _double2uint_[rn|rz|ru|rd]、_ _uint2double_[rn|rz|ru|rd])、double 与 64 位 int 转换(_ _double2ll_[rn|rz|ru|rd]、_ _ll2double_[rn|rz|ru|rd])和 double 与 64 位 unsigned int 转换(_ _double2ull_[rn|rz|ru|rd]、_ _ull2double_[rn|rz|ru|rd])等。另外还有 float 和 short(半精度)的转换(_ _half2float、_ _float2half)。

SM 中的特殊功能单元 SFU 实现了 6 种常用超越函数,支持单精度正弦、余弦、对数、指数、倒数和平方根倒数的计算,且性能非常好。相应的内置函数如表 15.2 所示。

表 15.2 SFU 内置函数

内置函数	运算	内置函数	运算
__cosf(x)	cosx	__log10f(x)	$\log_{10}x$
__exp10f(x)	10^x	__powf(x,y)	x^y
__expf(x)	e^x	__sinf(x)	sinx
__fdividef(x,y)	x/y	__sincosf(x,sptr,cptr)	*s=sinx *c=cosx
__logf(x)	lnx		
__log2f(x)	$\log_2x$	__tanf(x)	tanx

15.3 算术运算性能

对于 GPU 来说,除了算术运算功能支持外,还需要关注算术运算性能。本节从参考文献[6]摘取了 Tesla 架构 GPU 的部分实验数据。

参考文献[6]对基于 Tesla 架构的 GT280 GPU 进行测试,表 15.3 和表 15.4 分别统计了各种类型数据的算术运算及内置函数的延迟和吞吐量。基于表 15.3 和表 15.4,可以得到一些结论:①(Tesla 架构)整型 32 位乘法性能较差,比加法等运算慢,符合前文提及 32 位整数乘法需要 4 条指令;②整数运算中乘加操作没有融合到一起;③除法运算相对其他运算都比较弱。

表 15.3 算术运算的延迟和吞吐量

operation	type	latency(clocks)	Throughput(ops/clock)
add,sub,max,min	uint,int	24	7.9
mad	uint,int	120	1.4
mul	uint,int	96	1.7
div	uint	6.8	0.28
div	int	684	0.23
rem	uint	728	0.24
rem	int	784	0.2
and,or,xor,shl,shr	uint	24	7.9
add,sub,max,min	float	24	7.9
mad	float	24	7.9
mul	float	24	11.2
div	float	137	1.5
div	double	1366	0.063
mad	double	48	1
mul	double	48	1
add,sub,max,min	double	48	1

表 15.4 内置函数的延迟和吞吐量

operation	type	latency(clocks)	Throughput(ops/clock)
__umul24()	uint	24	7.9
__mul24()	int	24	7.9
__usad()	uint	24	7.9
__sad()	int	24	7.9
__umulhi()	uint	144	1
__mulhi()	int	180	0.77
__fadd_rn(),__fadd_rz()	float	24	7.9
__fmul_rn(),__fmul_rz()	float	26	10.4
__fdividef()	float	52	1.9
__dadd_rn	double	48	1
__powf()	float	75	1
__tanf()	float	98	0.67
__exp2f()	float	48	2
rsprt()	float	28	2
sqrt()	float	56	2
__log2f()	float	28	2
__expf(),__exp10f()	float	72	2
__logf(),__log10f()	float	52	2
__sinf(),__cosf()	float	48	2

表 15.3 和表 15.4 中的数据是基于 Tesla 架构的，现在早已过时，但却可以给 GPU 程序开发者一个基本的认识。在 GPU 开发过程中，算术运算更重要的是实现运算功能，只有在访存优化完成，计算成为瓶颈的情况下才会考虑算术运算优化。

下面是一段测试算术加法指令的计算延迟代码，以及在本文实验平台 Tesla K20c GPU 上的运行结果。执行结果显示，在 K20c GPU 中，单精度浮点运算的延迟为 9 clocks，相较 GT280 得到了大幅缩减。若要测试更多算术运算延迟，可基于该代码进行扩展测试，亦可见参考文献[6]的 microbenchmark。

```
#define DATATYPE float
__global__ void latency_add(double *time,DATATYPE *out,int its)
{
    DATATYPE a=1.1;
    DATATYPE b=0.1;
    double time_tmp=0.0;
    unsigned int start_time=0, stop_time=0;
    for(int i=0;i<its;i++)
    {
        __syncthreads();
```

```
        start_time=clock();
        for(int j=0;j<128;j++)
        {
            a=a+b;b=b+a;
        }
        stop_time=clock();
        time_tmp+=(stop_time-start_time);
    }
    out[0]=a+b;
    time_tmp=time_tmp/128.0/its/2.0;
    time[0]=time_tmp;
}
int main()
{
    double *h_time;
    h_time=(double *)malloc(sizeof(double));
    double *d_time;
    cudaMalloc((void**)&d_time,sizeof(double));
    DATATYPE *d_out,*h_out;
    h_out=(DATATYPE *)malloc(sizeof(DATATYPE));
    cudaMalloc((void**)&d_out,sizeof(DATATYPE));
    int its=30;
    latency_add<<<1,1>>>(d_time,d_out,its);
    cudaMemcpy(h_out,d_out,sizeof(DATATYPE),cudaMemcpyDeviceToHost);
    cudaMemcpy(h_time,d_time,sizeof(double),cudaMemcpyDeviceToHost);
    printf("b+=a:\t%f\n",h_time[0]);
    cudaFree(d_time);
    cudaFree(d_out);
    free(h_out);
    free(h_time);
    return 0;
}
[fangmq@cn18%yhstar fmq_check]$./latency_add
b+=a:      9.015104
```

15.4 分支处理

GPU 的 SIMT 执行模式注定了其分支处理不如 CPU 直接。GPU 中一个 warp 的 32 个线程同时执行相同的指令,因此当遇到条件分支语句时,也要串行逐一执行各分支。图 15.1 展示了 warp 遇到的两种分支情况,图 15.1(a)显示 warp 的所有线程处于同一个分支内,此时 warp 直接执行该分支;图 15.1(b)显示 warp 的线程处于不同分支,不同分支需要串行执行,不符合分支判断的线程空闲。显然图 15.1(a)的执行方式效率更高、性能更好,程序员应尽量在编程时令 warp 的线程处于同一分支。

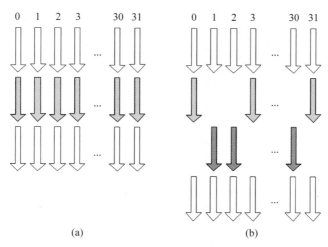

图 15.1　warp 分支处理

另外,参考文献[6]通过 microbenchmark 深入分析了 GPU 分支处理。分支处理包括分支和汇聚两部分,分支时相同 warp 不同分支串行执行,而不同的 warp 可以同时执行;汇聚点位于分支处,只有一个分支执行到汇聚点后,下一个分支才能开始执行。另外,参考文献[6]中还测试了分支处理顺序,结果显示分支处理顺序与线程 ID 号无关,仅与 if 和 else 顺序相关,其中 else 分支优先执行,if 分支后执行。

下面是笔者设计的 GPU 分支探索代码,用来探索 GPU 分支的处理机制。主要思想是给 warp 内的线程的变量赋不同值,利用 if 和 else 分支进行计时和运算,根据时间结果分析 GPU 的分支处理机制。其中,根据 warp 内的线程变量赋不同值的不同,将代码看成两个版本。

(1) if_else 版本,赋值为"int x=threadIdx.x;"。

(2) if_else_1 版本,赋值为"int x=31-threadIdx.x;"。

根据这两个版本的结果,可以分析得出一些有益的结论。根据在本文实验平台的 K20c GPU 上的运行结果,可以总结得出相应结论,在其他架构 GPU 上可能由于分支策略不同需要再设计新的代码,本书就不赘述了。

```
__global__ void if_else(unsigned int *time,DATATYPE *out)
{
    DATATYPE a=1.1;
    DATATYPE b=0.1;
    unsigned int s_time=0;
    __syncthreads();
    int x=threadIdx.x;       //if_else_1:      int x=31-threadIdx.x;
    if(x==0)     {s_time=clock();a+=b;b+=a;}
```

```
        else if(x==1)     {s_time=clock();a+=b;b+=a;}
        else if(x==2)     {s_time=clock();a+=b;b+=a;}
        else if(x==3)     {s_time=clock();a+=b;b+=a;}
        else if(x==4)     {s_time=clock();a+=b;b+=a;}
        else if(x==5)     {s_time=clock();a+=b;b+=a;}
        else if(x==6)     {s_time=clock();a+=b;b+=a;}
        else if(x==7)     {s_time=clock();a+=b;b+=a;}
        else if(x==8)     {s_time=clock();a+=b;b+=a;}
        else if(x==9)     {s_time=clock();a+=b;b+=a;}
        else if(x==10)    {s_time=clock();a+=b;b+=a;}
        else if(x==11)    {s_time=clock();a+=b;b+=a;}
        else if(x==12)    {s_time=clock();a+=b;b+=a;}
        else if(x==13)    {s_time=clock();a+=b;b+=a;}
        else if(x==14)    {s_time=clock();a+=b;b+=a;}
        else if(x==15)    {s_time=clock();a+=b;b+=a;}
        else if(x==16)    {s_time=clock();a+=b;b+=a;}
        else if(x==17)    {s_time=clock();a+=b;b+=a;}
        else if(x==18)    {s_time=clock();a+=b;b+=a;}
        else if(x==19)    {s_time=clock();a+=b;b+=a;}
        else if(x==20)    {s_time=clock();a+=b;b+=a;}
        else if(x==21)    {s_time=clock();a+=b;b+=a;}
        else if(x==22)    {s_time=clock();a+=b;b+=a;}
        else if(x==23)    {s_time=clock();a+=b;b+=a;}
        else if(x==24)    {s_time=clock();a+=b;b+=a;}
        else if(x==25)    {s_time=clock();a+=b;b+=a;}
        else if(x==26)    {s_time=clock();a+=b;b+=a;}
        else if(x==27)    {s_time=clock();a+=b;b+=a;}
        else if(x==28)    {s_time=clock();a+=b;b+=a;}
        else if(x==29)    {s_time=clock();a+=b;b+=a;}
        else if(x==30)    {s_time=clock();a+=b;b+=a;}
        else    {s_time=clock();a+=b;b+=a;}
    out[0]=a+b;
    time[threadIdx.x]=s_time;
}
int main()
{    unsigned int *h_time;
    h_time=(unsigned int *)malloc(sizeof(int) * 32);
    unsigned int *d_time;
    cudaMalloc((void **)&d_time,sizeof(int) * 32);
    DATATYPE *d_out,*h_out;
    h_out=(DATATYPE *)malloc(sizeof(DATATYPE));
    cudaMalloc((void **)&d_out,sizeof(DATATYPE));
    int its=30;
```

```
    if_else<<<1,32>>>(d_time,d_out);
    cudaMemcpy(h_out,d_out,sizeof(DATATYPE),cudaMemcpyDeviceToHost);
    cudaMemcpy(h_time,d_time,sizeof(int) * 32,cudaMemcpyDeviceToHost);
    for(int i=0;i<32;i++)
    {
        printf("%d\t%d\n",i,h_time[i]);
    }
    cudaFree(d_time);
    cudaFree(d_out);
    free(h_out);
    free(h_time);
    return 0;
}
```

编译并在 K20c GPU 上执行上述代码,执行结果比较冗长,在此省略,选取其中一组测试结果记录在表 15.5 中。表中数据已按时间顺序重新排列,从结果数据可以总结出以下结论:①所有的线程执行的分支都不在同一时刻,说明 warp 内分支处理过程确实如图 15.1 所示,串行执行各个分支;②if_else 和 if_else_1 两个版本的分支执行顺序是一致的,说明分支处理顺序与 if 和 else 顺序无关,仅与线程 ID 号相关,且按 0,1,2,3,16,17,18,19,4,5,6,7,20,21,22,23,8,9,10,11,24,25,26,27,12,13,14,15,28,29,30,31 的顺序执行。该结论与参考文献[6]相悖,可能的原因是不同架构产品的分支处理策略不同,笔者手头没有相应的 GPU 做验证,故不好下结论,有条件的读者可以进一步验证。

表 15.5 K20c 分支机制

if_else		if_else_1	
threadID	clock	threadID	clock
0	18812355	0	19163294
1	18812419	1	19163360
2	18812483	2	19163422
3	18812547	3	19163486
16	18812611	16	19163550
17	18812675	17	19163614
18	18812739	18	19163678
19	18812803	19	19163742
4	18812867	4	19163806
5	18812931	5	19163870
6	18812995	6	19163934
7	18813059	7	19163998
20	18813123	20	19164062
21	18813187	21	19164126
22	18813251	22	19164190
23	18813315	23	19164254

续表

if_else		if_else_1	
threadID	clock	threadID	clock
8	18813379	8	19164318
9	18813443	9	19164382
10	18813507	10	19164446
11	18813571	11	19164510
24	18813635	24	19164574
25	18813699	25	19164638
26	18813763	26	19164702
27	18813827	27	19164766
12	18813891	12	19164830
13	18813955	13	19164894
14	18814019	14	19164958
15	18814083	15	19165022
28	18814147	28	19165086
29	18814211	29	19165150
30	18814275	30	19165214
31	18814339	31	19165262

15.5 同步与测时

15.5.1 同步

同步操作可以实现共享的存储单元(例如共享存储、全局存储)的数据交互,其作用对象是 warp 而非具体的 thread。GPU 仅支持 block 中的所有 warp 同步。

最常用的同步语句是__syncthreads(),该内置函数的功能是等待线程块内所有线程到达后继续执行,一般用来维持线程块内共享存储数据的一致性。除此之外,还有类似的同步指令,其作用范围有所差异。表 15.6 描述了各同步函数及功能。

表 15.6 同步函数及其功能

同 步 函 数	功　　能
__syncthreads()	等待到所有由线程引起的共享存储访问对线程块内所有线程可见
__threadfence()	等待到所有由线程引起的共享存储和全局存储访问对线程块内所有线程可见
__threadfence_block()	等待到所有由线程引起的共享存储和全局存储访问对线程块内所有线程、设备中所有访问全局内存的线程可见
__threadfence_system()	等待到所有由线程引起的共享存储和全局存储访问对线程块内所有线程、设备中所有访问全局内存的线程、主机内访问锁页主机内存的线程可见

15.5.2 测时

GPU 上提供了 clock() 测时函数,使用方法与标准 C 语言的 clock() 函数类似。参考文献[6]利用 microbenchmark 测试 GTX280 得出以下结论。

(1) clock() 寄存器的数据每次左移一位,即半个着色器时钟频率增加 1。
(2) 两个非依赖的 clock() 需要 28 个时钟周期。
(3) 每个 TPC 内 clock 寄存器共享(非全局,亦非 SM)。

15.6 数学函数

CUDA C 提供了类 C 语言的丰富数学函数,表 15.7 罗列了数学函数及其最大 ULP 误差,其中,[f]表示带 f 为单精度数学函数,不带 f 为双精度数学函数。

表 15.7 数学函数及其最大 ULP 误差

函数	运算	ULP 误差	
		float	double
x+y	x+y	0#	0
x*y	x*y	0#	0
x/y	x/y	2@	0
1/x	1/x	1@	0
rsqrt[f](x)	平方根倒数	2	1
sqrt[f](x)	平方根	3@	0
cbrt[f](x)	立方根	1	1
rcbrt[f](x)	立方根倒数	2	1
hypot[f](x,y)	三角形斜边长	3	2
exp[f](x)	e 的指数	2	1
exp2[f](x)	2 的指数	2	1
exp10[f](x)	10 的指数	2	1
expm1[f](x)	e 的指数减 1	1	1
log[f](x)	自然(e)对数	1	1
log2[f](x)	以 2 为底的对数	3	1
log10[f](x)	以 10 为底的对数	3	1
log1p[f](x)	x+1 的自然对数	2	1
sin[f](x)	正弦	2	1
cos[f](x)	余弦	2	1
tan[f](x)	正切	4	2
sincos[f](x,s,c)	*s=sin(x); *c=cos(x)	2	1
sinpi[f](x)	sin(πx)	2	1
cospi[f](x)	cos(πx)	2	1

续表

函数	运算	ULP 误差	
		float	double
sincospi[f](x,s,c)	*s=sin(πx);*c=cos(πx)	2	1
asin[f](x)	反正弦	4	2
acos[f](x)	反余弦	3	2
atan[f](x)	反正切	2	2
atan2[f](y,x)	y/x 的反正切	3	2
sinh[f](x)	双曲正弦	3	1
cosh[f](x)	双曲余弦	2	1
tanh[f](x)	双曲正切	2	1
asinh[f](x)	反双曲正弦	3	2
acosh[f](x)	反双曲余弦	4	2
atanh[f](x)	反双曲正切	3	2
pow[f](x,y)	x 的 y 次方	8	2
erf[f](x)	误差函数	2	2
erfc[f](x)	误差函数的补	6	4
erfinv[f](x)	反误差函数	3	5
erfcinv[f](x)	反误差函数的补	4	6
erfcx[f](x)	缩放的误差函数	6	3
normcdf[f](x)	标准累计分布	6	5
normcdfinv[f](x)	反标准累计分布	5	7
lgamma[f](x)	伽玛函数的对数	6 $	4 $
tgamma[f](x)	真伽玛函数	11	8
fma[f](x,y,z)	x*y+z	0	0
frexp[f](x,exp)	小数部分	0	0
ldexp[f](x,exp)	缩放为 2 的幂次	0	0
scalbn[f](x,n)	对 x 缩放为 2^n 倍	0	0
scalbln[f](x,l)	对 x 缩放为 2^n 倍	0	0
logb[f](x)	得到指数	0	0
ilogb[f](x)	取指数	0	0
j0[f](x)	$n=0$ 的第一类贝塞尔函数	9%	7%
j1[f](x)	$n=1$ 的第一类贝塞尔函数	9%	7%
jn[f](x)	第一类贝塞尔函数	*	*
y0[f](x)	$n=0$ 的第二类贝塞尔函数	9%	7%
y1[f](x)	$n=1$ 的第二类贝塞尔函数	9%	7%
yn[f](x)	第二类贝塞尔函数	**	**
fmod[f](x,y)	浮点数取余	0	0
remainder[f](x,y)	余数	0	0

续表

函　　数	运　　算	ULP 误差	
		float	double
remquo[f](x,y,iptr)	余数,同时返回商	0	0
modf[f](x,iptr)	分割整数和小数部分	0	0
fdim[f](x,y)	正差值	0	0
trunc[f](x)	截断(向原点取整)	0	0
round[f](x)	最近整数取整	0	0
rint[f](x)	最近整数取整	0	0
nearbyint[f](x)	取整	0	0
ceil[f](x)	向上取整	0	0
floor[f](x)	向下取整	0	0
lrint[f](x)	向 long 转换	0	0
lround[f](x)	向 long 转换	0	0
llrint[f](x)	向 long long 转换	0	0
llround[f](x)	向 long long 转换	0	0

♯：在 SM 1.X 硬件上,加和乘合并为乘加指令,会因为中间尾数截断而降低精度。

@：在 SM 2.X 以后的硬件,若指定编译选项—prec-div=true 可降低误差为 0。

$：单精度在区间($-10.001, -2.264$)上误差为 6;双精度在区间($-11.001, -2.2637$)上误差为 4。

%：单精度,$|X|<8$ 时误差为 9,否则最大绝对误差为 2.2×10^{-6};双精度,$|X|<8$ 时误差为 7,否则最大绝对误差为 5×10^{-12}。

*：当 $n=128$,最大绝对误差为 2.2×10^{-6}(float)和 5×10^{-12}(double)。

**：单精度,$|X|<n$ 时误差为 $ceil(2+2.5n)$,否则最大绝对误差为 2.2×10^{-6};双精度,$|X|>1.5n$ 时,最大绝对误差为 5×10^{-12}。

15.7　warp 与 block 原语

15.7.1　warp 原语

warp 是 GPU 程序执行的基本单位,NVIDIA 提供了作用范围仅为 warp 的特殊指令,称为 warp 原语。warp 原语主要包括投票和洗牌两类。

投票是一条能够评估条件并广播 1 位结果给每个 thread 的指令。最早在 SM 1.2 中引入 VOTE 指令,例如,_ _any()函数在 warp 中任一 thread 判断为真(即返回 1),_ _all()函数只有在 warp 所有线程返回为真(即返回 1)。SM 2.X 后添加了对全部判断的支持,例如,函数_ _ballot()评估所有 thread 条件,返回 32 位值,每一位代表相应 thread 的判断结果。

```
//sm_12_atomic_functions.h
extern __device__ __device_builtin__ int          __any(int cond);
extern __device__ __device_builtin__ int          __all(int cond);
extern __device__ __device_builtin__ unsigned int __ballot(int);
```

洗牌允许 warp 内 thread 间不经共享存储进行数据交换。洗牌无法节省时间,但可以减少共享存储的访问,可避免 bank conflict 等问题。洗牌操作返回源线程的 var 值,源线程的计算各不相同,其中,__shfl()源线程为 srcLane 值,__shfl_up()将调用者线程减去 delta 为源线程,__shfl_down()将调用者线程加上 delta 为源线程,__shfl_xor()用调用者线程与 laneMask 按位异或计算得到源线程。

```
//sm_30_intrinsics.h
static __device__ __inline__ int __shfl(int var, int srcLane, int width=
warpSize);
static __device__ __inline__ int __shfl_up(int var, unsigned int delta, int
width=warpSize);
static __device__ __inline__ int __shfl_down(int var, unsigned int delta, int
width=warpSize);
static __device__ __inline__ int __shfl_xor(int var, int laneMask, int width=
warpSize);
static __device__ __inline__ float __shfl(float var, int srcLane, int width=
warpSize);
static __device__ __inline__ float __shfl_up(float var, unsigned int delta,
int width=warpSize);
static __device__ __inline__ float __shfl_down(float var, unsigned int delta,
int width=warpSize);
static __device__ __inline__ float __shfl_xor(float var, int laneMask, int
width=warpSize);
```

15.7.2 block 原语

在 block 层次,除了利用__syncthreads()函数实现同步操作外,SM 2.X 之后的硬件还提供了几个特殊功能同步指令,能聚合 block 内 thread 信息。其中,__syncthreads_count()评估一个判断,并返回判断为真的 thread 数量,__syncthreads_and()返回 block 内所有 thread 输入值的与运算,__syncthreads_or()返回 block 内所有 thread 输入值的或运算。

```
//sm_20_intrinsics.h
extern __device__ __device_builtin__ int __syncthreads_count(int);
extern __device__ __device_builtin__ int __syncthreads_and(int);
extern __device__ __device_builtin__ int __syncthreads_or(int);
```

15.8 kernel 启动、线程切换和循环处理

一般情况下，GPU 编程会涉及大规模循环处理，当循环量达到一定规模后无法直接处理，此时需要引入一些方法，例如，kernel 内部循环、启动大量 block、循环启动 kernel 函数等。尽管这 3 种模式在 GPU 中的开销（即 kernel 内循环开销、GPU 执行线程切换开销、kernel 函数启动开销）都非常小，但仍然存在差异，本节将通过实验量化测评这 3 类开销。

kernel 函数内循环开销：该开销的产生是由于启动 kernel 的线程数量不足以一次性处理整个问题，需要利用循环跳步的方法，完成全部问题的求解。而循环跳步过程将产生 kernel 函数内循环开销。

kernel 函数执行线程调度开销：GPU 的硬件核心数量是有限的，而启动 kernel 函数设定的线程总数量一般远大于硬件核心数量，GPU 通过线程切换和调度来执行 kernel 函数。此时有了线程切换开销的概念。

kernel 函数循环启动开销：kernel 函数启动开销是每个 kernel 函数启动必需的开销。循环启动 kernel 函数是解决大规模并行问题的重要途径之一，每次启动一个 kernel 函数处理其中一部分问题，通过循环启动 kernel 函数的方法完成全部运算。简单地说是将大问题分解为小问题，这种方法常用于 Geforce 系列显卡没有关闭运行时限制的情况，能有效避免 kernel 函数执行超时导致黑屏退出，提升程序通用性，11.5 节有相应的实例介绍。

下面设计了一个纯计算问题来测试 3 种方法的开销。设定一个二维矩阵，大小为 8192×8192，计算过程不涉及访存，为了避免代码被编译器直接优化，最后判断输出。利用上述 3 种方法实现该问题，下面展示了相应的 CUDA C 代码。实验结果显示 GPU 执行线程调度开销最小，kernel 内循环开销次之，kernel 函数启动开销最大。故编程开发时，可优先采用"线程切换"方法。

```
__global__ void loop_on_gpu(DATATYPE *array_a,int n)
{
    unsigned int bidx=blockIdx.x;
    unsigned int tidx=threadIdx.x;
    DATATYPE i=0;
    for(bidx=blockIdx.x;bidx<n;bidx+=gridDim.x)
    {
        for(tidx=threadIdx.x;tidx<n;tidx+=blockDim.x)
        {
            i=bidx+tidx;
        }
    }
    if((blockIdx.x==0)&&(threadIdx.x==n-1))
```

```
        {
            array_a[0]=i;
        }
}
__global__ void thread_change_on_gpu(DATATYPE *array_a,int n)
{
    unsigned int idx=blockIdx.x*blockDim.x+threadIdx.x;
    unsigned int bidx=idx/n;
    unsigned int tidx=idx%n;
    DATATYPE i;
    i=bidx+tidx;
    if(idx==n)
    {
        array_a[0]=i;
    }
}


__global__ void kernel_on_gpu(DATATYPE *array_a,int n,int p,int q)
{
    unsigned int bidx=p*512+blockIdx.x;
    unsigned int tidx=q*512+threadIdx.x;
    DATATYPE i;
    i=bidx+tidx;
    if((bidx==0)&&(tidx==n-1))
    {
        array_a[0]=i;
    }
}
int  main()
{
    struct timeval t1, t2, t3, t4;
    int i=0,j=0;
    int n=8192;
    int nn=(n+512-1)/512;
    DATATYPE *d_array_a;
    cudaMalloc((void**)&d_array_a,sizeof(DATATYPE));
    loop_on_gpu<<<512,512>>>(d_array_a,n);
    thread_change_on_gpu<<<nn*n,512>>>(d_array_a,n);
                                    //排除第一次 kernel 函数启动开销
    cudaThreadSynchronize();
    gettimeofday(&t1, NULL);
    loop_on_gpu<<<512,512>>>(d_array_a,n);
    cudaThreadSynchronize();
```

```
    gettimeofday(&t2, NULL);
    thread_change_on_gpu<<<nn * n,512>>>(d_array_a,n);
    cudaThreadSynchronize();
    gettimeofday(&t3, NULL);
    for(i=0;i<nn;i++)
    {
        for(j=0;j<nn;j++)
        {
            kernel_on_gpu<<<512,512>>>(d_array_a,n,i,j);
        }
    }
    cudaThreadSynchronize();
    gettimeofday(&t4, NULL);
    printf("loop_on_gpu: \t %f us\n",(t2.tv_sec-t1.tv_sec) * 1000.0 * 1000.0+
(t2.tv_usec-t1.tv_usec));
    printf("thread_change_on_gpu: \t %f us\n", ((t3.tv_sec-t2.tv_sec) * 1000.0 *
1000.0+ (t3.tv_usec-t2.tv_usec)));
    printf("one kernel call time:%f us \n", ((t4.tv_sec-t3.tv_sec) * 1000.0 *
1000.0+ (t4.tv_usec-t3.tv_usec)));
    return 0;
}
[fangmq@cn18%yhstar test_in_book]$  ./loop_thread_kernel_1
loop_on_gpu:    517.000000 us
thread_change_on_gpu:    3.000000 us
one kernel call time:2167.000000 us
```

上述结果只能说明理想问题中循环开销、线程切换开销和 kernel 启动开销的情况，真实问题涉及 GPU 存储访问，时间开销将集中在访存中，合理的访存模式和访存顺序将真正影响 kernel 函数的性能，需要考虑的因素也更加复杂。在 17.6 节的表 17.2 中，实验结果展示了这样一个实例，虽然线程切换开销小于线程内循环开销，但在访存密集型的复制操作中，2D 线程配置的循环版本明显比非循环版本（线程切换）性能要好。

第16章 GPU 存储体系

GPU 提供了层次式的存储体系(参见 3.4 节)，以辅助程序员优化程序性能，各存储单元的使用依赖于算法的访存模式和存储单元特性。GPU 存储单元包括全局存储、纹理存储、常量存储、共享存储、局部存储、寄存器等。另外，CPU 端(主机端)存储类型(页锁定存储和可分页存储)、CPU 与 GPU 通信接口和通信方式都会影响 GPU 程序执行的性能。本章逐点详细阐述各存储单元的特性和用法。

16.1 概述

各存储单元在 GPU 中的层次与位置的详细介绍可参阅 3.4 节，在此不再复述。表 16.1 给出了 GPU 存储单元的基本信息，包括位置、cache、访存类型和作用范围，其中，寄存器、共享存储位于片上，而常量存储、纹理存储拥有片上高速缓存，常量存储与纹理存储是只读存储。

表 16.1 GPU 存储基本信息

memory	location	cache	access	scope
register	on chip	N/A	R/W	thread
local	off chip	no	R/W	thread
shared	on chip	N/A	R/W	block
constant	off chip	yes	R	grid
global	off chip	yes	R/W	grid
texture	off chip	yes	R	grid

下面以 Tesla K20c GPU 为例测试了各存储单元的访存延迟情况。

测试数据的构造：除了寄存器，其他存储单元均支持数组存储。由于编译时可能包含诸如流水线、执行与访存掩藏等优化，因此需要设计一种依赖关系以避免这些编译器优化对访存延迟的影响。这里采用了一种固定跳步的方法(指针追逐方法)，即声明一个 int 数组，索引号为 index 对应的数据值为 index+step，访问时用本次读取的元素值作为下一次读取的索引。下面是该数组构造和访问的伪代码。

```
for i=1 to len
    array[i]=(i+step)%len;
for i=1 to n
    p=array[p];
```

寄存器访存延迟测试方法构造：寄存器也可以声明数组，但上述访问模式（p＝array[p]）将导致数组无法声明在寄存器中（详见 16.3 节），因此寄存器的访存延迟测定方法需要重新构造。简单的访问将直接被编译器优化。这里设计了下面的访问过程，每个数据的访问依赖前一个数据的赋值，最后将 r 写到全局存储，若无输出将被编译器直接优化。下面是测试寄存器访存延迟的 kernel 函数。

```
repeat128(r=p;p=q;q=x;x=y;y=z;z=r;)
out[0]=r;
__global__ void test_register_latency(double *time,DATATYPE *out,int its)
{
    int p=3;
    int q=1;
    int r,x=2,y=5,z=7;
    double time_tmp=0.0;
    unsigned int start_time=0, stop_time=0;
    for(int i=0;i<its;i++)
    {
        __syncthreads();
        start_time=clock();
        repeat128(r=p;p=q;q=x;x=y;y=z;z=r;)
        stop_time=clock();
        time_tmp+=(stop_time-start_time);
    }
    time_tmp=time_tmp/128.0/its;
    out[0]=r;
    time[0]=time_tmp;
}
```

下面给出常量存储、共享存储、局部存储、纹理存储和全局存储的访存延迟测试代码。

```
__constant__ DATATYPE d_const_array[ARRAYLEN];
__global__ void test_const_latency(double *time,DATATYPE *out,int its)
{
    int p=0;
    double time_tmp=0.0;
    unsigned int start_time=0, stop_time=0;
    for(int i=0;i<its;i++)
```

```
    {
        __syncthreads();
        start_time=clock();
        repeat128(p=d_const_array[p];)
        stop_time=clock();
        time_tmp+=(stop_time-start_time);
    }
    time_tmp=time_tmp/128.0/its;
    out[1]=p;
    time[1]=time_tmp;
}
__global__ void test_shared_latency(double *time, DATATYPE *out, int its,
DATATYPE *array)
{
    __shared__ DATATYPE shared_array[ARRAYLEN];
    int i;
    for(i=0;i<ARRAYLEN;i++)
    {
        shared_array[i]=array[i];
    }
    int p=0;
    double time_tmp=0.0;
    unsigned int start_time=0, stop_time=0;
    for(int i=0;i<its;i++)
    {
        __syncthreads();
        start_time=clock();
        repeat128(p=shared_array[p];)
        stop_time=clock();
        time_tmp+=(stop_time-start_time);
    }
    time_tmp=time_tmp/128.0/its;
    out[2]=p;
    time[2]=time_tmp;
}
__global__ void test_local_latency(double *time, DATATYPE *out, int its,
DATATYPE *array)
{
    DATATYPE local_array[ARRAYLEN];
    int i;
    for(i=0;i<ARRAYLEN;i++)
    {
        local_array[i]=array[i];
```

```
    }
    int p=0;
    double time_tmp=0.0;
    unsigned int start_time=0, stop_time=0;
    for(int i=0;i<its;i++)
    {
        __syncthreads();
        start_time=clock();
        repeat128(p=local_array[p];)
            stop_time=clock();
        time_tmp+=(stop_time-start_time);
    }
    time_tmp=time_tmp/128.0/its;
    out[3]=p;
    time[3]=time_tmp;
}
__global__ void test_global_latency(double *time, DATATYPE *out, int its,
DATATYPE *array)
{
    int p=0;
    double time_tmp=0.0;
    unsigned int start_time=0, stop_time=0;
    for(int i=0;i<its;i++)
    {
        __syncthreads();
        start_time=clock();
        repeat128(p=array[p];)
        stop_time=clock();
        time_tmp+=(stop_time-start_time);
    }
    time_tmp=time_tmp/128.0/its;
    out[4]=p;
    time[4]=time_tmp;
}
texture<int,1,cudaReadModeElementType>texref;
__global__ void test_texture_latency(double *time,DATATYPE *out,int its)
{
    int p=0;
    double time_tmp=0.0;
    unsigned int start_time=0, stop_time=0;
    for(int i=0;i<its;i++)
    {
        __syncthreads();
```

```
            start_time=clock();
            repeat128(p=tex1Dfetch(texref,p);)
            stop_time=clock();
            time_tmp+=(stop_time-start_time);
        }
        time_tmp=time_tmp/128.0/its;
        out[5]=p;
        time[5]=time_tmp;
    }
        DATATYPE *d_array;
        cudaMalloc((void**)&d_array,sizeof(DATATYPE) * ARRAYLEN);
        cudaMemcpy(d_array,h_array,sizeof(DATATYPE) * ARRAYLEN,
        cudaMemcpyHostToDevice);
        cudaMemcpyToSymbol(d_const_array,h_array,sizeof(DATATYPE) * ARRAYLEN);
        cudaBindTexture(NULL,texref,d_array,ARRAYLEN);
        double *d_time;
        cudaMalloc((void**)&d_time,sizeof(double) * 5);
        DATATYPE *d_out,*h_out;
        h_out=(DATATYPE *)malloc(sizeof(DATATYPE) * 6);
        cudaMalloc((void**)&d_out,sizeof(DATATYPE) * 6);
        test_register_latency    <<<1,1>>>(d_time,d_out,its);
        test_const_latency       <<<1,1>>>(d_time,d_out,its);
        test_shared_latency      <<<1,1>>>(d_time,d_out,its,d_array);
        test_local_latency       <<<1,1>>>(d_time,d_out,its,d_array);
        test_global_latency      <<<1,1>>>(d_time,d_out,its,d_array);
        test_texture_latency     <<<1,1>>>(d_time,d_out,its);
        cudaMemcpy(h_out,d_out,sizeof(DATATYPE) * 6,cudaMemcpyDeviceToHost);
        cudaMemcpy(h_time,d_time,sizeof(double) * 6,cudaMemcpyDeviceToHost);
        printf("%d\t%f\t%f\t%f\t%f\t%f\t%f\n",step,h_time[0],h_time[1],h_time[2],
        h_time[3],h_time[4],h_time[5]);
        cudaUnbindTexture(texref);
        cudaFree(d_array);
        cudaFree(d_time);
        cudaFree(d_out);
```

设置数组长度为 2048，循环测试 step 从 1 到 1024 时 Tesla K20c GPU 各级存储单元的访存延迟，统计访存延迟平均值如表 16.2 所示。其中，寄存器延迟为 6 次赋值的时钟数，该值几乎可以忽略，寄存器的访存延迟最小；其次是共享存储，与寄存器一样位于片上，访存延迟约为 48clocks；常量存储和纹理存储位于显存，但都拥有各自的片上缓存，平均访存延迟相当，约为 110～115clocks，接近全局存储的一半；局部存储和全局存储位于显存，访存延迟亦相当，其中局部存储的访存延迟比全局存储的访存延迟稍小。

表 16.2　GPU 访存延迟平均值　　　　　　　　　　单位：clocks

memory	register	constant	shared	local	global	texture
latency	0.19	110.13	47.66	203.54	218.71	115.60

实验结果显示常量存储访存延迟与 step 数值相关。图 16.1 罗列了 step 为 1 到 32 的常量存储访问延迟，当 step 从 1 到 16 时，访存延迟逐渐增加，step 大于 16 时，访存延迟趋于稳定。可能是受片上常量缓存大小的限制（纹理存储虽然同样利用了缓存机制，但访存延迟基本稳定）。上述现象也为常量存储优化提供了一条思路，即在构造常量存储时，需要考虑数据的访问顺序，尽可能将连续访问的数据邻近存放。

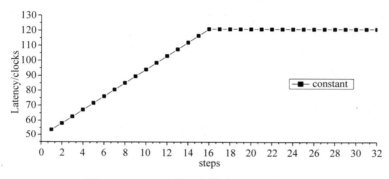

图 16.1　step 对常量存储访存延迟的影响

16.2　寄存器

寄存器是速度最快的存储单元，位于 GPU 芯片（SM）上，用于存储局部变量。每个 SM(SMX) 中有成千上万个 32 位寄存器（K20c 中单个 SMX 中有 65 535 个寄存器），kernel 函数启动时，这些寄存器被分配给指定线程。

有说法是寄存器的访存延迟是 1 clock，而这里的测试结果（16.1 节）远小于该值，可能原因是 GPU 程序执行时必然存在指令流水线，隐藏了寄存器的访存。当然寄存器不可能单独使用，必然伴随运算或赋值，能形成有效流水来掩盖访存延迟，故可以认为寄存器没有访存延迟。

寄存器除了能够存储普通的整型和浮点类型（双精度占两个寄存器）数据外，还支持内置数据类型，例如 char4、int4 和 float2 等。

在 Tesla K20c GPU 中，每个 SMX 拥有 65 536 个可用寄存器，block 的最大线程数量是 1024，SMX 可容纳的最大线程数量是 2048，因此平均每个线程的可用寄存器数量是 32~64 个，但不同 kernel 函数需要的寄存器数量也不相等，故又规定一个线程的最大寄存器数量是 256 个。

寄存器的使用量是影响占用率（occupancy）的重要因素，每个 SMX 的寄存器资源是有限的，若单个线程使用的寄存器数量越多，相同 SM 硬件资源能够运行的线程总量越少，GPU 的性能发挥越差。

查看 kernel 函数的寄存器数量方法有很多种，下面介绍 3 种方法。

（1）使用命令行选项"--ptxas-options= -v"来显示寄存器、共享存储、常量存储、本地存储和全局存储数量。下面展示一个实例，实例中 kernel 函数用了 3 个寄存器和 1048B 共享存储。

```
[fangmq@mn0%yhstar 12_memory]$nvcc -c register.cu --ptxas-options= -v
ptxas info    : 0 bytes gmem
ptxas info    : Compiling entry function '_Z13shared_accessPf' for 'sm_10'
ptxas info    : Used 3 registers, 1048 bytes smem
```

（2）使用 nvprof 检测 kernel 函数执行时使用的寄存器数量和共享存储数量。命令如下，具体运行实例和结果参见 11.4 节。

```
$nvprof --print-gpu-trace   ./cuda_2
```

（3）使用-keep 保留中间编译文件，查看.ptx 文件获知具体存储器使用情况。.ptx 文件中.reg 表示寄存器，.shared 表示共享存储器，.local 表示局部存储，.global 表示全局存储单元。下面展示这种方法的实例应用。

```
[fangmq@mn0%yhstar 12_memory]$nvcc -c register.cu -keep
[fangmq@mn0%yhstar 12_memory]$cat register.ptx
  ⋮
.entry _Z13shared_accessPf(
        .param .u64 __cudaparm__Z13shared_accessPf_a)
    {
    .reg .u32 %r<4>;
    .reg .u64 %rd<13>;
    .reg .f32 %f<4>;
    .shared .align 4 .b8 __cuda___cuda_local_var_14775_32_non_const_tmp8[1024];
    .loc    14    3    0
$LDWbegin__Z13shared_accessPf:
    mov.u64        %rd1, __cuda___cuda_local_var_14775_32_non_const_tmp8;
    .loc    14    6    0
    cvt.u64.u16    %rd2, %tid.x;
    cvt.u32.u64    %r1, %rd2;
    ld.param.u64   %rd3, [__cudaparm__Z13shared_accessPf_a];
    mul.lo.u64     %rd4, %rd2, 4;
    add.u64        %rd5, %rd3, %rd4;
    ld.global.f32  %f1, [%rd5+0];
    cvt.u64.u32    %rd6, %r1;
    mul.wide.u32   %rd7, %r1, 4;
    add.u64        %rd8, %rd1, %rd7;
```

```
        st.shared.f32    [%rd8+0], %f1;
        .loc    14    7    0
        bar.sync    0;
        .loc    14    8    0
        neg.s32          %r2, %r1;
        cvt.s64.s32      %rd9, %r2;
        mul.wide.s32     %rd10, %r2, 4;
        add.u64          %rd11, %rd1, %rd10;
        ld.shared.f32    %f2, [%rd11+1020];
        st.global.f32    [%rd5+0], %f2;
        .loc    14    9    0
        exit;
$LDWend__Z13shared_accessPf:
    } // _Z13shared_accessPf
```

利用这 3 种方法获得的存储单元使用数量存在差异,在以寄存器使用数量为参数判断程序性能时,前两种方法获得的寄存器数量相对精确(.ptx 文件仅是中间文件,还需要进一步编译生成可执行文件)。

16.3 局部存储

局部存储本身在硬件中没有特定的存储单元,而是从显存虚拟出来的地址空间。局部存储是为寄存器无法满足存储需求的情况设计的,主要用于存放单线程的大型数组和过多变量。局部存储是线程私有的,线程间相互不可见。由于 GPU 硬件本身没有局部存储单元,而是临时在显存(全局存储)申请的空间,故局部存储访存很慢(16.1 节测试结果显示,局部存储访存延迟与全局存储相当)。

尽管局部存储位于显存空间,但仍有大小限制,当申请的局部存储空间超过一定限制时,nvcc 在编译时会报错,错误信息如下:

```
ptxas error: Entry function '_Z7local_5Pf' uses too much local data (0x8000 bytes, 0x4000 max)
```

什么情况下变量(数组)使用局部存储呢?什么情况又使用寄存器?这取决于**程序复杂程度**、**空间大小**和**计算能力** 3 个因素,这三者影响了局部变量(数组)的声明位置。下面设计了一组实验分别探讨这 3 个要素如何影响寄存器和局部存储的使用(存储单元使用数量查看方法见 16.2 节)。

(1) 简单运算,一层循环。

```
#define NUMS 64
__global__ void local_1(float *a)
{
    float tmp[NUMS];
```

```
    int i;
    for(i=0;i<NUMS;i++)
    {
        tmp[i]=a[i];
    }
    for(i=0;i<NUMS;i++)
    {
        a[i]+=tmp[i];
    }
}
```

这种情况下,编译后查看存储单元使用报告,若不指定计算能力(默认使用 sm_10),NUMS≤=29 时,数组声明在寄存器上,否则数组声明在局部存储上;若指定计算能力为 3.5(即添加编译选项"-arch 'sm_35'"),NUMS 为 73。

当指定 sm_35 时,报告中未找到局部存储关键词 lmem,而是 stack frame。下面是使用不同计算能力的编译报告。

```
[fangmq@mn0%yhstar 12_memory]$nvcc -c local2.cu --ptxas-options=-v
ptxas info    : 0 bytes gmem
ptxas info    : Compiling entry function '_Z10local_1_30Pf' for 'sm_10'
ptxas info    : Used 4 registers, 24 bytes smem, 4 bytes cmem[1], 120 bytes lmem
ptxas info    : Compiling entry function '_Z10local_1_29Pf' for 'sm_10'
ptxas info    : Used 58 registers, 24 bytes smem
[fangmq@mn0%yhstar 12_memory]$nvcc -c local2.cu --ptxas-options=-v -arch 'sm_35'
ptxas info    : 0 bytes gmem
ptxas info    : Compiling entry function '_Z10local_1_73Pf' for 'sm_35'
ptxas info    : Function properties for _Z10local_1_73Pf
    0 bytes stack frame, 0 bytes spill stores, 0 bytes spill loads
ptxas info    : Used 31 registers, 328 bytes cmem[0]
ptxas info    : Compiling entry function '_Z10local_1_74Pf' for 'sm_35'
ptxas info    : Function properties for _Z10local_1_74Pf
    296 bytes stack frame, 0 bytes spill stores, 0 bytes spill loads
ptxas info    : Used 30 registers, 328 bytes cmem[0]
```

(2) 矩阵乘法,三层循环。

```
#define num_size 8
__global__ void local_2(float *a,float *b,float *c)
{
    float tmp_a[num_size * num_size];
    float temp;
```

```
    int i,j,k;
    for(i=0;i<num_size*num_size;i++)
    {
        tmp_a[i]=a[i];
    }
    for(i=0;i<num_size;i++)
    {
        for(j=0;j<num_size;j++)
        {
            temp=0.0;
            for(k=0;k<num_size;k++)
            {
                temp+=tmp_a[i*num_size+k]*b[k*num_size+j];
            }
            c[i*num_size+j]=temp;
        }
    }
}
```

测试结果显示,这种较为复杂的三层循环矩阵乘法运算中,默认使用 sm_10 编译时,num_size＝2 的数组才能存储在寄存器,而 num_size＞2 时,tmp_a 数组声明在 local memory;若使用 sm_35 编译,num_size＜＝6 时数组声明在寄存器,num_size＞6 时声明为 stack frame(局部存储)。

(3) 排序。

```
#define NUM 49
__global__ void local_3(float *a)
{
    float tmp[NUM];
    float minf=0.0,temp;
    int mind;
    int i,j;
    for(i=0;i<NUM;i++)
    {
        tmp[i]=a[i];
    }
    for(i=0;i<NUM;i++)
    {
        minf=tmp[i];
        mind=i;
        for(j=i;j<NUM;j++)
```

```
        {
            if(minf>tmp[j])
            {
                minf=tmp[j];
                mind=i;
            }
        }
        if(mind!=i)
        {
            temp=tmp[i];
            tmp[i]=tmp[mind];
            tmp[mind]=temp;
        }
    }
    a[0]=tmp[NUM-1];
}
```

存储声明报告显示,排序中不管 NUM 定义为何值,数组均被声明在局部存储器,而不是寄存器。

因为寄存器访存延迟远小于局部存储(详见 16.1 节),kernel 函数若涉及私有数组时,应尽量将其声明在寄存器,这是一个重要的优化原则。

16.4 共享存储器

共享存储位于 GPU 芯片上,访存延迟仅次于寄存器(16.1 节测得共享存储的访存延迟约为 48 clocks)。共享存储可以被一个 block 内的所有线程访问,可实现 block 内线程间的低开销通信。每个 SMX 上共享存储容量是有限的,因此共享内存的使用量将会影响 SMX 驻留的活动线程束数量,从而影响占用率(occupancy)。

SMX 的一级缓存和共享存储共用一个 64KB 高速存储(Maxwell 架构的共享存储单独存在)。SM 1.X 设备上 SM 中没有 L1 Cache(L1 Cache 位于 TPC),共享存储大小为 16KB;之后的硬件 SM 2.X 和 SM 3.X 上,高速存储器大小为 64KB,其中共享存储大小为 16KB 或 48KB,剩余为一级缓存;SM 3.X 还支持 32KB 共享存储和 32KB 一级缓存的分割方案。

16.4.1 共享存储使用

共享存储空间的开辟方法包括静态和动态两种。

(1) 静态共享存储:在 kernel 函数中利用 __shared__ 限定符声明具体尺寸的共享存储。例如,下面语句申请了一个 256 个 float 类型数据的数组。

```
__shared__ float array[256];
```

（2）动态共享存储：当共享存储空间不确定或变化时，需要动态申请共享存储空间。动态共享存储申请时，需要用 extern 关键字修饰_ _shared_ _限定符，kernel 函数中不指定具体共享存储空间的大小，而是在 kernel 函数启动时利用<<<>>>中的第 3 个参数来指定共享存储空间。

```
__global__ void kernel_shared(…)
{
    extern __shared__ float array[];
    ⋮
}
    kernel_shared<<<blocks,threads,sizeof(float) * 256>>>(…);
```

当应用需要多个共享存储数组时，静态共享存储只需多次声明即可，而动态共享存储只能声明一次，利用指针拆分共享存储空间。下面给出动态共享存储拆分的具体方法。

```
__global__ void kernel_shared(…)
{
    extern __shared__ float array[];
    float *array_a=array;
    float *array_b=&array[128];
    ⋮
}
```

同一 block 的 threads 可利用共享存储通信，即共享存储写后读时，需要同步，常用的同步指令有_ _syncthreads()，更多同步指令参见 15.5.1 节。

16.4.2　bank conflict

共享存储由交替排列的存储体（bank）构成，每个 bank 尺寸为 32b(4B)，如果线程束（warp）中多个线程同时访问同一存储体的不同地址将会引发存储体冲突（bank conflict）。一般来说，共享存储的 bank 数量和 warp 内线程数量是一致的，即 SM 1.X 有 16 组 bank，而 SM 2.X 和 SM 3.X 有 32 组 bank。

存储体冲突的后果：在 GPU 中，不同的 bank 组可以同时访存，若产生 bank conflict，多个线程要求访问同一个 bank，每次仅有一个线程能访问 bank，多个线程间需要串行访存，因而导致性能损失。

2D 数据访问时，假设单位数据尺寸为 4B（如 float），2D 数据的组织为 32×32，此时若一个 warp 访问 2D 数据的一行，每个线程访问不同 bank，所有线程同时访问，不存在 bank conflict；而如果一个 warp 访问 2D 数据的一列，所有线程访问同一组 bank，共产生 31 次 bank conflict，此时共享存储的访问是串行的。

针对这种情况，将 2D 数据声明为 32×33（SM 1.X 时为 16×17），即每行 33 个数据，2D 数据相同列的元素就位于不同 bank，同一列数据访存就避免了 bank conflict。

图 16.2 展示了 2D 共享存储 bank conflict 消除原理。共享存储有 32 组 bank,即图中 32 列。图 16.2(a)是 32×32 的 2D 数组在共享存储中的存储组织,数组中相同列号的数据存储在同一 bank 中,灰色标注的是列号为 0 的数据,如果一个 warp 所有线程访问 2D 数组中的 0 号列,即访问同一个 bank,将产生 bank conflict,导致 32 次访问需要串行进行。图 16.(b)中,2D 数组每行多填充了一个空元素(图中黑色带交叉纹理的网格),此时相同列号的元素被错开,而相同行号的元素位于不同的 bank 中,因此无论 warp 访问列数据还是行数据都不会产生 bank conflict。

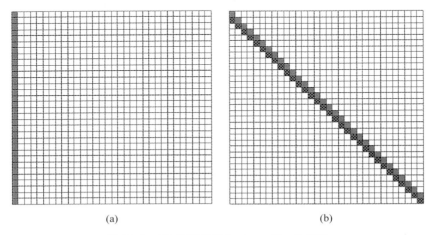

图 16.2　2D 共享存储 bank conflict 的消除

本书 9.4 节和 12.3.3 节均利用了避免 bank conflict 的方法来优化相应代码,并对比了避免 bank conflict 前后 kernel 函数的执行性能。

另外,当一个 warp 同时访问同一个共享存储单元时,如果访问类型是写,需要串行执行;如果访问类型是读,将触发广播操作。在计算能力 2.X 以上的设备中,多个线程访问同一个字的不同字节,硬件也只访存一次,不影响性能。

16.4.3　volatile 关键字

当某问题只需要 warp 内通信时,warp 是指令执行的基本单位,因此无须同步也能实现 warp 内不同线程通信。计算能力 2.X 之后,访问 __shared__ 声明的变量时,连续访问(没有同步语句)得到的数据可能是先前访问时缓存在寄存器中的数据,而不是到共享存储获取最新更新的值。对于计算能力 1.X 的设备,会直接读写共享存储,不会出现上述问题。

CUDA 提供了 volatile 关键字来解决这一问题。volatile 关键字可以将全局存储或共享存储中的变量声明为敏感变量,默认其他线程随时可能改变其值,因此每次运算都会读取新值,从而避免由于误读缓存数据引发的错误。下面给出了 volatile 关键字的声明和使用方法,具体的使用实例请参考 7.10 节。

```
__global__ void vector_dot_product_gpu_3_2(DATATYPE *a, DATATYPE *b, DATATYPE
*c_tmp, int n)
{
    extern __shared__ DATATYPE tmp[];
      ⋮
    volatile float *tmp_1=tmp;
      ⋮
}
```

16.4.4 共享存储原子操作

计算能力 1.1 提供了全局存储的原子操作,计算能力 1.2 提供了共享存储的原子操作。原子操作执行时,单个线程串行执行,其余线程循环等待,直到原子操作执行完毕,所有线程才继续执行后续代码。

下面是一个使用共享存储原子操作的例子。这个例子是基于 7.6 节 gpu_5 的代码进行修改的,删除了归约部分代码,修改为共享存储原子操作。代码中第一个原子操作的对象是共享存储,第二个原子操作的对象是全局存储。

```
__global__ void vector_dot_product_gpu_atom_atom(DATATYPE *a, DATATYPE *b,
DATATYPE *c, int n)
{
    __shared__ DATATYPE tmp[1];
    if((threadIdx.x==0)&&(blockIdx.x==0))
    {
        c[0]=0.0;
    }
    if(threadIdx.x==0)
    {
        tmp[0]=0.0;
    }
    const int t_n=blockDim.x * gridDim.x;
    int tid=blockIdx.x * blockDim.x+threadIdx.x;
    double temp=0.0;
    while(tid<n)
    {
        temp=__fadd_rn(temp,__fmul_rn(a[tid],b[tid]));
        tid+=t_n;
    }
    atomicAdd(tmp,temp);
    __syncthreads();
    if(threadIdx.x==0)
```

```
    {
        atomicAdd(c,tmp[0]);
    }
}
```

实验结果显示,处理 1024×1024 的向量数组时,修改后的双重原子操作的向量归约在<<<128,128>>>的线程维度最快,耗时 73.38μs;而 block 内使用归约,仅在最后使用全局存储原子操作的原始 gpu_5 版本耗时 57.50μs(见表 7.3)。实验结果说明原子操作不能盲目使用,例如 block 内归约运算不适合用共享存储原子操作替代。

16.5 常量存储

常量存储类于似局部存储,只是全局存储的虚拟地址,而不存在专门的常量存储单元。常量存储驻留在 GPU 板载内存上,存在一种特殊的只读缓存,该缓存提供了高速缓存和广播两个功能(亦可将常量存储理解为**常量缓存**)。这种特殊的缓存位于 SMX。常量存储的只读属性可以简化缓存管理,硬件无须管理复杂的写回策略。常量缓存启动的条件是同一 warp 所有线程同时访问相同的常量数据,若线程访问的数据位于不同地址则串行访问。

常量存储尺寸一般为 48KB,声明时使用_ _constant_ _限定符,常量内存声明的变量是全局变量,对 GPU 内所有 kernel 函数都可见。

常量存储中的数据从何而来?下面介绍几种来源。

(1) 定义时初始化。

```
_ _constant_ _ int d_array_m1[5]={4,1,5,3,4};
```

(2) 通过 cudaMemcpyToSymbol 函数传递数据。下面是常量存储的声明和传递实例,利用该方法可以将主机端 array_m2 数组数据传输到 d_array_m2 常量数组中。10.3 节有相应的使用实例。

```
#define msize 1024
_ _constant_ _ int d_array_m2[msize];
cudaMemcpyToSymbol(d_array_m2,array_m2,msize*sizeof(int),0,
cudaMemcpyHostToDevice);
cudaMemcpyFromSymbol(array_m2_1,d_array_m2,sizeof(int)*msize,0,
cudaMemcpyDeviceToHost);
```

(3) 利用 cudaGetSymbolAddress 函数获取常量存储的地址,然后在 kernel 函数中修改常量存储的值。下面实现了将常量数组每个元素加 1 的功能。需要注意,常量存储被访问时不要修改其值。

```
__global__ void constant_add(int *array_a)
{
    array_a[blockIdx.x * blockDim.x+threadIdx.x]++;
}
```

```
int *d_array_m2_1;
cudaGetSymbolAddress((void**)&d_array_m2_1,d_array_m2);
constant_add<<<4,256>>>(d_array_m2_1);
```

SM 2.X 之后的硬件利用了常量缓存的统一加载(load uniform，LDU)功能，对只读全局存储的一致访问做了优化。这种特殊优化需要同时具备以下 3 种属性：①只读(const 关键字声明)；②地址不依赖于线程 ID(一致访问)；③必须是全局存储。

利用第 10 章 1D 卷积运算进行测试，比较 gpu_1 和 gpu_2 并修改 gpu_1 的参数声明为 const，如此 gpu_1 中对数组 M 的访问符合只读全局存储一致访问条件，并分别采用 SM 1.0 和 SM 3.5 编译并执行，实验结果显示耗时跟以前一致，因此，笔者对 SM 2.X 后硬件支持只读全局存储的一致访问优化持怀疑态度。

16.6　全局存储

全局存储在某种意义上等同于 GPU 显存，kernel 函数通过全局存储读写显存。全局存储是 kernel 函数输入数据和写出结果的唯一来源(不包括常量存储、零拷贝)。由于显存位于 GPU 上，由内存控制器直接访问，因此访存带宽较大。17.2 节读取了 GPU 相关参数，得知 K20c 的访存时钟频率是 2600MHz，访存位宽为 320b。17.6 节实验测试了 GPU 访存带宽，K20c GPU 的访存带宽约为 150~160GB/s。

全局存储容量巨大，根据 GPU 型号不同各不相同。利用 cudaGetDeviceProperties() 函数得到参数结构体的 totalGlobalMem 字段可知 K20c 的全局存储为 5120MB(关闭 ECC 校验)或 4800MB(打开 ECC 校验)。

16.6.1　全局存储的使用

(1) 全局存储动态分配和释放的常用方法是 cudaMalloc() 和 cudaFree() 函数。两个函数在本书几乎所有实例章节均有应用，不再赘述。另外，cuda_runtime 还提供了对齐分配函数 cudaMallocPitch()，存储对齐有利于提高不规整数据的访问速率，在 8.4 节给出了对齐分配实例并通过实验对比分析了存储对齐的性能收益。动态全局存储的数据通信一般使用 cudaMemcpy() 函数。

```
extern __host__ cudaError_t CUDARTAPI cudaMalloc(void **devPtr, size_t size);
extern __host__ cudaError_t CUDARTAPI cudaFree(void *devPtr);
```

(2) 利用__device__限定符定义静态全局存储，静态共享存储只能在 kernel 函数或 device 函数访问，主机端没有指针故无法用 cudaMemcpy() 函数通信。本书提供两种静态全局存储与主机端通信方案,首先静态全局存储类似常量存储,可以用 cudaMemcpy-ToSymbol()和 cudaMemcpyFromSymbol()实现与主机端通信;另一种方法是利用数据复制,将数据转移到动态全局存储进行通信。

下面是一个静态全局存储的数据通信实例,给出了__device__限定符定义的静态全局存储的数据复制到动态全局存储的通信方法。

```
#define msize 1024
__device__ int d_array_m2[msize];
__global__ void global_init()
{
    d_array_m2[blockIdx.x*blockDim.x+threadIdx.x]=blockIdx.x+threadIdx.x;
}
__global__ void global_cpy(int *a)
{
    a[blockIdx.x*blockDim.x+threadIdx.x]=d_array_m2[blockIdx.x*blockDim.x
    +threadIdx.x];
}
```

```
int main()
{
    int *array_m2=(int *)malloc(sizeof(int)*msize);
    global_init<<<4,256>>>();
    cudaMemcpy(array_m2,d_array_m2,sizeof(int)*msize,cudaMemcpyDeviceToHost);
    vector_print(array_m2,10);

    int *d_array_m2_1;
    cudaMalloc((void**)&d_array_m2_1,sizeof(int)*msize);
    global_cpy<<<4,256>>>(d_array_m2_1);
    cudaMemcpy(array_m2,d_array_m2_1,sizeof(int)*msize,cudaMemcpyDeviceToHost);
    vector_print(array_m2,10);
    cudaFree(d_array_m2_1);
    free(array_m2);
    return 0;
}
```

编译执行上述代码,输出结果如下,实验结果验证了前文说法,即__device__限定符声明的全局存储无法直接用 cudaMemcpy()函数通信(第一行结果全为0),而将数据复制到动态全局存储可以实现通信(第二行输出结果正确)。

```
[fangmq@cn18%yhstar 12_memory]$./global
0  0  0  0  0  0  0  0  0  0
0  1  2  3  4  5  6  7  8  9
```

(3) kernel 函数中使用 malloc()和 free()函数分配和释放全局存储。自 Fermi 架构开始引入该功能,且 kernel 函数中 malloc() 的存储必须使用 free() 释放,用法与 C 语言类似,但要注意处理好不同线程间的逻辑索引关系。运行时每个线程执行各自的 malloc() 和 free() 函数。由于分配和释放均由线程执行,因此线程数量会直接影响存储分配和释放耗时,有兴趣的读者可以通过实验探索线程数量和耗时的关系。

(4) 全局存储的初始化。CUDA 提供了全局存储的初始化函数 cudaMenset(),类似的还有 cudaMenset2D()、cudaMemset3D() 和 cudaMemsetAsync() 等,详见 5.4.2 节。

(5) 全局存储的原子操作。全局存储的原子操作与 16.4.4 节共享存储原子操作类似,且在 7.6 节和 16.4.4 节实例中也用到全局存储的原子操作。分析实验结果可得出结论,应尽量避免或减少全局存储原子操作,可分割为 block 内的共享存储原子操作子任务。

16.6.2 全局存储的合并访问

从本质上说,全局存储的访问并非由线程完成的,而是通过内存控制器访问。一般一个内存控制器一次可访问 64b 数据,K20c 测得的访存带宽是 320b(见 17.2 节),共 5 个内存控制器。因此,全局存储访问时,只有合并访问才能充分发挥这些内存控制器的性能。

怎样的访问能够合并呢?首先合并访存的对象是 warp。一般情况下,合并访存需要满足以下几个条件。

(1) 每个线程访问的存储字长必须不小于 32b,当读取的存储字长为 8b(char) 或 16b(short) 时无法合并访存。

(2) warp 上线程对齐访问全局存储(即地址随线程 ID 偏移)。

(3) 每个 warp 中索引最小的线程访问的数据首地址必须是对齐的,访问数据是 32 位时要求 64 字节对齐,访问数据为 64 位时要求 128 字节对齐,访问数据为 128 位时要求 256 字节对齐。

伴随着 GPU 架构的发展,合并访存的条件正逐渐减少,早期 Tesla 架构 GPU 的合并访存非常严格。此后的 Fermi 架构和 Kepler 架构中,合并访存的必要条件不断弱化,例如,warp 索引最小线程访问数据的首地址对齐要求正在削弱,8.4 节矩阵乘法中,数据对齐对性能影响也非常小。

关于合并访存,笔者有个猜想。随着 GPU 的不断发展,未来合并访存的条件可能将不复存在,达到一种理想状态:warp 内访问的数据可以不连续,但位于一个全局存储片段中,而该片段可以被一个或多个内存控制器一次性访问,即能合并访存整个片段。

16.6.3 利用纹理缓存通道访问全局存储

在 16.5 节最后,使用常量缓存的统一加载功能来优化全局存储,但并未得到性能提升。本节考虑使用纹理缓存通道来加速全局存储访问。

在不绑定全局存储到纹理的前提下,CUDA 提供了两种纹理缓存通道访问全局存储的途径:①使用关键字 const restrict 修饰只读的全局存储;②利用 sm_32_intrinsics.h 头文件中定义的 __ldg() 内置函数(编译时需要指定计算能力 3.5)。

这里基于 6.4 节向量加法代码进行修改，分别利用这两种方法实现纹理缓存通道访问全局存储，下面是相关代码。

```
__global__ void vector_add_gpu_4(const DATATYPE * __restrict__ a, const DATATYPE * __restrict__ b, DATATYPE *c, int n)
{
    const int tidx=threadIdx.x;
    const int bidx=blockIdx.x;
    const int t_n=gridDim.x * blockDim.x;
    int tid=bidx * blockDim.x+tidx;
    while(tid<n)
    {
        c[tid]=a[tid]+b[tid];
        tid+=t_n;
    }
}
__global__ void vector_add_gpu_5(DATATYPE *a, DATATYPE *b, DATATYPE *c, int n)
{
    const int tidx=threadIdx.x;
    const int bidx=blockIdx.x;
    const int t_n=gridDim.x * blockDim.x;
    int tid=bidx * blockDim.x+tidx;
    while(tid<n)
    {
        c[tid]=__ldg(&a[tid])+__ldg(&b[tid]);
        tid+=t_n;
    }
}
    vector_add_gpu_4<<<blocknum,threadnum>>>(d_a,d_b,d_c,n);
    vector_add_gpu_5<<<blocknum,threadnum>>>(d_a,d_b,d_c,n);
```

此外，下面给出利用纹理存储的向量加法版本。

```
texture<float,1,cudaReadModeElementType>texref_a;
texture<float,1,cudaReadModeElementType>texref_b;
__global__ void vector_add_gpu_6(DATATYPE *c, int n)
{
    const int tidx=threadIdx.x;
    const int bidx=blockIdx.x;
    const int t_n=gridDim.x * blockDim.x;
    int tid=bidx * blockDim.x+tidx;
    while(tid<n)
    {
```

```
        c[tid]=tex1Dfetch(texref_a,tid)+tex1Dfetch(texref_b,tid);
        tid+=t_n;
    }
}
cudaMemcpy(d_a, a, sizeof(DATATYPE) * n, cudaMemcpyHostToDevice);
cudaMemcpy(d_b, b, sizeof(DATATYPE) * n, cudaMemcpyHostToDevice);
cudaBindTexture(NULL,texref_a,d_a,sizeof(DATATYPE) * n);
cudaBindTexture(NULL,texref_b,d_b,sizeof(DATATYPE) * n);
vector_add_gpu_6<<<blocknum,threadnum>>>(d_c,n);
cudaUnbindTexture(texref_a);
cudaUnbindTexture(texref_b);
cudaMemcpy(c, d_c, sizeof(DATATYPE) * n, cudaMemcpyDeviceToHost);
```

编译并执行上述代码，令向量长度为1024×1024×128，线程维度设置为<<<512，512>>>，统计 kernel 函数执行时间如表 16.3 所示。表中结果显示，利用纹理缓存通道访问全局存储的两个版本（gpu_4 和 gpu_5）、纹理存储版本（gpu_6）并不比全局存储访问（gpu_3）快。16.1 节测得的访存延迟是纹理存储比全局存储快近 1 倍，但纹理存储优化在实际应用中却没有起到正收益。可能的原因是纹理存储并不适用于该应用规则，常规 cache 能够起到较好的作用。

表 16.3　向量加法执行时间　　　　　　　　　　　　　单位：ms

	gpu_3	gpu_4	gpu_5	gpu_6	cublas
kernel	10.355	10.344	10.485	10.375	9.8553

16.7　纹理存储

纹理存储是 GPU 的重要特征之一，也是 GPU 编程优化的关键。纹理存储是从早期 GPU 用于纹理渲染发展而来的，具有缓存、纹理坐标缩放、地址映射、线性插值、类型转换等诸多功能。GPU 用于通用计算领域的主要功能是加速数据读取。

纹理存储涉及的相关概念非常广泛，包括主机内存、设备内存、CUDA 数组、纹理存储（1D、2D、3D、1D 分层纹理、2D 分层纹理）、纹理绑定、纹理拾取、归一化坐标、非归一化坐标、表面存储（1D、2D、3D）等。其中，主机内存和设备内存都已有描述，这里主机内存特指映射锁页存储，设备内存特指全局存储。

16.7.1　CUDA 数组

CUDA 数组专为纹理操作设计，位于显存池，且不能用指针访问，只能通过数组句柄和 1D、2D 或 3D 坐标访问。CUDA 数组寻址较为复杂，不详细介绍。

CUDA 数组的创建和销毁使用 cudaMallocArray()和 cudaFreeArray()函数，函数的详细介绍见 5.4.2 节。

CUDA 数组可分配大小是有限的，一维 CUDA 数组的最大宽度为 65 536，三维 CUDA 数组最大范围是 4096×4096×4096，详见 16.7.2 节。

CUDA 数组无法直接访问，必须绑定到纹理才能读数据，利用 cudaMemcpyToArray()、cudaMemcpyFromArray()、cudaMemcpy2DToArray() 和 cudaMemcpy2DFromArray() 等函数传输数据，若要通过 kernel 函数修改数据则需要绑定到表面存储。

16.7.2 纹理存储的操作和限制

纹理分为 1D 纹理、2D 纹理、3D 纹理、1D 分层纹理和 2D 分层纹理。

纹理无法单独存在，需要绑定到具体的存储单元，例如全局存储或 CUDA 数组。两者的使用方法是不同的。

1. 全局存储纹理操作

全局存储仅支持 1D 纹理和 2D 纹理的绑定和访问。

其中，1D 纹理绑定的是 cudaMalloc() 和 cudaFree() 分配和释放的全局存储，纹理绑定使用 cudaBindTexture() 函数，纹理访问使用 tex1Dfetch() 函数。相关的函数介绍请查阅 5.4.2 节和 5.4.7 节，16.6.2 节给出了 1D 纹理优化的向量加法运算并取得很好的优化效果。注意一维纹理的最大尺寸限制为 2^{27}。

2D 纹理绑定的全局存储空间必须是用 cudaMallocPitch() 函数分配的，利用 cudaBindTexture2D() 函数绑定到纹理，纹理访问函数为 tex2D()。注意该函数的索引必须为 float 类型。二维纹理的尺寸限制为 65536×65536。

```
static __inline__ __device__ float tex2D(texture<float, cudaTextureType2D,
cudaReadModeElementType>t, float x, float y);
//返回类型和纹理类型中的 float 可为其他任何类型,索引中的 float 类型不能改变
```

2. CUDA 数组的纹理操作

CUDA 数组支持所有的纹理类型。一维纹理绑定的 CUDA 数组使用 cudaMallocArray() 函数分配，其他如二维纹理、三维纹理等绑定的 CUDA 数组均由 cudaMalloc3DArray() 函数分配。CUDA 数组绑定到纹理均采用 cudaBindTextureToArray() 函数。

CUDA 数组绑定纹理后的访问函数是不同的：1D 纹理用 tex1D() 函数，2D 纹理用 tex2D() 函数，3D 纹理用 tex3D() 函数，1D 分层纹理用 tex1DLayered() 函数，2D 分层纹理用 tex2DLayered() 函数。

```
static __inline__ __device__ int tex1D(texture<int, cudaTextureType1D,
cudaReadModeElementType>t, float x);    //其中 int 型变量可以变换类型,下同
static __inline__ __device__ int tex2D(texture<int, cudaTextureType2D,
cudaReadModeElementType>t, float x, float y);
```

```
static __inline__ __device__ int tex3D(texture<int, cudaTextureType3D,
cudaReadModeElementType>t, float x, float y, float z);
static __inline__ __device__ int tex1DLayered(texture<int,
cudaTextureType1DLayered, cudaReadModeElementType>t, float x, int layer);
static __inline__ __device__ int tex2DLayered(texture<int,
cudaTextureType2DLayered, cudaReadModeElementType>t, float x, float y, int
layer);
```

3. 纹理的限制

GPU 的纹理并非无限的，使用时有一定的限制，而且各种维度的纹理尺寸限制是不同的。纹理的维度限制可以通过 cudaGetDeviceProperties() 函数获得的 cudaDeviceProp 结构体的参数获知。下面是一段获取纹理限制的 CUDA C 代码及其执行结果。

```c
#include<stdio.h>
#include<stdlib.h>
#include<cuda_runtime.h>
int main()
{
    int count,i;
    cudaGetDeviceCount(&count);
    if(count==0) {
        printf("There is no device.\n");
        return false;
    }
    printf("There are %d GPUs\n",count);
    for(i=0; i<count; i++) {
        cudaDeviceProp prop;
        if(cudaGetDeviceProperties(&prop, i)==cudaSuccess) {
            printf("GPU %d: \t %s\n",i,prop.name);
            printf("maxTexture1D\t:\t%d\n",prop.maxTexture1D);
            printf("maxTexture1DLinear\t:\t%d\n",prop.maxTexture1DLinear);
printf("maxTexture2D\t:\t%d,\t%d\n",prop.maxTexture2D[0],prop.maxTexture2D[1]);
printf("maxTexture2DLinear\t:\t%d,\t%d,\t%d\n",prop.maxTexture2DLinear[0],
prop.maxTexture2DLinear[1],prop.maxTexture2DLinear[2]);
printf("maxTexture2DGather\t:\t%d,\t%d\n",prop.maxTexture2DGather[0],prop.
maxTexture2DGather[1]);
printf(" maxTexture3D \t:\t%d,\t%d,\t%d\n", prop.maxTexture3D[0], prop.
maxTexture3D[1],prop.maxTexture3D[2]);
printf("maxTextureCubemap\t:\t%d\n",prop.maxTextureCubemap);
printf("maxTexture1DLayered\t:\t%d,\t%d\n", prop.maxTexture1DLayered[0],
prop.maxTexture1DLayered[1]);
```

```
printf("maxTexture2DLayered\t:\t%d,\t%d,\t%d\n",prop.maxTexture2DLayered
[0],prop.maxTexture2DLayered[1],prop.maxTexture2DLayered[2]);
printf("maxTextureCubemapLayered\t:\t%d,\t%d\n",prop.maxTextureCubemapLayered
[0],prop.maxTextureCubemapLayered[1]);
      }
      printf("\n\n\n");
   }
   return 0;
}
```

```
[fangmq@cn18%yhstar texture]$./texture_limit
There are 2 GPUs
GPU 0:        Tesla K20c
maxTexture1D:                65536
maxTexture1DLinear:          134217728
maxTexture2D:                65536,    65536
maxTexture2DLinear:          65000,    65000,    1048544
maxTexture2DGather:          16384,    16384
maxTexture3D:                4096,     4096,     4096
maxTextureCubemap:           16384
maxTexture1DLayered:         16384,    2048
maxTexture2DLayered:         16384,    16384,    2048
maxTextureCubemapLayered:16384,        2046
    ⋮
```

16.7.3 读取模式、纹理坐标、滤波模式和寻址模式

1. 读取模式

纹理访问有两种读取模式，分别是 cudaReadModeElementType 和 cudaReadMode-NormalizedFloat，cudaReadModeElementType 读取到的是指定的数据类型，而 cudaReadModeNormalizedFloat 模式读取的任何数据都会转换为 float 类型。

```
enum __device_builtin__ cudaTextureReadMode
{
    cudaReadModeElementType=0,      /**<Read texture as specified element type */
    cudaReadModeNormalizedFloat=1   /**<Read texture as normalized float */
};
```

2. 非归一化坐标和归一化坐标

除了全局存储的 1D 纹理存储的读取函数 tex1Dfetch() 的索引为整数外，其他纹理存储读取函数的索引均为 float 类型，这一点区别于其他存储单元。正是如此，纹理坐标有非归一化坐标和归一化坐标两种形式（见图 16.3）。

非归一化坐标：类似数学常用的坐标，范围为[0,max]。非归一化坐标可能导致部

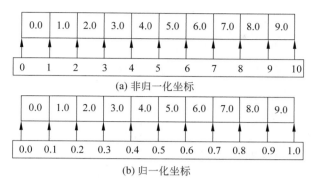

图 16.3　非归一化坐标和归一化坐标

分纹理操作不可用，比如重叠和镜像寻址。

归一化坐标：将坐标按比例缩放至 [0,1] 范围内。

3. 滤波模式

纹理操作还会涉及滤波模式，纹理结构体提供了相应的参数设置，用 texture.filterMode 配置纹理的滤波模式。

```
enum __device_builtin__ cudaTextureFilterMode
{
    cudaFilterModePoint=0,      /**<Point filter mode*/
    cudaFilterModeLinear=1      /**<Linear filter mode*/
};
```

滤波模式有点滤波和线性滤波，默认为点滤波，根据浮点坐标点返回纹理值（向下取值）；而线性滤波获取两个相邻的纹理值，并以纹理坐标为权值进行线性插值。例如，在图 16.3(a) 的非归一化坐标中，点滤波地址为 4.5 的取值为 4.0，而线性滤波的取值为 4.5。

4. 寻址模式

寻址模式是指定处理超出范围的纹理坐标的方式。寻址模式主要有 4 种，分别是夹取寻址、边界寻址、镜像寻址和线程束寻址（重叠寻址）（见图 16.4）。寻址模式利用纹理结构体的 texture.addressMode[0/1/2] 配置。

```
enum __device_builtin__ cudaTextureAddressMode
{
    cudaAddressModeWrap=0,      /**<Wrapping address mode*/
    cudaAddressModeClamp=1,     /**<Clamp to edge address mode*/
    cudaAddressModeMirror=2,    /**<Mirror address mode*/
    cudaAddressModeBorder=3     /**<Border address mode*/
};
```

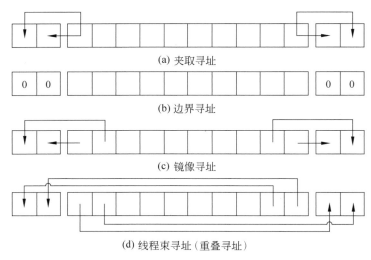

图 16.4 纹理寻址模式

夹取寻址：超出地址范围的取地址最大值上的值或最小值上的值。
边界寻址：超出地址范围的取值均为 0。
镜像寻址：以边界为镜面，按超出距离计算对应的值。
线程束寻址（重叠寻址）：根据超出地址范围距离从头开始计算相应的值。

16.7.4 表面存储

CUDA 数组无法直接读写，必须通过纹理绑定 CUDA 数组才能读取其中的值，若要修改其值则只能通过 cudaMemcpyToArray() 等函数传输数据。表面存储的提出为 CUDA 数组的读写操作提供了支持。表面存储是自 Fermi 架构开始引入的。

区别于纹理存储拥有独立的纹理缓存，表面存储使用与全局存储类似的 L2 Cache。不同的表面存储维度有不同的读写函数，例如，1D 表面读函数 surf1Dread() 和 1D 表面写函数 surf1Dwrite()。下面罗列了表面读写函数，其中表面存储的数据类型是可变的，使用时根据需要选取。

```
template<> __inline__ __device__ int surf1Dread(surface<void,
cudaSurfaceType1D>surf, int x, enum cudaSurfaceBoundaryMode mode)
template<> __inline__ __device__ int surf2Dread(surface<void,
cudaSurfaceType2D>surf, int x, int y, enum cudaSurfaceBoundaryMode mode)
template<> __inline__ __device__ int surf3Dread(surface<void,
cudaSurfaceType3D>surf, int x, int y, int z, enum cudaSurfaceBoundaryMode mode)
template<> __inline__ __device__ int surf1DLayeredread(surface<void,
cudaSurfaceType1DLayered>surf, int x, int layer, enum cudaSurfaceBoundaryMode
mode)
```

```
template<>__inline____device__ int surf2DLayeredread(surface<void,
cudaSurfaceType2DLayered>surf, int x, int y, int layer,
enum cudaSurfaceBoundaryMode mode)
```

```
static__inline____device__ void surf1Dwrite(int val, surface<void,
cudaSurfaceType1D>surf, int x, enum cudaSurfaceBoundaryMode
mode=cudaBoundaryModeTrap)
static__inline____device__ void surf2Dwrite(int val, surface<void,
cudaSurfaceType2D>surf, int x, int y, enum cudaSurfaceBoundaryMode
mode=cudaBoundaryModeTrap)
static__inline____device__ void surf3Dwrite(int val, surface<void,
cudaSurfaceType3D>surf, int x, int y, int z, enum cudaSurfaceBoundaryMode
mode=cudaBoundaryModeTrap)
static__inline____device__ void surf1DLayeredwrite(int val, surface<void,
cudaSurfaceType1DLayered>surf, int x, int layer, enum cudaSurfaceBoundaryMode
mode=cudaBoundaryModeTrap)
static__inline____device__ void surf2DLayeredwrite(int val, surface<void,
cudaSurfaceType2DLayered>surf, int x, int y, int layer, enum
cudaSurfaceBoundaryMode mode=cudaBoundaryModeTrap)
```

表面存储的定义和绑定方法如下：

```
surface<int,1>surf1d;
    ⋮
cudaArray *array;
    ⋮
cudaBindSurfaceToArray(surf1d,array);
```

```
template<class T, int dim=1>
struct __device_builtin_surface_type__ surface: public surfaceReference
{
    __host__ surface(void)
    {
      channelDesc=cudaCreateChannelDesc<T>();
    }
    __host__ surface(struct cudaChannelFormatDesc desc)
    {
      channelDesc=desc;
    }
};
extern __host__ cudaError_t CUDARTAPI cudaBindSurfaceToArray(const struct
surfaceReference *surfref, const struct cudaArray *array, const struct
cudaChannelFormatDesc *desc);
```

与纹理类似，表面存储也是有限的，且不同维度的表面存储限制也不相同。表面的维度限制可以通过 cudaGetDeviceProperties()函数获得的 cudaDeviceProp 结构体中的参数测得。下面是一段获取表面存储限制的 CUDA C 代码及其执行结果。

```
#include<stdio.h>
#include<stdlib.h>
#include<cuda_runtime.h>
int main()
{
    int count,i;
    cudaGetDeviceCount(&count);
    if(count==0) {
        printf("There is no device.\n");
        return false;
    }
    printf("There are %d GPUs\n",count);
    for(i=0; i<count; i++) {
        cudaDeviceProp prop;
        if(cudaGetDeviceProperties(&prop, i)==cudaSuccess) {
            printf("GPU %d: \t %s\n",i,prop.name);
            printf("maxSurface1D\t:\t%d\n",prop.maxSurface1D);
printf("maxSurface2D\t:\t%d,\t%d\n",prop.maxSurface2D[0],prop.maxSurface2D[1]);
printf("maxSurface3D \t: \t%d, \t%d, \t%d \n", prop. maxSurface3D [0], prop. maxSurface3D[1],prop.maxSurface3D[2]);
printf("maxSurfaceCubemap\t:\t%d\n",prop.maxSurfaceCubemap);
printf("maxSurface1DLayered \t: \t%d, \t%d \n", prop. maxSurface1DLayered [0], prop.maxSurface1DLayered[1]);
printf("maxSurface2DLayered\t:\t%d,\t%d,\t%d\n", prop.maxSurface2DLayered[0],prop.maxSurface2DLayered[1],prop.maxSurface2DLayered[2]);
printf("maxSurfaceCubemapLayered\t:\t%d,\t%d\n",prop.maxSurfaceCubemapLayered[0],prop.maxSurfaceCubemapLayered[1]);
printf("surfaceAlignment\t:\t%d\n",prop.surfaceAlignment);
        }
        printf("\n\n\n");
    }
    return 0;
}
```

```
[fangmq@cn18%yhstar texture]$./surface_limit
There are 2 GPUs
GPU 0:       Tesla K20c
maxSurface1D:        65536
maxSurface2D:        65536,    32768
maxSurface3D:        65536,    32768,    2048
```

```
maxSurfaceCubemap:             32768
maxSurface1DLayered:           65536,   2048
maxSurface2DLayered:           65536,   32768,   2048
maxSurfaceCubemapLayered:      32768,   2046
surfaceAlignment:              512
```

16.8 主机端内存

CUDA 编程涉及的主机端存储主要有两种,分别是可分页内存(pageable memory)和页锁定内存(pinned memory)。

可分页内存利用 malloc() 和 free() 函数分配和释放,该内存是可换页的,即内存页可能被换出到磁盘。可分页内存无法使用 DMA。普通 C 程序中使用的存储就是这种可分页内存。

页锁定内存利用 cudaHostAlloc() 和 cudaFreeHost() 函数分配和释放,该内存一直位于内存空间中,不会被换出到磁盘。页锁定内存支持 DMA 访问,支持与 GPU 的异步通信。但由于内存空间有限,若页锁定内存分配过多将导致内存空间不足。当然,当前服务器或工作站中,一般主机端内存已达到 64GB,甚至 512GB,相对于 GPU 的显存空间(K20c GPU 中 4800MB),还是很充足的。

页锁定内存的另一种申请和释放方法是利用 cudaHostRegister() 函数将可分页内存注册为页锁定空间,用 cudaHostUnregister() 函数解除注册。

页锁定内存的另一个重要优势是传输速率。一般情况下,页锁定内存的传输速率是可分页内存的两倍左右(也存在一些特殊情况,详见 17.7 节)。既然页锁定内存明显优于可分页内存,那么使用页锁定内存后程序性能是否一定比可分页内存性能更好呢?

这里设计了一组实验探索上述问题,首先编程实现可分页内存和页锁定内存的 3 种分配和释放方法。

```
void pagable_memory_test(DATATYPE *a0,DATATYPE *a,DATATYPE *d_a,int n)
{
    a=(DATATYPE *)malloc(sizeof(DATATYPE) * n);
    for(int i=0;i<n;i++)
    {
        a[i]=a0[i];
    }
    cudaMemcpy(d_a, a, sizeof(DATATYPE) * n, cudaMemcpyHostToDevice);
    cudaMemcpy(a, d_a, sizeof(DATATYPE) * n, cudaMemcpyDeviceToHost);
    free(a);
}
void pinned_memory_test(DATATYPE *a0,DATATYPE *b,DATATYPE *d_a,int n)
```

```
{
    cudaHostAlloc((void**) &b, sizeof(DATATYPE) * n,cudaHostAllocDefault);
    for(int i=0;i<n;i++)
    {
        b[i]=a0[i];
    }
    cudaMemcpy(d_a, b, sizeof(DATATYPE) * n, cudaMemcpyHostToDevice);
    cudaMemcpy(b, d_a, sizeof(DATATYPE) * n, cudaMemcpyDeviceToHost);
    cudaFreeHost(b);
}
void pinned_register_memory_test(DATATYPE *a0,DATATYPE *c,DATATYPE *d_a,int n)
{
    c=(DATATYPE * )malloc(sizeof(DATATYPE) * n);
    cudaHostRegister(c,sizeof(DATATYPE) * n,0);
    for(int i=0;i<n;i++)
    {
        c[i]=a0[i];
    }
    cudaMemcpy(d_a, c, sizeof(DATATYPE) * n, cudaMemcpyHostToDevice);
    cudaMemcpy(c, d_a, sizeof(DATATYPE) * n, cudaMemcpyDeviceToHost);
    cudaHostUnregister(c);
    free(c);
}
```

编译并执行上述代码，测试各步骤时间，统计时间结果如表 16.4 所示。其中，页锁定 1 指代利用 cudaHostAlloc() 函数申请的页锁定内存，页锁定 2 指代 cudaHostRegister() 函数注册的页锁定内存。

表 16.4 可分页内存和页锁定内存对比　　　　　　　　单位：ms

1G	分配	注册	CPU 复制	H2D	D2H	解除	释放	all_time
可分页	0.01	0.00	442.55	328.98	369.20	0.00	2.01	1142.73
页锁定 1	469.13	0.00	237.44	170.06	160.33	0.00	165.57	1202.52
页锁定 2	0.07	243.21	235.51	169.72	160.34	45.93	1.63	856.41
2G	分配	注册	CPU 复制	H2D	D2H	解除	释放	all_time
可分页	0.01	0.00	894.57	828.07	731.91	0.00	3.93	2458.49
页锁定 1	936.87	0.00	472.43	339.67	320.84	0.00	329.96	2399.76
页锁定 2	0.03	483.80	479.31	339.58	320.81	91.50	3.97	1719.00

表 16.4 中的结果显示：

(1) 对比表中 CPU 复制、H2D、D2H 几列数据，显然页锁定内存在 CPU 访问和 CPU/GPU 通信中有巨大优势，比可分页内存快了近 1 倍。

注：事实上页锁定内存的 CPU 访存性能并不优于可分页内存，这里之所以可分页内

存的 CPU 访问耗时更长,是由于 malloc 函数执行时不分配存储空间,而是在第 1 次使用存储的分配,故其分配存储的时间累加到了 CPU 访问中。为了避免该影响,可在 malloc 函数后增加两个 memset 函数用来排除(两倍的)memset 开销。具体实验测试和结果不在此展开,有兴趣的读者可实践测试。

(2) 对比分配、注册、解除、释放几列数据,cudaHostAlloc( ) 函数分配空间和 cudaFreeHost( ) 函数释放空间非常耗时,而 cudaHostRegister( ) 函数和 cudaHost-Unregister( )函数耗时较少。

(3) 从总时间看,用 cudaHostAlloc( )函数分配页锁定内存的方法并不比可分页内存快,甚至更慢(测试中有 CPU 复制、H2D 和 D2H 3 个部分来抵消页锁定内存分配和释放开销,若具体应用仅需一次通信,性能将更差)。

(4) 从总时间看,用 cudaHostRegister( )函数注册页锁定内存的方法是性能最好的。故编程时宜采用注册页锁定内存的方法。

16.9 零拷贝操作

用 cudaHostAlloc( )函数分配页锁定内存或 cudaHostRegister( )函数注册页锁定内存时,可将 flags 设置为 cudaHostAllocMapped,再通过 cudaHostGetDevicePointer( )函数可获得在设备端指向主机端内存空间的指针,此时 kernel 函数可直接访问位于主机端的页锁定存储(称为映射锁页内存)。这种无须通信直接访问主机端数据的操作称为零拷贝(zerocopy)操作。

```
extern __host__ cudaError_t CUDARTAPI cudaHostGetDevicePointer(void ** pDevice, 
void *pHost, unsigned int flags);
```

11.3.3 节探索了 4 种零拷贝实现来优化曼德博罗特集计算,并总结了零拷贝的数条使用原则:①零拷贝使用时内存访问必须是连续的、对齐的;②4B 数据尺寸适合使用零拷贝,而 3B 数据尺寸不适合使用零拷贝;③零拷贝隐含计算与通信异步执行;④零拷贝访存要求严格,访存模式兼容性差,且一旦访存模式不兼容将导致极大的性能瓶颈。

第17章 GPU 关键性能测评

17.1 GPU 性能测评概述

GPU 作为一个相对陌生的设备,读者需要了解它的哪些信息呢? 本章将从 GPU 的基本参数、浮点运算性能、通信带宽、访存带宽和测评软件等角度展开测评,试图为读者展现一个具体全面的 GPU 设备。

GPU 相关参数:GPU 作为一类高性能计算设备,其设备型号、数量、计算核心数、各级存储容量、计算能力、时钟频率、grid 和 block 网格维度、ECC 错误校验等参数都是 GPU 程序开发的关注点。

GPU 浮点计算能力:浮点计算能力是科学计算关注的重点,是衡量 GPU 协处理器性能的最重要指标。

GPU 访存带宽:访存带宽指计算核心对内存(显存)和各级缓存(存储)的读写能力。任何计算都需要内存存储数据,计算时从内存读取数据,最终将结果写回内存,因此访存带宽非常重要。而由于存储墙的限制,存储的发展无法匹配计算核心发展的摩尔定律,因此访存带宽往往是决定程序真实执行性能的最重要的因素。

GPU 通信带宽:GPU 是协处理器,与 CPU 端存储是分离的,故 GPU 并行计算时,必须先将 CPU 端的代码(程序)和数据传输到 GPU, GPU 才能执行 kernel 函数。此时涉及 CPU 与 GPU 通信,其中通信接口 PCI-E 的版本和性能会直接影响通信带宽,此外,主机端内存分页方式(页锁定和可分页)也将直接影响通信带宽。

GPU-Z、CUDA-Z 等软件可用于测试 GPU 性能,分辨 GPU 真伪。其中,GPU-Z 可获取 GPU 的硬件参数,包括对 CUDA、OpenCL、Direct X 等技术的支持;CUDA-Z 提供了 GPU 的详细参数信息,还提供通信带宽、浮点运算能力等重要性能参数。结合这两个软件的运行结果数据(见图 17.1),对比相关数据,就能判断 GPU 显卡的真实情况,是否被二级厂商"裁剪",了解 GPU 显卡性能强弱。显卡的显存大小往往是商家宣传的主要参数,但事实上,决定显卡性能的重要因素是浮点运算能力、通信带宽和访存带宽。

第 17 章 GPU 关键性能测评

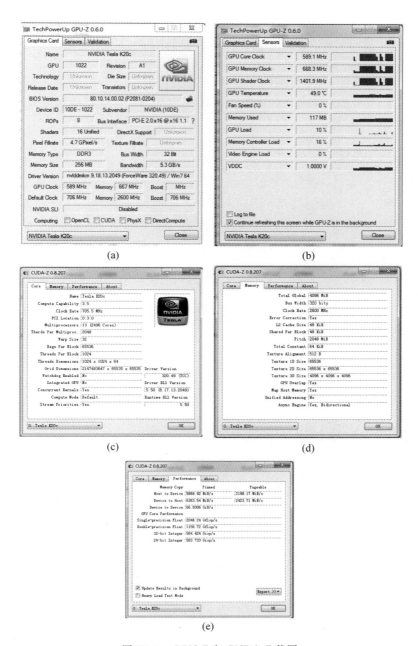

图 17.1 GPU-Z 和 CUDA-Z 截图

NVVP(nvprof)：NVIDIA visual profiler 是 CUDA 程序性能测试工具。主要用来测试 CUDA 程序中与 GPU 相关的函数时间，包括 kernel 函数、通信函数、存储分配和释放函数等。NVVP 包含图形版本（NVIDIA visual profiler）和命令行版本（nvprof）。

17.2 GPU 参数获取

除了采用辅助软件 GPU-Z、CUDA-Z 的方法外，通过 CUDA 编程也能获取 GPU 详细参数信息。本节提供几个常用的函数来辅助 CUDA 编程，并解释获取 GPU 详细参数的过程。

17.2.1 GPU 选择

编写 CUDA 程序时，程序员可能想知道当前系统中 GPU 设备的数量、GPU 型号，以及多 GPU 中哪个 GPU 性能最强？下面利用一个函数可完成这些功能：输出平台 GPU 数量和型号，自动选择性能最高的 GPU 并设置为运算 GPU。

```
bool get_best_gpu() {
    int count,maxid;
    int i;
    double core_performance[10],maxflops;
    double memory_performance[10];
    cudaGetDeviceCount(&count);
    if(count==0) {
        printf("There is no device.\n");
        return false;
    }
    printf("There are %d GPUs\n",count);
    for(i=0; i<count; i++) {
        cudaDeviceProp prop;
        if(cudaGetDeviceProperties(&prop, i)==cudaSuccess) {
            printf("GPU %d: \t %s\n",i,prop.name);
    core_performance[i]=(float)prop.clockRate * (float)prop.
multiProcessorCount * 192.0/1000.0/1000.0;
    memory_performance[i]=(float)prop.memoryClockRate * (float)prop.
memoryBusWidth * 1000.0/8.0/1024.0/1024.0/1024.0;
        }
    }
    maxflops=0.0;
    for(i=0;i<count;i++)
    {
        if(maxflops<core_performance[i])
        {
            maxflops=core_performance[i];
            maxid=i;
        }
```

```
        }
        cudaSetDevice(maxid);
        printf("GPU %d is the best device and is set!\n",maxid);
        return true;
}
```

上述代码在本书实验平台上的执行结果为

```
[fangmq@cn18%yhstar test_in_book]$./get_best_gpu_of_all_gpus
There are 2 GPUs
GPU 0:                Tesla K20c
GPU 1:                Tesla K20c
GPU 0 is the best device and is set!
```

17.2.2 详细设备参数获取

设备参数函数 cudaGetDeviceProperties() 获得的参数结构体 cudaDeviceProp 为用户提供全面的 GPU 参数信息（详见 5.4.1 节）。下面分别展示了设备参数函数 cudaGetDeviceProperties() 及其使用方法。

```
extern __host__ cudaError_t CUDARTAPI cudaGetDeviceProperties(struct
cudaDeviceProp *prop, int device);
cudaDeviceProp prop;
cudaGetDeviceProperties(&prop,i);     //i 为设备号
```

下面是本书实验平台上执行 deviceQuery（CUDA Toolkit Samples）程序的输出结果：

```
[fangmq@cn18%yhstar deviceQuery]$./deviceQuery
./deviceQuery Starting…
CUDA Device Query (Runtime API) version (CUDART static linking)
Detected 2 CUDA Capable device(s)

Device 0: "Tesla K20c"
    CUDA Driver Version/Runtime Version            5.5/5.5
    CUDA Capability Major/Minor version number:    3.5
    Total amount of global memory:                 5120 MBytes (5368512512 bytes)
    (13) Multiprocessors, (192) CUDA Cores/MP:     2496 CUDA Cores
    GPU Clock rate:                                706 MHz (0.71 GHz)
    Memory Clock rate:                             2600 Mhz
    Memory Bus Width:                              320-bit
```

```
       L2 Cache Size:                                      1310720 bytes
       Maximum Texture Dimension Size(x,y,z)               1D= (65536),
                                                           2D= (65536, 65536),
                                                           3D= (4096, 4096, 4096)
       Maximum Layered 1D Texture Size, (num) layers       1D= (16384), 2048 layers
       Maximum Layered 2D Texture Size, (num) layers       2D= (16384, 16384), 2048 layers
       Total amount of constant memory:                    65536 bytes
       Total amount of shared memory per block:            49152 bytes
       Total number of registers available per block: 65536
       Warp size:                                          32
       Maximum number of threads per multiprocessor:       2048
       Maximum number of threads per block:                1024
       Max dimension size of a thread block (x,y,z): (1024, 1024, 64)
       Max dimension size of a grid size    (x,y,z): (2147483647, 65535, 65535)
       Maximum memory pitch:                               2147483647 bytes
       Texture alignment:                                  512 bytes
       Concurrent copy and kernel execution:               Yes with 2 copy engine(s)
       Run time limit on kernels:                          No
       Integrated GPU sharing Host Memory:                 No
       Support host page-locked memory mapping:            Yes
       Alignment requirement for Surfaces:                 Yes
       Device has ECC support:                             Disabled
       Device supports Unified Addressing (UVA):           Yes
       Device PCI Bus ID/PCI location ID:                  3/0
       Compute Mode:
          < Default (multiple host threads can use::cudaSetDevice() with device
             simultaneously)>
  …(GPU1 参数省略)
  >Peer access from Tesla K20c (GPU0) ->Tesla K20c (GPU1): Yes
  >Peer access from Tesla K20c (GPU1) ->Tesla K20c (GPU0): Yes

  deviceQuery, CUDA Driver=CUDART, CUDA Driver Version=5.5, CUDA Runtime Version
  =5.5, NumDevs=2, Device0=Tesla K20c, Device1=Tesla K20c
  Result=PASS
```

另外，在 CUDA 安装目录的 doc 文件夹有 CUDA_Toolkit_Reference_Manual.pdf 文件，该文件提供了详细的 GPU 函数信息。

17.3 精确测时方法汇总

要正确测评程序性能，精确的测时函数必不可少。本节总结了多种常用测时方法，并分析其优缺点。

17.3.1　clock 测时

clock()函数是最通用的测时函数,可以跨平台(Windows、Linux)。该方法的测时精度为秒级。以下两种情况不适合采用 clock 方法测时:①Linux 下测试 openMP 程序执行时间;②时间小于 1s 时误差较大。

注意:该方法还可用于 kernel 函数中。

```
#include<time.h>
    clock_t t0,t1;
    t0=clock();
    //测时区间
    t1=clock();
    printf("runtime=%f(s)",(float)(t1-t0)/CLOCKS_PER_SEC);
```

17.3.2　gettimeofday 测时

测时函数 gettimeofday()适用于 Linux 环境,是使用最广泛、测时最准确的方法之一。该方法的精度为毫秒级。缺点是不兼容 Windows 平台。本书采纳的程序执行时间数据均用该方法测得。

```
#include<sys/time.h>
double dtime()
{
    double tseconds=0.0;
    struct timeval mytime;
    gettimeofday(&mytime,(struct timezone *)0);
    tseconds=(double)(mytime.tv_sec+mytime.tv_usec*1.0e-6);
    return tseconds;
}
    double tstart,tstop,ttime;
    tstart=dtime();
    …//测时区间
    tstop=dtime();
    ttime=tstop-tstart;
    printf("runtime=%f(s)\n",ttime);
```

17.3.3　CUDA 事件测时

CUDA 事件测时利用 CUDA 运行时函数来实现测时功能,能精确到毫秒级。优势是通用性强(适用于多种系统平台,适用于多种 CUDA 版本),仅需引入 CUDA 运行时函数即可。缺点是测时比较烦琐,需要 CUDA 环境。

```
#include<cuda_runtime.h>
    cudaEvent_t time_start,time_stop;
    cudaEventCreate(&time_start);
    cudaEventCreate(&time_stop);
    cudaEventRecord(time_start,0);
    cudaEventSynchronize(time_start);
    …//测时区间
    cudaEventRecord(time_stop,0);
    cudaEventSynchronize(time_stop);
    float elapsedTime;
    cudaEventElapsedTime(&elapsedTime,time_start,time_stop);
    printf("runtime=%f ms \n",elapsedTime);
```

17.3.4 cutil 库函数测时

cutil 库函数也提供了测时功能。优点是使用简单；缺点是需要 cutil 库，而从 CUDA 5.0 版本开始官方默认不再提供 cutil 函数库，通用性受到限制。

```
#include<cutil.h>
#include<cutil_inline.h>
    unsigned int timer=0;
    cutilCheckError(cutCreateTimer(&timer));
    cutilCheckError(cutStartTimer(timer));
    //测时区域
    cutilCheckError(cutStopTimer(timer));
    printf(" runtime=%f(ms) \n", cutGetTimerValue(timer));
```

建议读者在特定实例（如矩阵乘法）中实践上述测时方法，对比实验结果，总结和分析各种测时方法的优缺点。

17.4 GPU 预热与启动开销

本节将探讨的 GPU 预热和启动开销往往会被忽略，但又真实影响 GPU 实测性能。

GPU 预热：GPU 第一次被 kernel 函数调用时，存在启动时间（预热开销），该值往往较大，前文的一些实践中已发现这个问题。本节将利用 CUDA C 代码量化该值。

kernel 启动开销：每个 kernel 函数启动都需要时间，即 kernel 启动开销。

下面给出了测试 GPU 预热时间和 kernel 函数启动开销的 CUDA C 代码，编译执行并测试 GPU 预热时间和 kernel 启动开销。实验结果显示，GPU 预热时间约为 3s，平均 kernel 函数启动开销为 $6\mu s$ 左右。

```
__global__ void nothing_gpu(void){}
int main()
{
    struct timeval t1, t2, t3, t4;
    int i=0;
    gettimeofday(&t1, NULL);
    nothing_gpu<<<512, 512>>>();
    cudaThreadSynchronize();
    gettimeofday(&t2, NULL);
    for(i=0;i<1000;i++)
    {
        nothing_gpu<<<512, 512>>>();
    }
    cudaThreadSynchronize();
    gettimeofday(&t3, NULL);
    nothing_gpu<<<512, 512>>>();
    gettimeofday(&t4, NULL);

    printf("First call for kernel on GPU time: %f µs\n",(t2.tv_sec-t1.tv_sec) * 1000.0 * 1000.0+(t2.tv_usec-t1.tv_usec));
    printf("Average time for kernel calls:%f µs\n", ((t3.tv_sec-t2.tv_sec) * 1000.0 * 1000.0+(t3.tv_usec-t2.tv_usec))/1000.0);
    printf("one kernel call time:%f µs\n", ((t4.tv_sec-t3.tv_sec) * 1000.0 * 1000.0+ (t4.tv_usec-t3.tv_usec)));
    return 0;
}
```

```
[fangmq@cn18%yhstar test_in_book]$./first_kernel_and_average_kernel_O3
First call for kernel on GPU time: 3238591.000000µs
Average time for kernel calls:6.267000µs
one kernel call time:6.000000 µs
```

17.5　GPU 浮点运算能力

浮点运算能力和访存带宽是 GPU 最重要的性能指标,直接关系 GPU 运算性能。NVIDIA 发布新 GPU 时,会公布详细的性能数据。但该峰值性能往往连理想测试都未必能够达到,本节通过编程测试获得 GPU 浮点运算的理想性能。

下面是测试 GPU 浮点运算能力的 CUDA C 代码,其中核心运算是乘加运算。

```
#define DATATYPE float
#define REP 512
__global__ void get_gflops_on_gpu(DATATYPE *a,DATATYPE x,DATATYPE *b,DATATYPE y,int n)
```

```c
{
    int idx=blockIdx.x * blockDim.x+threadIdx.x;
    int idy=blockIdx.y * blockDim.y+threadIdx.y;
    DATATYPE temp=a[idy * n+idx];
    DATATYPE temp1=b[idy * n+idx];
#pragma unroll
    for(int i=0;i<REP;i++)
    {
        temp+=temp * x;
        temp1+=temp1 * y;
    }
    a[idy * n+idx]=temp;
    b[idy * n+idx]=temp1;
}
    dim3 blocks(dimb,dimb,1);
    dim3 threads(dimt,dimt,1);
    gettimeofday(&t1, NULL);
    get_gflops_on_gpu<<<blocks,threads>>>(d_a,1.10,d_b,0.99,n);
    cudaThreadSynchronize();
    gettimeofday(&t2, NULL);
    double timeused= (t2.tv_sec-t1.tv_sec) * 1000.0 * 1000.0+ (t2.tv_usec-t1.tv_usec);
    double gflops= (double)(dimb * dimb * dimt * dimt * 2.0 * REP * 2.0)/timeused/1000.0;
    printf("<(%d,%d),(%d,%d)>\t %f Gflops\n",dimb,dimb,dimt,dimt,gflops);
```

分别测试单精度浮点运算、双精度浮点运算和整数运算性能,统计结果如表 17.1 所示。

表 17.1　Tesla K20c GPU 峰值计算能力　　　　　　　　单位:Gflops

blocks	threads	float	double	int
(16,16)	(16,16)	1636.80	972.59	530.50
(32,32)	(16,16)	2088.99	1106.95	566.92
(64,64)	(16,16)	2218.47	1141.67	573.12
(16,16)	(32,32)	2080.90	1069.46	558.66
(32,32)	(32,32)	2216.19	1113.84	565.42
(64,64)	(32,32)	2245.73	1126.40	574.73

表 17.1 中数据显示,Tesla K20c GPU 的单精度浮点运算峰值性能可达 2245.73Gflops,双精度浮点运算峰值性能可达 1141.67Gflops,整型运算性能为 574.43Gflops。测试结果与 CUDA-Z 结果基本一致,但是并未达到该款 GPU 的峰值性能,而在 8.8.3 节分析中,利用 CUBLAS 库函数实现的矩阵乘法运算性能达到 2659Gflops。

17.6 GPU 访存带宽

存储墙是计算机性能的重要瓶颈之一,仅有浮点运算能力还不能实现数据处理,访存带宽才是影响真实运算性能的关键。本节通过编程测试 GPU 的 global memory 访存带宽。

本文分别设计了 1D、2D、2D 循环 3 种维度的访存带宽测试 kernel 函数,并分别测试不同线程维度参数的访存带宽,下面是相应的 CUDA C 代码,每次读出和写入算两次访存。此外利用 cudaMemcpy() 函数的内存复制作为对比测试。

```
__global__ void memory_copy_on_global0(DATATYPE *in, DATATYPE *out,long n)
{
    int idx=blockIdx.x * blockDim.x+threadIdx.x;
    int nn=blockDim.x * gridDim.x;
    for(;idx<n;idx+=nn)
    {
        out[idx]=in[idx];
    }
}
__global__ void memory_copy_on_global_2d_0(DATATYPE *in, DATATYPE *out,int ni,
long n)
{
    int idx=blockIdx.x * blockDim.x+threadIdx.x;
    int idy=blockIdx.y * blockDim.y+threadIdx.y;
    int id=idy * ni+idx;
    out[id]=in[id];
}
__global__ void memory_copy_on_global_2d_3(DATATYPE *in, DATATYPE *out,int ni,
long n)
{
    int bidx,bidy,idx,idy,id;
    int nn=ni/blockDim.x;
    for(bidy=blockIdx.y;bidy<nn;bidy+=gridDim.y)
    {
        idy=bidy * blockDim.y+threadIdx.y;
        for(bidx=blockIdx.x;bidx<nn;bidx+=gridDim.x)
        {
            idx=bidx * blockDim.x+threadIdx.x;
            id=idy * ni+idx;
            out[id]=in[id];
        }
```

```
    }
  }
    memory_copy_on_global0<<<blocknum,threadnum>>>(d_in,d_out,n*n*256);
    memory_copy_on_global_2d_0<<<blocks,threads>>>(d_in,d_out,n*16,n*n*
256);
    memory_copy_on_global_2d_3<<<blocks1,threads1>>>(d_in,d_out,n*16,n*n
*256);
    cudaMemcpy(d_out,d_in,sizeof(DATATYPE)*n*n*256,cudaMemcpyDeviceToDevice);
```

分别测试不同维度情况下的 GPU 访存带宽（float 类型数据），统计如表 17.2 所示。表中数据显示，利用 2D 循环维度时，线程维度为<<<(64,64),(32,32)>>>时，K20c 的访存带宽达到 151.49GB/s，cudaMemcpy() 函数测得的带宽可视为 GPU 的最大访存带宽（K20c 约为 160GB/s）。

表 17.2　GPU 访存（global memory）带宽（float）

1D		2D			2D 循环			Memcpy (D2D)
grid	GB/s	grid		GB/s	grid		GB/s	
<256,256>	137.80	(1024,1024)	(16,16)	137.96	(16,16)	(16,16)	96.89	159.53
<1024,256>	123.88	(1024,1024)	(16,16)	138.34	(32,32)	(16,16)	113.18	159.72
<256,1024>	120.23	(512,512)	(32,32)	116.10	(16,16)	(32,32)	149.28	160.05
<1024,1024>	145.32	(512,512)	(32,32)	115.59	(32,32)	(32,32)	147.65	160.15
<512,512>	121.22	(512,512)	(32,32)	115.89	(64,64)	(32,32)	151.49	159.74

若将数据类型由 float 型改为 double 型，几乎所有的线程维度配置均能达到近 150GB/s，其中，2D 维度<<<(1024,1024),(16,16)>>>获得最佳访存带宽 157.10GB/s。同理，测试 int 类型数据的访存带宽，与 float 类型基本一致，不再赘述。

测试 char 类型数据的访存带宽，统计结果如表 17.3 所示。表中数据显示，用普通方式访问 char 类型时访存带宽较小，最高仅达 61.55GB/s。可能的原因是 global memory 访问没有合并访存。解决方法是组合访问。

表 17.3　GPU 访存（global memory）带宽（char）

1D		2D			2D 循环			Memcpy (D2D)
grid	GB/s	grid		GB/s	grid		GB/s	
<256,256>	49.30	(1024,1024)	(16,16)	45.84	(16,16)	(16,16)	48.90	157.43
<1024,256>	61.03	(1024,1024)	(16,16)	45.96	(32,32)	(16,16)	55.56	159.29
<256,1024>	59.08	(512,512)	(32,32)	36.28	(16,16)	(32,32)	54.12	158.93
<1024,1024>	63.02	(512,512)	(32,32)	36.31	(32,32)	(32,32)	52.76	159.34
<512,512>	60.01	(512,512)	(32,32)	36.30	(64,64)	(32,32)	61.55	158.78

为了提高 char 类型访存带宽，对 char 类型数据组合访问，即将 4 个 char 类型数据合并成一个结构体（即 CUDA 提供内置变量类型 char4），然后测试访存带宽，结果显示访存带宽为 148.60GB/s。

表 17.2 的实验结果还显示了一个现象：密集访存情况下，2D 线程配置的循环模式比非循环模式性能要好，2D 循环模式访存带宽可达 151GB/s，而 2D 非循环模式（线程切换）仅能达到 138GB/s，结合 15.8 的结论（线程切换开销较线程内循环开销少），说明了以访存为主的应用中，需要更加关注访存。

17.7 GPU 通信带宽

GPU 存储与 CPU 内存相互独立，若要 GPU 计算，必须先将数据传输到 GPU 显存中。CPU 与 GPU 通过 PCI-E 接口通信，本书实验平台上 GPU 的 PCI-E 接口是 16 通道的，PCI-E 版本一般为 2.0，该版本每个通道带宽约为 500MB/s，由于数据包的额外开销，8GB/s 理论带宽上能达到 6GB/s。

影响 CPU 和 GPU 通信速率的因素除了 PCI-E 通道数、版本外，还有 CPU 端内存使用方式。CPU 端内存有可分页（pagable）内存和页锁定（pinned）内存，区别是操作系统是否将分页交换到硬盘。显然不存在分页交换的页锁定内存访问更快；但受内存总量限制，无法分配超过内存大小的页锁定内存，而且可能影响其他进程对内存的使用。

关于可分页内存和页锁定内存已在 16.8 节详细阐述。可分页内存的分配和释放采用 C 标准函数 malloc() 和 free()。页锁定内存的分配和释放需要采用 CUDA 提供的函数 cudaHostAlloc() 和 cudaFreeHost()。CUDA 还提供了注册页锁定存储的方法，采用 C 标准函数 malloc() 分配空间，用 cudaHostRegister() 函数将可分页存储注册为页锁定存储，利用 cudaHostUnregister() 解除注册。

16.8 节实验数据显示页锁定内存比可分页内存通信性能好。那么页锁定内存和可分页内存在不同数据规模时的通信性能如何？从主机端到设备端通信和设备端到主机端通信有何差别？本节通过实验来量化测评。

NVIDIA Tesla K20c GPU 平台上，设置数据规模从 1B 到 1GB 不等，分别测试页锁定内存和可分页内存主机端到设备端通信和设备端到主机端通信时间，结果统计如图 17.2 所示。

图 17.2 中数据显示：

(1) 当数据规模大于 8KB 时，页锁定内存的通信性能明显比可分页内存的通信性能好。

(2) 当数据规模小于 8KB 时（见图 17.2(b)），两种类型的通信性能相当。

(3) 从数据规模大于 1MB 的部分结果可以看出，页锁定内存的设备端到主机端通信速率比主机端到设备端通信速率更高，而可分页内存的通信速率恰好相反，主机端到设备

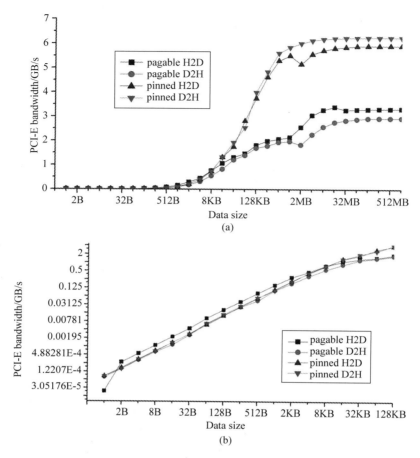

图 17.2 数据量和传输速率关系

端的通信速率优于设备端到主机端的通信速率。

（4）从图中趋势可以看出，当数据量小于 512KB 时，通信速率与数据规模正相关；而数据规模达到 512KB 后，通信速率提升较为平缓；数据规模达到 8MB 后通信速率基本达到最大通信带宽并趋于稳定。

17.8 NVIDIA Visual Profiler

NVIDIA Visual Profiler(NVVP)是 GPU 程序性能分析工具，提供了 kernel 函数时间统计、通信时间统计、带宽计算、时间轴信息和各种程序性能指标分析等功能。下面分别介绍 Windows 和 Linux(命令行)两种模式的使用方法。

Windows 下的 NVVP 用法如下，执行后时间轴界面如图 17.3 所示(Linux 图形界面下的 NVVP 性能分析工具的用法和 Windows 下的相同)。

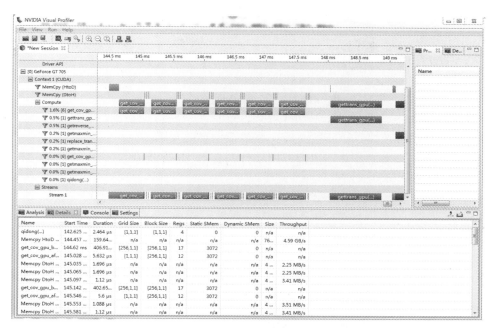

图 17.3　Windows 下的 NVVP 分析结果截图

位置	开始菜单/NVIDIA Corporation/CUDA Toolkit/（或者直接查找 CUDA Toolkit 安装目录下的子目录）
用法	File->New Session （File：选择可执行文件 exe） （Working Directory：工作目录） （Arguments：命令行参数） （Environment：环境变量）
具体分析	参考 NVIDIA 提供的相关文档

Linux 命令行 NVVP 用法：在命令行使用 nvprof 命令分析 GPU 程序性能。第 7 章有 nvprof 命令使用实例，用法如下：

```
$ nvprof ./vector_dot_product_cublas
$ nvprof ./程序名
```

此时显示的信息并不完整，仅统计了总体信息，若需要每个 kernel 函数或通信的具体时间信息和起始时间信息，需要添加选项 "——print-gpu-trace"。本书 11.4 节利用该方法分析计算与通信是否真正重叠。

```
$ nvprof --print-gpu-trace ./程序名
$ nvprof --print-gpu-trace ./cuda_2
```

附录 D 给出了 nvprof 命令的帮助信息。

17.9 程序性能对比约定

最优串行程序约定：访存连续、最内层循环向量化。

kernel 函数性能对比约定：kernel 函数执行若需要通信，应将通信时间统计到 GPU 并行时间。因为通信时间是 GPU 并行产生的最主要的开销。

多 GPU 获取设备并预热产生的开销应计入 GPU 并行开销。

第18章 CPUs 和 GPUs 协同

前面主要阐述了如何将 C 语言代码改为 CUDA C 代码,如何逐步优化 kernel 函数,分析性能影响因素等,本章将着重介绍如何利用多核 CPU 和多 GPU 进行协同计算。为方便理解,本章基于第 8 章的矩阵乘法,利用本文实验平台(2 个 8 核 E5-2670 CPU 和 2 个 Tesla K20c GPU)开展协同优化研究。

18.1 协同优化基点

利用多 CPU 和多 GPU 协同计算矩阵乘,问题规模必须够大,单独的 CPU 或 GPU 设备计算时间超出预期,否则协同计算就意义不大了。为了简化问题,减少复杂的不同规模问题探讨,设定矩阵乘法规模为$(16384 \times 16384) \times (16384 \times 16384)$。

本节还设定 CPU 并行矩阵乘基点和 GPU 并行矩阵乘基点,基点在一定程度上代表单独设备上的性能优化极值。本章讨论中,即便有更好的优化策略或方法,也不再继续优化 kernel 函数或 CPU 多核代码,而是将重点放在异构系统多个计算设备的协同上。

18.1.1 CPU 并行矩阵乘基点

8.1.4 节探讨了最优串行矩阵乘,其中转置优化的矩阵乘法效率最高,本章在此基础上开发了 OpenMP 多线程并行版本,作为 CPU 并行矩阵乘基点,具体代码如下:

```
void matmult(DATATYPE *a,DATATYPE *b, DATATYPE *c, int m, int n,
int l)
{//a:m 行 l 列,b:l 行 n 列的转置,c:m 行 n 列
    int i, j, k;
    double temp;
    DATATYPE *b1=(DATATYPE *) malloc(sizeof(DATATYPE ) * l * n);
#pragma omp parallel for private(j)
```

```
    for(i=0;i<n;i++)
    {
        for(j=0;j<l;j++)
        {
            b1[i * l+j]=b[j * n+i];
        }
    }
#pragma omp parallel for private(j,k,temp)
    for(i=0; i<m; i++) {
        for(j=0; j<n; j++) {
            temp=0;
            for(k=0; k<l; k++) {
                temp+=a[i * l+k] * b1[j * n+k];
            }
            c[i * n+j]=temp;
        }
    }
    free(b1);
}
```

编译并执行 CPU 并行矩阵乘法,执行结果如下。该结果将用于协同计算的任务划分。

```
$./cpu
16384     194325.999022     (ms)     42.155965     Gflops
```

18.1.2　GPU 并行矩阵乘基点

GPU 并行矩阵乘基点选择 8.8.1 节 gpu_4_1 版本矩阵乘代码,详细代码请查阅相应章节。下面是执行结果。该结果将用于协同计算的任务划分。

```
./gpu
16384    32    34401.619911    (ms)    238.128321    Gflops
```

18.2　CPU/GPU 协同

要想利用 CPU 与 GPU 协同计算,首先要分割计算任务,利用上节测得的结果,将计算任务 n 划分为 GPU 任务 gpu_n 和 CPU 任务 cpu_n,计算公式见代码(见图 18.1)。其次需要重新计算 GPU 启动 kernel 函数的线程(块)维度。Kernel 函数启动后异步执行,故可在 kernel 启动函数后直接添加 CPU 并行计算代码,即可实现 CPU 与 GPU 的协同运算。

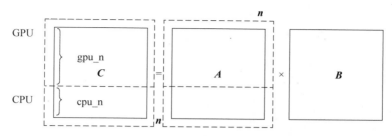

图 18.1 矩阵乘法 CPU/GPU 任务分割

下面是相应的 CPU/GPU 异构协同计算矩阵乘代码。

```
#define threadnx 32
   ⋮
   int gpu_gflops=238;
   int cpu_gflops=42;
   int gpu_n=(gpu_gflops*n/(gpu_gflops+cpu_gflops)+threadnx-1)/threadnx*
threadnx;
   int cpu_n=n-gpu_n;
   ⋮
   int bx=(n+threadnx-1)/threadnx;
   dim3 blocks(bx,((gpu_n+threadnx-1)/threadnx));
   dim3 threads(threadnx, threadnx);
   ⋮
   matrix_multiplication_gpu_4_1<<<blocks,threads>>>(d_a,pitch_a/sizeof
(DATATYPE),d_b,pitch_b/sizeof(DATATYPE),d_c1,pitch_c/sizeof(DATATYPE),n);
   matmult(&a[gpu_n*n],b,&c1[gpu_n*n],cpu_n,n,n);
   ⋮
```

理论上，CPU 与 GPU 协同计算非常简单，但实践时却没这么容易。从代码逻辑上看，上述代码没有任何问题，但无法编译，原因如下。

（1）nvcc 编译器不支持 OpenMP 代码编译。

（2）icc(gcc) 编译器不支持 CUDA 内置符号和内置函数的编译。

因此需要对上述代码进行封装，将包含 CUDA 内置符号和内置函数的部分代码封装在 cu 文件中用 nvcc 编译(kernel.cu)；将包含 OpenMP 的代码封装在 cpp 文件中用 icc 或 gcc 编译(cpukernel.cpp)；然后在最外层用 cpp 文件调用。用 icc 或 gcc 编译，需要添加相应的 nvcc 头文件和库文件(cpu_gpu.cpp)。

```
/*****************************************************************/
/* cpukernel.cpp                                                 */
/*****************************************************************/
#include<stdlib.h>
#include<stdio.h>
#include<omp.h>
```

```
#define DATATYPE float
void matmult(DATATYPE *a,DATATYPE *b, DATATYPE *c, int m, int n,int l) {//a:m行l
列,b:l行n列的转置,c:m行n列
    int i, j, k;
    double temp;
    DATATYPE *b1=(DATATYPE * ) malloc(sizeof(DATATYPE) * l * n);
#pragma omp parallel for private(j)
    for(i=0;i<n;i++)
    {
        for(j=0;j<l;j++)
        {
            b1[i * l+j]=b[j * n+i];
        }
    }
#pragma omp parallel for private(j,k,temp)
    for(i=0; i<m; i++) {
        for(j=0; j<n; j++) {
            temp=0;
            for(k=0; k<l; k++) {
                temp+=a[i * l+k] * b1[j * n+k];
            }
            c[i * n+j]=temp;
        }
    }
    free(b1);
}
```

```
/********************************************************************/
/* kernel.cu                                                        */
/********************************************************************/
#include<cuda_runtime.h>
#define threadnx 32
#define DATATYPE float
__global__ void matrix_multiplication_gpu_4_1(const DATATYPE *a, size_t lda,
const DATATYPE *b, size_t ldb, DATATYPE *c, size_t ldc, int n){
    __shared__ DATATYPE matA[threadnx][threadnx];
    __shared__ DATATYPE matB[threadnx][threadnx];
    const int tidc=threadIdx.x;
    const int tidr=threadIdx.y;
    const int bidc=blockIdx.x * threadnx;
    const int bidr=blockIdx.y * threadnx;
    int i, j;
    float results=0.0;
    float comp=0.0;
```

```
        float t;
        for(j=0; j<n; j+=threadnx)
        {
            matA[tidr][tidc]=a[(tidr+bidr) * lda+tidc+j];
            matB[tidr][tidc]=b[(tidr+j) * ldb+tidc+bidc];
            __syncthreads();
            for(i=0; i<threadnx; i++)
            {
                comp -=matA[tidr][i] * matB[i][tidc];
                t=results-comp;
                comp=(t-results)+comp;
                results=t;
            }
            __syncthreads();
        }
        if(tidr+bidr<n && tidc+bidc<n)
        {
            c[(tidr+bidr) * ldc+tidc+bidc]=results;
        }
}
void matmult(DATATYPE *a,DATATYPE *b, DATATYPE *c, int m, int n,int l);
void cpu_gpu_matrix_multiplication (DATATYPE *,DATATYPE * b, DATATYPE * c1,
DATATYPE *d_a,DATATYPE *d_b,DATATYPE *d_c1, size_t pitch_a,size_t pitch_b,size_
t pitch_c,dim3 blocks,dim3 threads,int gpu_n,int cpu_n,int n)
{
    cudaMemcpy2D(d_a, pitch_a, a, sizeof(DATATYPE) * n, sizeof(DATATYPE) * n,
    gpu_n, cudaMemcpyHostToDevice);
    cudaMemcpy2D(d_b, pitch_b, b, sizeof(DATATYPE) * n, sizeof(DATATYPE) * n,
    n, cudaMemcpyHostToDevice);
    matrix_multiplication_gpu_4_1<<<blocks,threads>>>(d_a,pitch_a/sizeof
    (DATATYPE),d_b,pitch_b/sizeof(DATATYPE),d_c1,pitch_c/sizeof(DATATYPE),
    n);
    matmult(&a[gpu_n * n],b,&c1[gpu_n * n],cpu_n,n,n);
    cudaMemcpy2D(c1, sizeof(DATATYPE) * n, d_c1, pitch_c, sizeof(DATATYPE) *
n, gpu_n, cudaMemcpyDeviceToHost);
}
/*********************************************************************/
/* cpu_gpu.cpp                                                       */
/*********************************************************************/
#include<stdlib.h>
#include<stdio.h>
#define DATATYPE float
#include<cuda_runtime.h>
```

```c
#include<omp.h>
#define threadnx 32
#include<sys/time.h>
#define TIMETESTEVENT
double dtime()
{
    double tseconds=0.0;
    struct timeval mytime;
    gettimeofday(&mytime,(struct timezone *)0);
    tseconds=(double)(mytime.tv_sec+mytime.tv_usec*1.0e-6);
    return tseconds;
}
void cpu_gpu_matrix_multiplication(DATATYPE *a,DATATYPE *b,DATATYPE *c1,
DATATYPE *d_a,DATATYPE *d_b,DATATYPE *d_c1,size_t pitch_a,size_t pitch_b,size_
t pitch_c,dim3 blocks,dim3 threads,int gpu_n,int cpu_n,int n);
int main()
{
    int n=16384;
    DATATYPE *a,*b,*c1;
    a=(DATATYPE *)malloc(sizeof(DATATYPE) * n * n);
    b=(DATATYPE *)malloc(sizeof(DATATYPE) * n * n);
    c1=(DATATYPE *)malloc(sizeof(DATATYPE) * n * n);
    FILE *dat_a,*dat_b;
    dat_a=fopen("a.dat","rb");
    fread(a,sizeof(DATATYPE),n*n,dat_a);
    fclose(dat_a);
    dat_b=fopen("b.dat","rb");
    fread(b,sizeof(DATATYPE),n*n,dat_b);
    fclose(dat_b);
    int gpu_gflops=238;
    int cpu_gflops=42;
    int gpu_n=(gpu_gflops * n/(gpu_gflops+cpu_gflops)+threadnx-1)/threadnx *
    threadnx;
    int cpu_n=n-gpu_n;
    DATATYPE *d_a,*d_b,*d_c1;
    size_t pitch_a, pitch_b, pitch_c;
    cudaMallocPitch((void **) &d_a, &pitch_a, sizeof(DATATYPE) * n, gpu_n);
    cudaMallocPitch((void **) &d_b, &pitch_b, sizeof(DATATYPE) * n, n);
    cudaMallocPitch((void **) &d_c1, &pitch_c, sizeof(DATATYPE) * n, gpu_n);
    int bx=(n+threadnx-1)/threadnx;
    dim3 blocks(bx, ((gpu_n+threadnx-1)/threadnx));
    dim3 threads(threadnx, threadnx);
#ifdef TIMETESTEVENT
```

```
    double time_0,time_1;
    time_0=dtime();
#endif
cpu_gpu_matrix_multiplication(a,b,c1,d_a,d_b,d_c1,pitch_a,pitch_b,pitch_c,
blocks,threads,gpu_n,cpu_n,n);
#ifdef TIMETESTEVENT
    time_1=dtime();
    printf("%d\t %d\t %f\t(ms)\t%f\tGflops\n",n,threadnx,(time_1-time_0) *
1000,(float) ((float)n * (float)n * (float)n * 2.0f)/1024.0f/1024.0f/1024.0f/
(time_1-time_0));
#endif
    cudaFree(d_a);
    cudaFree(d_b);
    cudaFree(d_c1);
    FILE *dat_c;
    dat_c=fopen("c_cpu_gpu.dat","wb");
    fwrite(c1,sizeof(DATATYPE),n * n,dat_c);
    fclose(dat_c);
    free(a);
    free(b);
    free(c1);
    return 0;
}
```

编译命令和执行结果如下。显然，CPU 与 GPU 协同运算的性能比单独的 CPU 与 GPU 运算性能都好。

```
[fangmq@mn0%yhstar cpu_gpu]$icc -openmp -c cpukernel.cpp
[fangmq@mn0%yhstar cpu_gpu]$nvcc cpukernel.o kernel.cu -c
[fangmq@mn0%yhstar cpu_gpu]$icc -openmp cpukernel.o kernel.o cpu_gpu.cpp -o
cpu_gpu -I/vol/home/fangmq/gpu_cuda/cuda/cuda_5_5/include -L/vol/home/
fangmq/gpu_cuda/cuda/cuda_5_5/lib64 -lcudart
[fangmq@cn18%yhstar cpu_gpu]$./cpu_gpu
16384      32      30728.428125     (ms)      266.593526      Gflops
```

18.3 多 GPU 协同

本节探讨多 GPU 协同计算，本文的实验平台装备了 2 个 Tesla K20c GPU，本节的任务是同时利用这两个 GPU 运算矩阵乘法。根据不同的环境，本节给出 3 种实现方式，分别是 CUDA 版本、OpenMP＋CUDA 和 MPI＋CUDA。

OpenMP 和 MPI 编程知识请参考相关文献资料。

18.3.1 CUDA 版本

CUDA 版本的思想是利用 kernel 函数启动的异步特性，与 18.2 节 CPU 与 GPU 协同计算类似。利用 cudaSetDevice() 函数实现不同 GPU 设备的切换。由于两个 GPU 都是 Tesla K20c GPU，任务直接对半划分。下面是详细代码。

```
cudaSetDevice(0);
DATATYPE *d_a,*d_b,*d_c;
size_t pitch_a, pitch_b, pitch_c;
cudaMallocPitch((void **) &d_a, &pitch_a, sizeof(DATATYPE) * n, n/2);
cudaMallocPitch((void **) &d_b, &pitch_b, sizeof(DATATYPE) * n, n);
cudaMallocPitch((void **) &d_c, &pitch_c, sizeof(DATATYPE) * n, n/2);
cudaSetDevice(1);
DATATYPE *d_a1,*d_b1,*d_c1;
size_t pitch_a1, pitch_b1, pitch_c1;
cudaMallocPitch((void **) &d_a1, &pitch_a1, sizeof(DATATYPE) * n, n/2);
cudaMallocPitch((void **) &d_b1, &pitch_b1, sizeof(DATATYPE) * n, n);
cudaMallocPitch((void **) &d_c1, &pitch_c1, sizeof(DATATYPE) * n, n/2);
int bx=(n+threadnx-1)/threadnx;
dim3 blocks(bx, bx/2);
dim3 threads(threadnx, threadnx);
cudaSetDevice(0);
cudaMemcpy2D(d_a, pitch_a, a, sizeof(DATATYPE) * n, sizeof(DATATYPE) * n,
n/2, cudaMemcpyHostToDevice);
cudaMemcpy2D(d_b, pitch_b, b, sizeof(DATATYPE) * n, sizeof(DATATYPE) * n,
n, cudaMemcpyHostToDevice);
cudaSetDevice(1);
cudaMemcpy2D(d_a1, pitch_a, &a[n * n/2], sizeof(DATATYPE) * n, sizeof
(DATATYPE) * n, n/2, cudaMemcpyHostToDevice);
cudaMemcpy2D(d_b1, pitch_b, b, sizeof(DATATYPE) * n, sizeof(DATATYPE) * n,
n, cudaMemcpyHostToDevice);
cudaSetDevice(0);
matrix_multiplication_gpu_4_1<<<blocks,threads>>>(d_a,pitch_a/sizeof
(DATATYPE),d_b,pitch_b/sizeof(DATATYPE),d_c,pitch_c/sizeof(DATATYPE),n);
cudaSetDevice(1);
matrix_multiplication_gpu_4_1<<<blocks,threads>>>(d_a1,pitch_a/sizeof
(DATATYPE),d_b1,pitch_b/sizeof(DATATYPE),d_c1,pitch_c/sizeof(DATATYPE),
n);
cudaSetDevice(0);
cudaMemcpy2D(c1, sizeof(DATATYPE) * n, d_c, pitch_c, sizeof(DATATYPE) * n,
n/2, cudaMemcpyDeviceToHost);
cudaSetDevice(1);
```

```
    cudaMemcpy2D(&c1[n * n/2], sizeof(DATATYPE) * n, d_c1, pitch_c, sizeof
    (DATATYPE) * n, n/2, cudaMemcpyDeviceToHost);
    cudaFree(d_a);
    cudaFree(d_b);
    cudaFree(d_c);
    cudaFree(d_a1);
    cudaFree(d_b1);
    cudaFree(d_c1);
    ⋮
```

编译并执行上述代码,为了更好地分析代码的执行情况,利用 nvprof 命令执行上述程序,结果如下。由结果可知,kernel 函数的执行是重叠的,其他部分(通信)是串行执行的。此时浮点计算性能达到 459Gflops(已包括通信时间),接近单 GPU 的两倍。

```
[fangmq@cn18%yhstar gpu_gpu]$nvprof --print-gpu-trace ./gpu
==14873==NVPROF is profiling process 14873, command: ./gpu
16384     32      17832.106829    (ms)      459.396081    Gflops
==14873==Profiling application: ./gpu
==14873==Profiling result:
 Start   Duration      Grid Size     Block Size     Regs*      SSMem*       DSMem*
Size   Throughput      Device     Context      Stream     Name
666.39ms  169.96ms   -       -       -      -       536.87MB   3.1588GB/s   Tesla K20c (0)    1
    2   [CUDA memcpy HtoD]
836.69ms  340.27ms   -       -       -      -       1.0737GB   3.1555GB/s   Tesla K20c (0)    1
    2   [CUDA memcpy HtoD]
1.17724s  168.92ms   -       -       -      -       536.87MB   3.1783GB/s   Tesla K20c (1)    2
    2   [CUDA memcpy HtoD]
1.34649s  339.55ms   -       -       -      -       1.0737GB   3.1622GB/s   Tesla K20c (1)    2
    2   [CUDA memcpy HtoD]
1.68650s  16.2643s  (512 256 1)   (32 32 1)    24   8.1920KB    0B      -       -    Tesla
K20c (0)   1    2   matrix_multiplication_gpu_4_1(float const *, unsigned long,
float const *, unsigned long, float *, unsigned long, int) [183]
1.68697s  16.2674s  (512 256 1)   (32 32 1)    24   8.1920KB    0B      -       -    Tesla
K20c (1)   2    2   matrix_multiplication_gpu_4_1(float const *, unsigned long,
float const *, unsigned long, float *, unsigned long, int) [193]
17.9508s  274.03ms   -       -       -      -       536.87MB   1.9592GB/s   Tesla K20c (0)    1
    2   [CUDA memcpy DtoH]
18.2258s  271.47ms   -       -       -      -       536.87MB   1.9777GB/s   Tesla K20c (1)    2
    2   [CUDA memcpy DtoH]

Regs: Number of registers used per CUDA thread.
SSMem: Static shared memory allocated per CUDA block.
DSMem: Dynamic shared memory allocated per CUDA block.
```

18.3.2 OpenMP+CUDA

18.3.1 节 CUDA 版本的不足之处是通信函数串行执行，且程序中利用 cudaSetDevice()函数不断切换 GPU。本节利用 2 个 OpenMP 线程来分别控制两个 GPU 完成矩阵乘法运算。

下面是 OpenMP+CUDA 混合并行代码。代码中 GPU 空间分配部分亦能并行，但为了与其他版本对比，仅统计 kernel 函数计算和数据通信的时间。

```
/******************************************************************/
/* kernel.cu                                                      */
/******************************************************************/
#include<stdlib.h>
#include<stdio.h>
#define DATATYPE float
#include<cuda_runtime.h>
#define threadnx 32
__global__ void matrix_multiplication_gpu_4_1(const DATATYPE *a, size_t lda,
const DATATYPE *b, size_t ldb, DATATYPE *c, size_t ldc, int n){ ⋮ }
void gpu_gpu_matrix_multiplication(DATATYPE * a, DATATYPE * b, DATATYPE * c1,
DATATYPE *d_a, DATATYPE *d_b, DATATYPE *d_c1, size_t pitch_a,size_t pitch_b,size_
t pitch_c,dim3 blocks,dim3 threads, int n)
{
    int gpuid;
    cudaGetDevice(&gpuid);
    if(gpuid==0)
    {
        cudaMemcpy2D(d_a, pitch_a, a, sizeof(DATATYPE) * n, sizeof(DATATYPE) *
        n, n/2, cudaMemcpyHostToDevice);
    }
    else if(gpuid==1)
    {
        cudaMemcpy2D(d_a, pitch_a, &a[n * n/2], sizeof(DATATYPE) * n, sizeof
        (DATATYPE) * n, n/2, cudaMemcpyHostToDevice);
    }
    cudaMemcpy2D(d_b, pitch_b, b, sizeof(DATATYPE) * n, sizeof(DATATYPE) * n,
    n, cudaMemcpyHostToDevice);
    matrix_multiplication_gpu_4_1<<<blocks,threads>>>(d_a,pitch_a/sizeof
    (DATATYPE),d_b,pitch_b/sizeof(DATATYPE),d_c1,pitch_c/sizeof(DATATYPE),n);
    if(gpuid==0)
    {
        cudaMemcpy2D(c1, sizeof(DATATYPE) * n, d_c1, pitch_c, sizeof(DATATYPE)
        * n, n/2, cudaMemcpyDeviceToHost);
```

```
        }
        else if(gpuid==1)
        {
            cudaMemcpy2D(&c1[n*n/2], sizeof(DATATYPE) * n, d_c1, pitch_c, sizeof
            (DATATYPE) * n, n/2, cudaMemcpyDeviceToHost);
        }
    }
}
/*********************************************************************/
/* omp_cuda.cpp                                                      */
/*********************************************************************/
#include<stdlib.h>
#include<stdio.h>
#define DATATYPE float
#include<omp.h>
#include<cuda_runtime.h>
    ⋮
    cudaSetDevice(0);
    DATATYPE *d_a,*d_b,*d_c;
    size_t pitch_a, pitch_b, pitch_c;
    cudaMallocPitch((void **) &d_a, &pitch_a, sizeof(DATATYPE) * n, n/2);
    cudaMallocPitch((void **) &d_b, &pitch_b, sizeof(DATATYPE) * n, n);
    cudaMallocPitch((void **) &d_c, &pitch_c, sizeof(DATATYPE) * n, n/2);
    cudaSetDevice(1);
    DATATYPE *d_a1,*d_b1,*d_c1;
    size_t pitch_a1, pitch_b1, pitch_c1;
    cudaMallocPitch((void **) &d_a1, &pitch_a1, sizeof(DATATYPE) * n, n/2);
    cudaMallocPitch((void **) &d_b1, &pitch_b1, sizeof(DATATYPE) * n, n);
    cudaMallocPitch((void **) &d_c1, &pitch_c1, sizeof(DATATYPE) * n, n/2);
    int bx= (n+threadnx-1)/threadnx;
    dim3 blocks(bx, bx/2);
    dim3 threads(threadnx, threadnx);
#pragma omp parallel num_threads(2)
    {
        int cpuid=omp_get_thread_num();
        if(cpuid==0)
        {
            cudaSetDevice(0);
gpu_gpu_matrix_multiplication(a,b,c1,d_a,d_b,d_c,pitch_a,pitch_b,pitch_c,
blocks,threads,n);
        }
        else if(cpuid==1)
        {
            cudaSetDevice(1);
```

```
            gpu_gpu_matrix_multiplication(a,b,c1,d_a1,d_b1,d_c1,pitch_a,pitch_b,pitch_c,
            blocks,threads,n);
                    }
                }
            ⋮
```

编译方法与 13.3.1 节类似，详细编译命令和 nvprof 执行结果如下。结果显示，kernel 函数和通信函数的执行已重叠。性能亦达到 477 Gflops。

```
[fangmq@mn0%yhstar omp_cuda]$ nvcc -c kernel.cu
[fangmq@mn0%yhstar omp_cuda]$ icc -openmp kernel.o omp_cuda.cpp -o omp_cuda -
I/vol/home/fangmq/gpu_cuda/cuda/cuda_5_5/include - L/vol/home/fangmq/gpu_
cuda/cuda/cuda_5_5/lib64 -lcudart
[fangmq@cn18%yhstar gpu_gpu]$ nvprof --print-gpu-trace ./omp_cuda_gpu
==21805==NVPROF is profiling process 21805, command: ./omp_cuda_gpu
16384    32    17153.040171    (ms)    477.582978    Gflops
==21805==Profiling application: ./omp_cuda_gpu
==21805==Profiling result:
   Start  Duration       Grid Size   Block Size     Regs*    SSMem*       DSMem*
Size   Throughput    Device     Context    Stream    Name
481.52ms  166.30ms  -    -    -    -    536.87MB   3.2284GB/s   Tesla K20c (0)   1
   2   [CUDA memcpy HtoD]
481.67ms  203.79ms  -    -    -    -    536.87MB   2.6344GB/s   Tesla K20c (1)   2
   2   [CUDA memcpy HtoD]
648.14ms  322.07ms  -    -    -    -    1.0737GB   3.3339GB/s   Tesla K20c (0)   1
   2   [CUDA memcpy HtoD]
685.90ms  436.12ms  -    -    -    -    1.0737GB   2.4620GB/s   Tesla K20c (1)   2
   2   [CUDA memcpy HtoD]
970.66ms  16.2636s   (512 256 1)   (32 32 1)   24   8.1920KB   0B   -   -   Tesla
K20c (0)   1   2   matrix_multiplication_gpu_4_1(float const*, unsigned long,
float const *, unsigned long, float *, unsigned long, int) [184]
1.12239s  16.2680s   (512 256 1)   (32 32 1)   24   8.1920KB   0B   -   -   Tesla
K20c (1)   2   2   matrix_multiplication_gpu_4_1(float const*, unsigned long,
float const *, unsigned long, float *, unsigned long, int) [194]
17.2342s  255.72ms  -    -    -    -    536.87MB   2.0994GB/s   Tesla K20c (0)   1
   2   [CUDA memcpy DtoH]
17.3904s  240.99ms  -    -    -    -    536.87MB   2.2278GB/s   Tesla K20c (1)   2
   2   [CUDA memcpy DtoH]

Regs: Number of registers used per CUDA thread.
SSMem: Static shared memory allocated per CUDA block.
DSMem: Dynamic shared memory allocated per CUDA block.
```

18.3.3 MPI+CUDA

若两个 GPU 位于不同节点,上述两种方法就不可行,此时需要引入 MPI 与 CUDA 混合编程,实现多节点上 GPU 的协同计算。笔者没有多节点多 GPU 环境,这里仅在单节点两个 GPU 上模拟执行。

下面给出 MPI+CUDA 混合编程矩阵乘法的详细代码。代码中,两个 MPI 进程分别读取文件,运算后分别写文件,其中,fseek()函数用来定位文件位置。

```
/****************************************************************/
/* kernel.cu                                                    */
/****************************************************************/
#include<cuda_runtime.h>
#define threadnx 32
#define DATATYPE float
__global__ void matrix_multiplication_gpu_4_1(const DATATYPE *a, size_t lda,
const DATATYPE *b, size_t ldb, DATATYPE *c, size_t ldc, int n){ … }
void gpu_gpu_matrix_multiplication(DATATYPE * a, DATATYPE * b, DATATYPE * c1,
DATATYPE *d_a,DATATYPE *d_b,DATATYPE *d_c1,size_t pitch_a,size_t pitch_b,size_
t pitch_c,dim3 blocks,dim3 threads, int n)
{
    cudaMemcpy2D(d_a, pitch_a, a, sizeof(DATATYPE) * n, sizeof(DATATYPE) * n,
    n/2, cudaMemcpyHostToDevice);
    cudaMemcpy2D(d_b, pitch_b, b, sizeof(DATATYPE) * n, sizeof(DATATYPE) * n,
    n, cudaMemcpyHostToDevice);
    matrix_multiplication_gpu_4_1<<<blocks,threads>>>(d_a,pitch_a/sizeof
    (DATATYPE),d_b,pitch_b/sizeof(DATATYPE),d_c1,pitch_c/sizeof(DATATYPE),
    n);
    cudaMemcpy2D(c1, sizeof(DATATYPE) * n, d_c1, pitch_c, sizeof(DATATYPE) *
    n, n/2, cudaMemcpyDeviceToHost);
}
/****************************************************************/
/* mpi_cuda.cpp                                                 */
/****************************************************************/
#include<stdlib.h>
#include<stdio.h>
#define DATATYPE float
#include<mpi.h>
#include<cuda_runtime.h>
#define threadnx 32
#include<sys/time.h>
#define TIMETESTEVENT
double dtime()
```

```c
{
    double tseconds=0.0;
    struct timeval mytime;
    gettimeofday(&mytime,(struct timezone *)0);
    tseconds=(double)(mytime.tv_sec+mytime.tv_usec*1.0e-6);
    return tseconds;
}
void gpu_gpu_matrix_multiplication(DATATYPE *a,DATATYPE *b,DATATYPE *c1,
DATATYPE *d_a,DATATYPE *d_b,DATATYPE *d_c1,size_t pitch_a,size_t pitch_b,size_t pitch_c,dim3 blocks,dim3 threads, int n);
int main(int argc, char *argv[])
{
    MPI_Init(&argc,&argv);
    int rank,size;
    MPI_Status status;
    MPI_Comm_rank(MPI_COMM_WORLD,&rank);
    MPI_Comm_size(MPI_COMM_WORLD,&size);
    int n=16384;
    DATATYPE *a,*b,*c1;
    a=(DATATYPE *)malloc(sizeof(DATATYPE) * n * n/2);
    b=(DATATYPE *)malloc(sizeof(DATATYPE) * n * n);
    c1=(DATATYPE *)malloc(sizeof(DATATYPE) * n * n/2);
    FILE *dat_a,*dat_b;
    dat_a=fopen("a.dat","rb");
    fseek(dat_a,sizeof(DATATYPE) * rank * n * n/2,SEEK_SET);
    fread(a,sizeof(DATATYPE),n * n/2,dat_a);
    fclose(dat_a);
    dat_b=fopen("b.dat","rb");
    fread(b,sizeof(DATATYPE),n * n,dat_b);
    fclose(dat_b);
    cudaSetDevice(rank);
    DATATYPE *d_a,*d_b,*d_c;
    size_t pitch_a, pitch_b, pitch_c;
    cudaMallocPitch((void **) &d_a, &pitch_a, sizeof(DATATYPE) * n, n/2);
    cudaMallocPitch((void **) &d_b, &pitch_b, sizeof(DATATYPE) * n, n);
    cudaMallocPitch((void **) &d_c, &pitch_c, sizeof(DATATYPE) * n, n/2);
    int bx=(n+threadnx-1)/threadnx;
    dim3 blocks(bx, bx/2);
    dim3 threads(threadnx, threadnx);
    MPI_Barrier(MPI_COMM_WORLD);
#ifdef TIMETESTEVENT
```

```
    double time_0,time_1;
    time_0=dtime();
#endif
    gpu_gpu_matrix_multiplication(a,b,c1,d_a,d_b,d_c,pitch_a,pitch_b,pitch_
    c,blocks,threads,n);
    MPI_Barrier(MPI_COMM_WORLD);
#ifdef TIMETESTEVENT
    time_1=dtime();
    if(rank==0)
    {
    printf("%d\t %d\t %f\t (ms)\t%f\tGflops\n",n,threadnx,(time_1-time_0) *
    1000,(float)((float)n * (float)n * (float)n * 2.0f)/1024.0f/1024.0f/1024.0f/
    (time_1-time_0));
    }
#endif
    cudaFree(d_a);
    cudaFree(d_b);
    cudaFree(d_c);
    FILE *dat_c;
    dat_c=fopen("c_mpi_cuda.dat","wb");
    fseek(dat_c,sizeof(DATATYPE) * rank * n * n/2,SEEK_SET);
    fwrite(c1,sizeof(DATATYPE),n * n/2,dat_c);
    fclose(dat_c);
    MPI_Barrier(MPI_COMM_WORLD);
    free(a);
    free(b);
    free(c1);
    MPI_Finalize();
    return 0;
}
```

下面是 MPI+CUDA 代码的编译命令和执行结果,在两个 GPU 上获得了 470 Gflops 的浮点运算性能,仅比 OpenMP+CUDA 稍差。

```
[fangmq@mn0%yhstar mpi_cuda]$ nvcc -c kernel.cu
[fangmq@mn0%yhstar mpi_cuda]$ mpicc kernel.o mpi_cuda.cpp -o mpi_cuda -I/vol/
home/fangmq/gpu_cuda/cuda/cuda_5_5/include -L/vol/home/fangmq/gpu_cuda/
cuda/cuda_5_5/lib64 -lcudart
[fangmq@mn0%yhstar gpu_gpu]$ yhrun -n 2 -p gpu ./mpi_cuda_gpu
16384    32    17393.656969    (ms)    470.976288    Gflops
```

18.4 CPUs/GPUs 协同

18.2 节利用 CPU 和 GPU 协同计算,18.3 节利用 GPU 和 GPU 协同,本节利用多 CPU 和多 GPU 协同计算。本质上是综合 18.2 节和 18.3 节内容,但在编程实现时并不是简单平移。下面分 3 小节阐述。

18.4.1 CUDA 版本

逻辑上只需在 18.3.1 节的 CUDA 版本的 kernel 函数启动语句后添加 OpenMP 多线程并行执行区域即可实现 CPUs 和 GPUs 的协同计算。

```
    ⋮
    cudaSetDevice(0);
    matrix_multiplication_gpu_4_1<<<blocks,threads>>>(d_a,pitch_a/sizeof
(DATATYPE),d_b,pitch_b/sizeof(DATATYPE),d_c,pitch_c/sizeof(DATATYPE),n);
    cudaSetDevice(1);
    matrix_multiplication_gpu_4_1<<<blocks,threads>>>(d_a1,pitch_a/sizeof
(DATATYPE),d_b1,pitch_b/sizeof(DATATYPE),d_c1,pitch_c/sizeof(DATATYPE),n);
    matmult(&a[gpu_n*n*2],b,&c1[gpu_n*n*2],cpu_n,n,n);
    ⋮
```

但由于 nvcc 和 icc 编译器不兼容,需用 3 个文件封装上述代码,详细代码如下。其中,根据 CPU、GPU 数量重新做了任务分割。

```
/****************************************************************/
/* cpukernel.cpp                                                */
/****************************************************************/
#include<stdlib.h>
#include<stdio.h>
#include<omp.h>
#define DATATYPE float
void matmult(DATATYPE *a,DATATYPE *b, DATATYPE *c, int m, int n,int l) {//a:m 行 l
列,b:l 行 n 列的转置,c:m 行 n 列
    int i, j, k;
    double temp;
    DATATYPE *b1=(DATATYPE *) malloc(sizeof(DATATYPE) * l * n);
#pragma omp parallel for private(j)
    for(i=0;i<n;i++)
    {
        for(j=0;j<l;j++)
        {
            b1[i*l+j]=b[j*n+i];
```

```
            }
        }
#pragma omp parallel for private(j,k,temp)
    for(i=0; i<m; i++) {
        for(j=0; j<n; j++) {
            temp=0;
            for(k=0; k<l; k++) {
                temp+=a[i * l+k] * b1[j * n+k];
            }
            c[i * n+j]=temp;
        }
    }
    free(b1);
}
```

```
/**********************************************************************/
/* kernel.cu                                                          */
/**********************************************************************/
#include<cuda_runtime.h>
#define threadnx 32
#define DATATYPE float
__global__ void matrix_multiplication_gpu_4_1(const DATATYPE *a, size_t lda,
const DATATYPE *b, size_t ldb, DATATYPE *c, size_t ldc, int n){
    __shared__ DATATYPE matA[threadnx][threadnx];
    __shared__ DATATYPE matB[threadnx][threadnx];
    const int tidc=threadIdx.x;
    const int tidr=threadIdx.y;
    const int bidc=blockIdx.x * threadnx;
    const int bidr=blockIdx.y * threadnx;
    int i, j;
    float results=0.0;
    float comp=0.0;
    float t;
    for(j=0; j<n; j+=threadnx)
    {
        matA[tidr][tidc]=a[(tidr+bidr) * lda+tidc+j];
        matB[tidr][tidc]=b[(tidr+j) * ldb+tidc+bidc];
        __syncthreads();
        for(i=0; i<threadnx; i++)
        {
            comp -=matA[tidr][i] * matB[i][tidc];
            t=results-comp;
            comp=(t-results)+comp;
            results=t;
```

```
            }
            __syncthreads();
        }
        if(tidr+bidr<n && tidc+bidc<n)
        {
            c[(tidr+bidr)*ldc+tidc+bidc]=results;
        }
}
void matmult(DATATYPE *a,DATATYPE *b, DATATYPE *c, int m, int n,int l);
void cpu_gpu_matrix_multiplication(DATATYPE * a, DATATYPE * b, DATATYPE * c1,
DATATYPE *d_a, DATATYPE *d_b, DATATYPE *d_c, DATATYPE *d_a1, DATATYPE *d_b1,
DATATYPE *d_c1, size_t pitch_a,size_t pitch_b,size_t pitch_c,dim3 blocks,dim3
threads, int gpu_n,int cpu_n,int n)
{
    cudaSetDevice(0);
    cudaMemcpy2D(d_a, pitch_a, a, sizeof(DATATYPE) * n, sizeof(DATATYPE) * n,
    gpu_n, cudaMemcpyHostToDevice);
    cudaMemcpy2D(d_b, pitch_b, b, sizeof(DATATYPE) * n, sizeof(DATATYPE) * n,
    n, cudaMemcpyHostToDevice);
    cudaSetDevice(1);
    cudaMemcpy2D(d_a1, pitch_a, &a[n * gpu_n], sizeof(DATATYPE) * n, sizeof
    (DATATYPE) * n, gpu_n, cudaMemcpyHostToDevice);
    cudaMemcpy2D(d_b1, pitch_b, b, sizeof(DATATYPE) * n, sizeof(DATATYPE) * n,
    n, cudaMemcpyHostToDevice);
    cudaSetDevice(0);
    matrix_multiplication_gpu_4_1<<<blocks,threads>>>(d_a,pitch_a/sizeof
    (DATATYPE),d_b,pitch_b/sizeof(DATATYPE),d_c,pitch_c/sizeof(DATATYPE),n);
    cudaSetDevice(1);
    matrix_multiplication_gpu_4_1<<<blocks,threads>>>(d_a1,pitch_a/sizeof
    (DATATYPE),d_b1,pitch_b/sizeof(DATATYPE),d_c1,pitch_c/sizeof(DATATYPE),
    n);
    matmult(&a[gpu_n*n*2],b,&c1[gpu_n*n*2],cpu_n,n,n);
    cudaSetDevice(0);
    cudaMemcpy2D(c1, sizeof(DATATYPE) * n, d_c, pitch_c, sizeof(DATATYPE) * n,
    gpu_n, cudaMemcpyDeviceToHost);
    cudaSetDevice(1);
    cudaMemcpy2D(&c1[n * gpu_n], sizeof(DATATYPE) * n, d_c1, pitch_c, sizeof
    (DATATYPE) * n, gpu_n, cudaMemcpyDeviceToHost);
}
/******************************************************************/
/* gpu.cpp                                                        */
/******************************************************************/
#include<stdlib.h>
```

```c
#include<stdio.h>
#define DATATYPE float
//#define PRINTNEED
#include<omp.h>
#include<cuda_runtime.h>
#define threadnx 32
#include<sys/time.h>
#define TIMETESTEVENT
double dtime()
{
    double tseconds=0.0;
    struct timeval mytime;
    gettimeofday(&mytime,(struct timezone *)0);
    tseconds=(double)(mytime.tv_sec+mytime.tv_usec*1.0e-6);
    return tseconds;
}
void cpu_gpu_matrix_multiplication(DATATYPE *a,DATATYPE *b,DATATYPE *c1,
DATATYPE *d_a,DATATYPE *d_b,DATATYPE *d_c,DATATYPE *d_a1,DATATYPE *d_b1,
DATATYPE *d_c1,size_t pitch_a,size_t pitch_b,size_t pitch_c,dim3 blocks,dim3
threads,int gpu_n,int cpu_n,int n);

int main()
{
    int n=16384;
    DATATYPE *a,*b,*c1;
    a=(DATATYPE *)malloc(sizeof(DATATYPE)*n*n);
    b=(DATATYPE *)malloc(sizeof(DATATYPE)*n*n);
    c1=(DATATYPE *)malloc(sizeof(DATATYPE)*n*n);
    FILE *dat_a,*dat_b;
    dat_a=fopen("a.dat","rb");
    fread(a,sizeof(DATATYPE),n*n,dat_a);
    fclose(dat_a);
    dat_b=fopen("b.dat","rb");
    fread(b,sizeof(DATATYPE),n*n,dat_b);
    fclose(dat_b);
    int gpu_gflops=238;
    int cpu_gflops=42;
    int gpu_n=(gpu_gflops*n/(gpu_gflops*2+cpu_gflops)+threadnx-1)/threadnx
    *threadnx;
    int cpu_n=n-gpu_n*2;
    printf("gpu_n=%d\tcpu_n=%d\n",gpu_n,cpu_n);
    cudaSetDevice(0);
    DATATYPE *d_a,*d_b,*d_c;
```

```
        size_t pitch_a, pitch_b, pitch_c;
        cudaMallocPitch((void **) &d_a, &pitch_a, sizeof(DATATYPE) * n, gpu_n);
        cudaMallocPitch((void **) &d_b, &pitch_b, sizeof(DATATYPE) * n, n);
        cudaMallocPitch((void **) &d_c, &pitch_c, sizeof(DATATYPE) * n, gpu_n);
        cudaSetDevice(1);
        DATATYPE *d_a1,*d_b1,*d_c1;
        size_t pitch_a1, pitch_b1, pitch_c1;
        cudaMallocPitch((void **) &d_a1, &pitch_a1, sizeof(DATATYPE) * n, gpu_n);
        cudaMallocPitch((void **) &d_b1, &pitch_b1, sizeof(DATATYPE) * n, n);
        cudaMallocPitch((void **) &d_c1, &pitch_c1, sizeof(DATATYPE) * n, gpu_n);
        int bx=(n+threadnx-1)/threadnx;
        dim3 blocks(bx, (gpu_n/threadnx));
        dim3 threads(threadnx, threadnx);
#ifdef TIMETESTEVENT
        double time_0,time_1;
        time_0=dtime();
#endif
        cpu_gpu_matrix_multiplication(a,b,c1,d_a,d_b,d_c,d_a1,d_b1,d_c1,pitch_a,
        pitch_b,pitch_c,blocks,threads,gpu_n,cpu_n,n);
#ifdef TIMETESTEVENT
        time_1=dtime();
        printf("%d\t %d\t %f\t(ms) \t%f\tGflops\n",n,threadnx,(time_1-time_0) *
        1000,(float)((float)n * (float)n * (float)n * 2.0f)/1024.0f/1024.0f/1024.0f/
        (time_1-time_0));
#endif
        cudaFree(d_a);
        cudaFree(d_b);
        cudaFree(d_c);
        cudaFree(d_a1);
        cudaFree(d_b1);
        cudaFree(d_c1);
        FILE *dat_c;
        dat_c=fopen("c_gpu.dat","wb");
        fwrite(c1,sizeof(DATATYPE),n * n,dat_c);
        fclose(dat_c);
        free(a);
        free(b);
        free(c1);
        return 0;
}
```

编译并执行上述代码,实验结果如下,浮点运算性能为 475 Gflops,比原来的 459 Gflops 好。但仍不及 18.3.2 节 OpenMP+CUDA 版本,原因是多核 CPU 计算获得的收益无法(或刚好)弥补 GPU 的通信开销。

```
[fangmq@cn18%yhstar cpus_gpus]$./gpu_run
gpu_n=7552    cpu_n=1280
16384    32    17220.758915    (ms)    475.704935    Gflops
```

18.4.2 OpenMP+CUDA

本节基于 18.3.2 节 OpenMP+CUDA 代码,增加 CPU 并行运算。由于 18.1.1 节 CPU 基点代码中的转置和矩阵乘法是严格独立的,因此只能采取嵌套并行的手段。下面给出了修改后的 OpenMP+CUDA 代码,其中 omp_set_nested(1)指定了允许嵌套并行。

```c
/******************************************************************/
/* kernel.cu                                                    */
/******************************************************************/
#include<stdlib.h>
#include<stdio.h>
#define DATATYPE float
#include<cuda_runtime.h>
#define threadnx 32
__global__ void matrix_multiplication_gpu_4_1(const DATATYPE *a, size_t lda,
const DATATYPE *b, size_t ldb, DATATYPE *c, size_t ldc, int n){
    __shared__ DATATYPE matA[threadnx][threadnx];
    __shared__ DATATYPE matB[threadnx][threadnx];
    const int tidc=threadIdx.x;
    const int tidr=threadIdx.y;
    const int bidc=blockIdx.x *threadnx;
    const int bidr=blockIdx.y *threadnx;
    int i, j;
    float results=0.0;
    float comp=0.0;
    float t;
    for(j=0; j<n; j+=threadnx)
    {
        matA[tidr][tidc]=a[(tidr+bidr) * lda+tidc+j];
        matB[tidr][tidc]=b[(tidr+j) * ldb+tidc+bidc];
        __syncthreads();
        for(i=0; i<threadnx; i++)
        {
            comp -=matA[tidr][i] * matB[i][tidc];
            t=results-comp;
            comp=(t-results)+comp;
```

```
            results=t;
        }
        __syncthreads();
    }
    if(tidr+bidr<n && tidc+bidc<n)
    {
        c[(tidr+bidr) * ldc+tidc+bidc]=results;
    }
}
void gpu_gpu_matrix_multiplication(DATATYPE *a,DATATYPE *b,DATATYPE *c1,
DATATYPE *d_a,DATATYPE *d_b,DATATYPE *d_c1, size_t pitch_a,size_t pitch_b,size_
t pitch_c,dim3 blocks,dim3 threads, int gpu_n,int n)
{
    int gpuid;
    cudaGetDevice(&gpuid);
    if(gpuid==0)
    {
        cudaMemcpy2D(d_a, pitch_a, a, sizeof(DATATYPE) * n, sizeof(DATATYPE) *
        n, gpu_n, cudaMemcpyHostToDevice);
    }
    else if(gpuid==1)
    {
        cudaMemcpy2D(d_a, pitch_a, &a[n * gpu_n], sizeof(DATATYPE) * n, sizeof
        (DATATYPE) * n, gpu_n, cudaMemcpyHostToDevice);
    }
    cudaMemcpy2D(d_b, pitch_b, b, sizeof(DATATYPE) * n, sizeof(DATATYPE) * n,
n, cudaMemcpyHostToDevice);

    matrix_multiplication_gpu_4_1<<<blocks,threads>>>(d_a,pitch_a/sizeof
    (DATATYPE),d_b,pitch_b/sizeof(DATATYPE),d_c1,pitch_c/sizeof(DATATYPE),
    n);

    if(gpuid==0)
    {
        cudaMemcpy2D(c1, sizeof(DATATYPE) * n, d_c1, pitch_c, sizeof(DATATYPE)
        * n, gpu_n, cudaMemcpyDeviceToHost);
    }
    else if(gpuid==1)
    {
        cudaMemcpy2D(&c1[n * gpu_n], sizeof(DATATYPE) * n, d_c1, pitch_c,
        sizeof(DATATYPE) * n, gpu_n, cudaMemcpyDeviceToHost);
    }
}
```

```cpp
/***********************************************************************/
/* omp_cuda.cpp                                                      */
/***********************************************************************/
#include<stdlib.h>
#include<stdio.h>
#define DATATYPE float
//#define PRINTNEED
#include<omp.h>
#include<cuda_runtime.h>
#define threadnx 32
#include<sys/time.h>
#define TIMETESTEVENT
double dtime()
{
    double tseconds=0.0;
    struct timeval mytime;
    gettimeofday(&mytime,(struct timezone *)0);
    tseconds=(double)(mytime.tv_sec+mytime.tv_usec*1.0e-6);
    return tseconds;
}
void gpu_gpu_matrix_multiplication(DATATYPE *a,DATATYPE *b,DATATYPE *c1,
DATATYPE *d_a,DATATYPE *d_b,DATATYPE *d_c1,size_t pitch_a,size_t pitch_b,size_
t pitch_c,dim3 blocks,dim3 threads, int gpu_n,int n);
void matmult(DATATYPE *a,DATATYPE *b,DATATYPE *c, int m, int n,int l) {//a:m行l
列,b:l行n列的转置,c:m行n列
    int i, j, k;
    double temp;
    DATATYPE * b1=(DATATYPE *) malloc(sizeof(DATATYPE) * l * n);
#pragma omp parallel for private(j) num_threads(14)
    for(i=0;i<n;i++)
    {
        for(j=0;j<l;j++)
        {
            b1[i*l+j]=b[j*n+i];
        }
    }
#pragma omp parallel for private(j,k,temp) num_threads(14)
    for(i=0; i<m; i++) {
        for(j=0; j<n; j++) {
            temp=0;
            for(k=0; k<l; k++) {
```

```
                temp+=a[i*l+k]*b1[j*n+k];
            }
            c[i*n+j]=temp;
        }
    }
    free(b1);
}
int main()
{
    int n=16384;
    DATATYPE *a,*b,*c1;
    a=(DATATYPE *)malloc(sizeof(DATATYPE) * n * n);
    b=(DATATYPE *)malloc(sizeof(DATATYPE) * n * n);
    c1=(DATATYPE *)malloc(sizeof(DATATYPE) * n * n);
    FILE *dat_a,*dat_b;
    dat_a=fopen("a.dat","rb");
    fread(a,sizeof(DATATYPE),n*n,dat_a);
    fclose(dat_a);
    dat_b=fopen("b.dat","rb");
    fread(b,sizeof(DATATYPE),n*n,dat_b);
    fclose(dat_b);
    int gpu_gflops=238;
    int cpu_gflops=42;
    int gpu_n=(gpu_gflops*n/(gpu_gflops*2+cpu_gflops)+threadnx-1)/threadnx
    *threadnx;
    int cpu_n=n-gpu_n*2;
    printf("gpu_n=%d\tcpu_n=%d\n",gpu_n,cpu_n);
    cudaSetDevice(0);
    DATATYPE *d_a,*d_b,*d_c;
    size_t pitch_a, pitch_b, pitch_c;
    cudaMallocPitch((void **) &d_a, &pitch_a, sizeof(DATATYPE) * n, gpu_n);
    cudaMallocPitch((void **) &d_b, &pitch_b, sizeof(DATATYPE) * n, n);
    cudaMallocPitch((void **) &d_c, &pitch_c, sizeof(DATATYPE) * n, gpu_n);
    cudaSetDevice(1);
    DATATYPE *d_a1,*d_b1,*d_c1;
    size_t pitch_a1, pitch_b1, pitch_c1;
    cudaMallocPitch((void **) &d_a1, &pitch_a1, sizeof(DATATYPE) * n, gpu_n);
    cudaMallocPitch((void **) &d_b1, &pitch_b1, sizeof(DATATYPE) * n, n);
    cudaMallocPitch((void **) &d_c1, &pitch_c1, sizeof(DATATYPE) * n, gpu_n);
    int bx=(n+threadnx-1)/threadnx;
    dim3 blocks(bx, (gpu_n/threadnx));
```

```
        dim3 threads(threadnx,threadnx);
#ifdef TIMETESTEVENT
    double time_0,time_1;
    time_0=dtime();
#endif
    double temp;
    DATATYPE *b1=(DATATYPE *) malloc(sizeof(DATATYPE) * n * n);
    omp_set_nested(1);
#pragma omp parallel num_threads(3)
    {
        int i, j, k;
        int cpuid=omp_get_thread_num();
        if(cpuid==0)
        {
            cudaSetDevice(0);
gpu_gpu_matrix_multiplication(a,b,c1,d_a,d_b,d_c,pitch_a,pitch_b,pitch_c,
blocks,threads,gpu_n,n);
        }
        else if(cpuid==1)
        {
            cudaSetDevice(1);
gpu_gpu_matrix_multiplication(a,b,c1,d_a1,d_b1,d_c1,pitch_a,pitch_b,pitch_c,
blocks,threads,gpu_n,n);
        }
        else
        {
            matmult(&a[n * gpu_n * 2],b,&c1[n * gpu_n * 2],cpu_n,n,n);
        }
    }
    free(b1);
#ifdef TIMETESTEVENT
    time_1=dtime();
    printf("%d\t %d\t %f\t (ms)\t%f\tGflops\n",n,threadnx,(time_1-time_0) *
1000,(float)((float)n * (float)n * (float)n * 2.0f)/1024.0f/1024.0f/1024.0f/
(time_1-time_0));
#endif
    cudaFree(d_a);
    cudaFree(d_b);
    cudaFree(d_c);
    cudaFree(d_a1);
    cudaFree(d_b1);
```

```
    cudaFree(d_c1);
    FILE *dat_c;
    dat_c=fopen("c_omp_cuda.dat","wb");
    fwrite(c1,sizeof(DATATYPE),n*n,dat_c);
    fclose(dat_c);
    free(a);
    free(b);
    free(c1);
    return 0;
}
```

编译上述代码并执行,执行结果如下。可见增加了 CPU 计算后性能反而远远不如 18.3.2 节两个 GPU 的版本。可能的原因是嵌套的性能损失较大。

```
[fangmq@cn18%yhstar cpus_gpus]$./omp_cuda_run
gpu_n=7552    cpu_n=1280
16384    32    18549.577951    (ms)    441.627299    Gflops
```

18.4.3　MPI＋OpenMP＋CUDA

本节在 18.3.3 节 MPI＋CUDA 混合编程实现多 GPU 协同计算基础上,增加利用 OpenMP 实现的 CPU 多核并行计算,实现 MPI＋OpenMP＋CUDA 三者协同计算。下面给出了相应的代码,由于是在单节点双 GPU 上模拟执行,因此 OpenMP 的线程数量指定为节点总线程的一半(即 8 个线程)。

```
/******************************************************************/
/* cpukernel.cpp                                                  */
/******************************************************************/
#include<stdlib.h>
#include<stdio.h>
#include<omp.h>
#define DATATYPE float
void matmult(DATATYPE *a,DATATYPE *b, DATATYPE *c, int m, int n,int l) {//a:m行l
列,b:l 行 n 列的转置,c:m 行 n 列
    int i, j, k;
    double temp;
    DATATYPE *b1= (DATATYPE *) malloc(sizeof(DATATYPE) * l * n);
#pragma omp parallel for private(j) num_threads(8)
    for(i=0;i<n;i++)
```

```
            {
                for(j=0;j<l;j++)
                {
                    b1[i*l+j]=b[j*n+i];
                }
            }
#pragma omp parallel for private(j,k,temp) num_threads(8)
        for(i=0; i<m; i++) {
            for(j=0; j<n; j++) {
                temp=0;
                for(k=0; k<l; k++) {
                    temp+=a[i*l+k] * b1[j*n+k];
                }
                c[i*n+j]=temp;
            }
        }
        free(b1);
}
/************************************************************************/
/* kernel.cu                                                          */
/************************************************************************/
#include<cuda_runtime.h>
#define threadnx 32
#define DATATYPE float
__global__ void matrix_multiplication_gpu_4_1(const DATATYPE *a, size_t lda,
const DATATYPE *b, size_t ldb, DATATYPE*c, size_t ldc, int n){
    __shared__ DATATYPE matA[threadnx][threadnx];
    __shared__ DATATYPE matB[threadnx][threadnx];
    const int tidc=threadIdx.x;
    const int tidr=threadIdx.y;
    const int bidc=blockIdx.x * threadnx;
    const int bidr=blockIdx.y * threadnx;
    int i, j;
    float results=0.0;
    float comp=0.0;
    float t;
    for(j=0; j<n; j+=threadnx)
    {
        matA[tidr][tidc]=a[(tidr+bidr) * lda+tidc+j];
        matB[tidr][tidc]=b[(tidr+j) * ldb+tidc+bidc];
        __syncthreads();
```

```
            for(i=0; i<threadnx; i++)
            {
                comp-=matA[tidr][i]*matB[i][tidc];
                t=results-comp;
                comp=(t-results)+comp;
                results=t;
            }
            __syncthreads();
    }
    if(tidr+bidr<n && tidc+bidc<n)
    {
        c[(tidr+bidr)*ldc+tidc+bidc]=results;
    }
}
void matmult(DATATYPE *a,DATATYPE *b, DATATYPE *c, int m, int n,int l);
void gpu_gpu_matrix_multiplication(DATATYPE * a, DATATYPE * b, DATATYPE * c1,
DATATYPE *d_a,DATATYPE *d_b,DATATYPE *d_c1,size_t pitch_a,size_t pitch_b,size_
t pitch_c,dim3 blocks,dim3 threads, int gpu_n,int cpu_n,int n)
{
    cudaMemcpy2D(d_a, pitch_a, a, sizeof(DATATYPE) * n, sizeof(DATATYPE) * n,
      gpu_n, cudaMemcpyHostToDevice);
    cudaMemcpy2D(d_b, pitch_b, b, sizeof(DATATYPE) * n, sizeof(DATATYPE) * n,
      n, cudaMemcpyHostToDevice);
    matrix_multiplication_gpu_4_1<<<blocks,threads>>>(d_a,pitch_a/sizeof
      (DATATYPE),d_b,pitch_b/sizeof(DATATYPE),d_c1,pitch_c/sizeof(DATATYPE),
      n);
    matmult(&a[gpu_n*n],b,&c1[gpu_n*n],cpu_n,n,n);
    cudaMemcpy2D(c1, sizeof(DATATYPE) * n, d_c1, pitch_c, sizeof(DATATYPE) *
n, gpu_n, cudaMemcpyDeviceToHost);
}
/*******************************************************************/
/* mpi_omp_cuda.cpp                                                */
/*******************************************************************/
#include<stdlib.h>
#include<stdio.h>
#define DATATYPE float
//#define PRINTNEED
#include<mpi.h>
#include<cuda_runtime.h>
#define threadnx 32
#include<sys/time.h>
```

```
#define TIMETESTEVENT
double dtime()
{
    double tseconds=0.0;
    struct timeval mytime;
    gettimeofday(&mytime,(struct timezone *)0);
    tseconds=(double)(mytime.tv_sec+mytime.tv_usec * 1.0e-6);
    return tseconds;
}
void gpu_gpu_matrix_multiplication(DATATYPE *a,DATATYPE *b,DATATYPE *c1,
DATATYPE *d_a,DATATYPE *d_b,DATATYPE *d_c1,size_t pitch_a,size_t pitch_b,size_
t pitch_c,dim3 blocks,dim3 threads, int gpu_n,int cpu_n,int n);
int main(int argc, char *argv[])
{
    MPI_Init(&argc,&argv);
    int rank,size;
    MPI_Status status;
    MPI_Comm_rank(MPI_COMM_WORLD,&rank);
    MPI_Comm_size(MPI_COMM_WORLD,&size);
    int n=16384;
    DATATYPE *a,*b,*c1;
    a=(DATATYPE *)malloc(sizeof(DATATYPE) * n * n/2);
    b=(DATATYPE *)malloc(sizeof(DATATYPE) * n * n);
    c1=(DATATYPE *)malloc(sizeof(DATATYPE) * n * n/2);
    FILE *dat_a,*dat_b;
    dat_a=fopen("a.dat","rb");
    fseek(dat_a,sizeof(DATATYPE) * rank * n * n/2,SEEK_SET);
    fread(a,sizeof(DATATYPE),n * n/2,dat_a);
    fclose(dat_a);
    dat_b=fopen("b.dat","rb");
    fread(b,sizeof(DATATYPE),n * n,dat_b);
    fclose(dat_b);
    int gpu_gflops=238;
    int cpu_gflops=42;
    int gpu_n=(gpu_gflops * n/(gpu_gflops * 2+cpu_gflops)+threadnx-1)/threadnx
    *threadnx;
    int cpu_n=n/2-gpu_n;//half
    printf("rank=%d\tgpu_n=%d\tcpu_n=%d\n",rank,gpu_n,cpu_n);
    cudaSetDevice(rank);
    DATATYPE *d_a,*d_b,*d_c;
    size_t pitch_a, pitch_b, pitch_c;
```

```
        cudaMallocPitch((void **) &d_a, &pitch_a, sizeof(DATATYPE) * n, gpu_n);
        cudaMallocPitch((void **) &d_b, &pitch_b, sizeof(DATATYPE) * n, n);
        cudaMallocPitch((void **) &d_c, &pitch_c, sizeof(DATATYPE) * n, gpu_n);
        int bx= (n+threadnx-1)/threadnx;
        dim3 blocks(bx, gpu_n/threadnx);
        dim3 threads(threadnx, threadnx);
        MPI_Barrier(MPI_COMM_WORLD);
#ifdef TIMETESTEVENT
        double time_0,time_1;
        time_0=dtime();
#endif
        gpu_gpu_matrix_multiplication(a,b,c1,d_a,d_b,d_c,pitch_a,pitch_b,pitch_
        c,blocks,threads,gpu_n,cpu_n,n);
        MPI_Barrier(MPI_COMM_WORLD);
#ifdef TIMETESTEVENT
        time_1=dtime();
        if(rank==0)
        {
printf("%d\t %d\t %f\t (ms) \t%f\tGflops\n",n,threadnx,(time_1-time_0) * 1000,
(float)((float)n * (float)n * (float)n * 2.0f)/1024.0f/1024.0f/1024.0f/(time_1
-time_0));
        }
#endif
        cudaFree(d_a);
        cudaFree(d_b);
        cudaFree(d_c);
        FILE *dat_c;
        dat_c=fopen("c_mpi_cuda.dat","wb");
        fseek(dat_c,sizeof(DATATYPE) * rank * n * n/2,SEEK_SET);
        fwrite(c1,sizeof(DATATYPE),n * n/2,dat_c);
        fclose(dat_c);
        MPI_Barrier(MPI_COMM_WORLD);
        free(a);
        free(b);
        free(c1);
        MPI_Finalize();
        return 0;
}
```

上述代码的编译命令和执行结果如下。编译命令包含了 icc、nvcc、mpicc 3 种编译器的混合编译，mpicc 编译、链接时需要添加 nvcc 的头文件目录和库文件目录。执行结果显示，MPI+OpenMP+CUDA 程序获得了 481 Gflops 性能，为上述所有实验结果中的最大值。

```
$ icc -c cpukernel.cpp -openmp
$ nvcc -c kernel.cu
$ mpicc -openmp cpukernel.o kernel.o mpi_cuda.cpp -o mpi_cuda -I/vol/home/
fangmq/gpu_cuda/cuda/cuda_5_5/include -L/vol/home/fangmq/gpu_cuda/cuda/cuda
_5_5/lib64 -lcudart
[fangmq@mn0%yhstar cpus_gpus]$ yhrun -N 1 -n 2 -p gpu -c 8 ./mpi_cuda_run
rank=1    gpu_n=7552    cpu_n=640
rank=0    gpu_n=7552    cpu_n=640
16384    32    17024.771929    (ms)    481.181189    Gflops
```

注意：若采用 gcc 编译 cpukernel.cpp，-openmp 选项需改为-fopenmp，mpicc 编译报错时，可改为 mpicxx。

18.5 本章小结

本章以矩阵乘法为对象，研究了 CPUs 与 GPUs 协同运算，取得了较理想的效果，但仍存在一些性能缺陷，有待进一步研究（见表 18.1）。在协同计算中，CPU 和 GPU 的任务划分并不能简单地用浮点运算性能计算，还需考虑其他因素，比如 CPU 矩阵乘法的转置运算量、GPU 运算中的数据通信等。有兴趣的读者可深入研究，令 CPUs、GPUs 负载均衡。本章更大的意义是为读者的混合编程提供借鉴，从这个角度看继续调优已然不再重要。

表 18.1 CPUs 和 GPUs 协同矩阵乘性能对比

	版 本	Gflops
	CPU	42.16
	GPU	238.13
	CPU+GPU	266.59
2GPU	CUDA	459.40
2GPU	OMP+CUDA	477.58
2GPU	MPI+CUDA	470.98
CPU+2GPU	CUDA	475.70
CPU+2GPU	OMP+CUDA	441.63
CPU+2GPU	MPI+OMP+CUDA	481.18

此外，本章并未涉及计算与通信重叠，其原因是矩阵乘法计算的通信开销相对计算开销可以忽略。本书 11.4 节给出了计算和通信重叠优化的实例。

附　录

附录 A　判断法 1D 卷积代码
附录 B　曼德博罗特集的系列优化代码
附录 C　几种图像处理完整源码
附录 D　nvprof 帮助菜单
附录 E　NVCC 帮助菜单
附录 F　几种排序算法源代码

附录 A

判断法 1D 卷积代码

附录 A.1 判断法 1D 卷积 basic 版

```c
#include<stdlib.h>
#include<stdio.h>
#include<cuda_runtime.h>
#define m 5
#define block_width 512
void vector_init(DATATYPE *a,int n)
{
    for(int i=0;i<n;i++)
    {
        a[i]=((DATATYPE) rand()/0x7fffff+(DATATYPE) rand()/
        ((DATATYPE)0x7fffff * 0x7fffff))/1000;
    }
}
__global__ void convolution_1d_basic_kernel(float *array_m,
float *array_n,float *array_p,int n)
{
    int i=blockIdx.x * blockDim.x+threadIdx.x;
    float pvalue=0.0;
    int n_start_point=i-(m/2);
    for(int j=0;j<m;j++)
    {
        if((n_start_point+j>=0)&&(n_start_point+j<n))
        {
            pvalue+=array_n[n_start_point+j] * array_m[j];
        }
    }
    array_p[i]=pvalue;
}
⋮
    int m_2=m/2;
```

```
    float *array_m,*array_n,*array_p1;
    array_m=(float *)malloc(sizeof(float) * m);
    array_n=(float *)malloc(sizeof(float) * n);
    array_p1=(float *)malloc(sizeof(float) * n);
    vector_init(array_m,m);
    vector_init(array_n,n);
    int grid_width=n/block_width;
    float * d_array_m, * d_array_n, * d_array_p1;
    cudaMalloc((void **) &d_array_m, sizeof(float) * m);
    cudaMalloc((void **) &d_array_n, sizeof(float) * n);
    cudaMalloc((void **) &d_array_p1, sizeof(float) * n);
    cudaMemcpy(d_array_m, array_m, sizeof(float) * m, cudaMemcpyHostToDevice);
    cudaMemcpy(d_array_n, array_n, sizeof(float) * n, cudaMemcpyHostToDevice);
    convolution_1d_basic_kernel<<<grid_width,block_width>>>(d_array_m,d_
    array_n,d_array_p1,n);
    cudaMemcpy(array_p1,d_array_p1, sizeof(float) * n, cudaMemcpyDeviceToHost);
    cudaFree(d_array_m);
    cudaFree(d_array_n);
    cudaFree(d_array_p1);
    free(array_m);
    free(array_n);
    free(array_p1);
    ⋮
```

附录 A.2　判断法 1D 卷积 constant 版

```
#include<stdlib.h>
#include<stdio.h>
#include<cuda_runtime.h>
#define m 5
#define block_width 512
void vector_init(DATATYPE *a,int n)
{
    for(int i=0;i<n;i++)
    {
        a[i]=((DATATYPE) rand()/0x7ffff+(DATATYPE) rand()/((DATATYPE)0x7ffff
        * 0x7ffff))/1000;
    }
}
```

附录 A 判断法 1D 卷积代码

```
__constant__ float d_array_m1[m];
__global__ void convolution_1d_const_kernel(float *array_n,float *array_p,int n)
{
    int i=blockIdx.x*blockDim.x+threadIdx.x;
    float pvalue=0.0;
    int n_start_point=i-(m/2);
    for(int j=0;j<m;j++)
    {
        if((n_start_point+j>=0)&&(n_start_point+j<n))
        {
            pvalue+=array_n[n_start_point+j] * d_array_m1[j];
        }
    }
    array_p[i]=pvalue;
}
⋮
    float *array_m,*array_n,*array_p2;
    int m_2=m/2;
    array_m=(float *)malloc(sizeof(float) * m);
    array_n=(float *)malloc(sizeof(float) * n);
    array_p2=(float *)malloc(sizeof(float) * n);
    vector_init(array_m,m);
    vector_init(array_n,n);
    int grid_width=n/block_width;
    float *d_array_n,*d_array_p1;
    cudaMalloc((void **) &d_array_n, sizeof(float) * n);
    cudaMalloc((void **) &d_array_p1, sizeof(float) * n);
    cudaMemcpy(d_array_n, array_n, sizeof(float) * n, cudaMemcpyHostToDevice);
    cudaMemcpyToSymbol(d_array_m1,array_m,sizeof(float) * m);
    convolution_1d_const_kernel<<<grid_width,block_width>>>(d_array_n,d_array_p1,n);
    cudaMemcpy(array_p2,d_array_p1, sizeof(float) * n, cudaMemcpyDeviceToHost);
    cudaFree(d_array_n);
    cudaFree(d_array_p1);
    free(array_m);
    free(array_n);
    free(array_p2);
⋮
```

附录 A.3 判断法 1D 卷积 shared 版

```c
#include<stdlib.h>
#include<stdio.h>
#include<cuda_runtime.h>
#define m 5
#define block_width 512
void vector_init(DATATYPE *a,int n)
{
    for(int i=0;i<n;i++)
    {
        a[i]=((DATATYPE) rand()/0x7ffff+(DATATYPE) rand()/((DATATYPE)0x7ffff
        * 0x7ffff))/1000;
    }
}
__constant__ float d_array_m1[m];
#define mm (block_width+m-1)
__global__ void convolution_1d_const_shared_kernel(float *array_n,float *
array_p,int n)
{
    int i=blockIdx.x * blockDim.x+threadIdx.x;
    __shared__ float array_ns[mm];
    int m2=m/2;
    int halo_index_left=(blockIdx.x-1) * blockDim.x+threadIdx.x;
    if(threadIdx.x>=blockDim.x-m2)
    {
        array_ns[threadIdx.x-(blockDim.x-m2)]=(halo_index_left<0)? 0:array_n
        [halo_index_left];
    }
    array_ns[threadIdx.x+m2]=array_n[blockIdx.x * blockDim.x+threadIdx.x];

    int halo_index_right=(blockIdx.x+1) * blockDim.x+threadIdx.x;
    if(threadIdx.x<m2)
    {
        array_ns[m2+blockDim.x+threadIdx.x]=(halo_index_right>=m2)?0:array_
        n[halo_index_right];
    }
    __syncthreads();

    float pvalue=0.0;
    for(int j=0;j<m;j++)
```

```
    {
        pvalue+=array_ns[threadIdx.x+j] * d_array_m1[j];
    }
    array_p[i]=pvalue;
}
...
    float *array_m,*array_n,*array_p3;
    int m_2=m/2;
    array_m=(float *)malloc(sizeof(float) * m);
    array_n=(float *)malloc(sizeof(float) * n);
    array_p3=(float *)malloc(sizeof(float) * n);
    vector_init(array_m,m);
    vector_init(array_n,n);
    int grid_width=n/block_width;
    float *d_array_n,*d_array_p1;
    cudaMalloc((void **) &d_array_n, sizeof(float) * n);
    cudaMalloc((void **) &d_array_p1, sizeof(float) * n);
    cudaMemcpy(d_array_n, array_n, sizeof(float) * n, cudaMemcpyHostToDevice);
    cudaMemcpyToSymbol(d_array_m1,array_m,sizeof(float) * m);
    convolution_1d_const_shared_kernel<<<grid_width,block_width>>>(d_array_n,d_array_p1,n);
    cudaMemcpy(array_p3,d_array_p1, sizeof(float) * n, cudaMemcpyDeviceToHost);
    cudaFree(d_array_n);
    cudaFree(d_array_p1);
    free(array_m);
    free(array_n);
    free(array_p3);
...
```

附录 A.4 判断法 1D 卷积 cache 版

```
#include<stdlib.h>
#include<stdio.h>
#include<cuda_runtime.h>
#define m 5
#define block_width 512
void vector_init(DATATYPE *a,int n)
{
    for(int i=0;i<n;i++)
    {
        a[i]=((DATATYPE) rand()/0x7ffff+(DATATYPE) rand()/((DATATYPE)0x7ffff
        * 0x7ffff))/1000;
```

```
        }
    }
    __constant__ float d_array_m1[m];
    #define mm (block_width+m-1)
    __global__ void convolution_1d_const_shared_cache_kernel(float *array_n,
    float *array_p,int n)
    {
        int i=blockIdx.x * blockDim.x+threadIdx.x;
        __shared__ float array_ns[block_width];
        array_ns[threadIdx.x]=array_n[i];
        __syncthreads();

        int this_tile_start_point=blockIdx.x * blockDim.x;
        int nest_tile_start_point= (blockIdx.x+1) * blockDim.x;
        int n_start_point=i-(m/2);
        float pvalue=0.0;
        for(int j=0;j<m;j++)
        {
            int n_index=n_start_point+j;
            if(n_index>=0&&n_index<n)
            {
                if((n_index>=this_tile_start_point)&&(n_index<nest_tile_start_
                point))
                {
                    pvalue+=array_ns[threadIdx.x+j-(m/2)] * d_array_m1[j];
                }
                else
                {
                    pvalue+=array_n[n_index] * d_array_m1[j];
                }
            }
        }
        array_p[i]=pvalue;
    }
    ⋮
        float *array_m,*array_n,*array_p4;
        int m_2=m/2;
        array_m=(float *)malloc(sizeof(float) * m);
        array_n=(float *)malloc(sizeof(float) * n);
        array_p4=(float *)malloc(sizeof(float) * n);
        vector_init(array_m,m);
        vector_init(array_n,n);
        int grid_width=n/block_width;
```

```
    float *d_array_n,*d_array_p1;
    cudaMalloc((void **) &d_array_n, sizeof(float) * n);
    cudaMalloc((void **) &d_array_p1, sizeof(float) * n);
    cudaMemcpy(d_array_n, array_n, sizeof(float) * n, cudaMemcpyHostToDevice);
    cudaMemcpyToSymbol(d_array_m1,array_m,sizeof(float) * m);
    convolution_1d_const_shared_cache_kernel<<<grid_width,block_width>>>(d_array_n,d_array_p1,n);
    cudaMemcpy(array_p4,d_array_p1, sizeof(float) * n, cudaMemcpyDeviceToHost);
    cudaFree(d_array_n);
    cudaFree(d_array_p1);
    free(array_m);
    free(array_n);
    free(array_p4);
    ⋮
```

附录 B 曼德博罗特集的系列优化代码

附录 B.1 完整版串行 C 代码

```c
#include<stdlib.h>
#include<stdio.h>
#define TIMETESTEVENT
#include<sys/time.h>
double dtime()
{
    double tseconds=0.0;
    struct timeval mytime;
    gettimeofday(&mytime,(struct timezone *)0);
    tseconds= (double)(mytime.tv_sec+mytime.tv_usec * 1.0e-6);
    return tseconds;
}


#include "bmpimage.h"   //见附录 C.1
#define MAX_ITE 255
typedef struct _LI_RGB
{
    unsigned char b,g,r;
}LI_RGB;
void ComputeColor(double x, double y, LI_RGB *color)
{
    double zx, zy;
    double tempx, tempy;
    int count;

    count=0;
    zx=0;
    zy=0;

    while((count<MAX_ITE) && ((zx * zx+zy * zy)<4.0))
```

```
        {
            tempx=zx * zx-zy * zy+x;
            tempy=2.0 * zx * zy+y;
            zx=tempx;
            zy=tempy;
            count++;
        }

    (*color).b=(count<=8 ? (count * 16-1): (unsigned char)((count-8)/120 * 128
    +127));
    (*color).g=(count<=32 ? (count * 8-1): 255);
    (*color).r=(count<=16 ? (count * 16-1): (unsigned char)((count-16)/239 *
    128+127));
}

int main_test(int n)
{
    int i, j;
    int bmpflag;
    LI_RGB *pRGB;
    double x0, x1, y0, y1;
    //int n;
    double temp, dx, dy;
    double x, y;

    x0=-2;
    x1=2;
    y0=-2;
    y1=2;
    //n=1024;
    bmpflag=1;

    pRGB=(LI_RGB *) malloc (n * n * sizeof(LI_RGB));

    if(x1<x0)
    {
        temp=x0;
        x0=x1;
        x1=temp;
    }
    if(y1<y0)
    {
```

```c
            temp=y0;
            y0=y1;
            y1=temp;
        }

        //计算坐标差值
    dx=(x1-x0)/n;
    dy=(y1-y0)/n;

#ifdef TIMETESTEVENT
    double time_0,time_1;
    time_0=dtime();
#endif


    for(j=0; j<n; j++)
    {
        y=y1-j*dy;
        for(i=0; i<n; i++)
        {
            x=x0+i*dx;
            ComputeColor(x, y, &pRGB[j*n+i]);
        }
    }
#ifdef TIMETESTEVENT
    time_1=dtime();
    printf("n=\t%d\t time=\t%f\n",n,(time_1-time_0)*1000);
#endif

    if(n<=1024)
    {
        FILE *outfile;
        outfile=BmpWriteHeader("mandelbrot_serial.bmp",n,n,24);
        unsigned char *rgb=(unsigned char *)malloc(sizeof(char)*3*n*n);
        for(int i=0;i<n*n;i++)
        {
            rgb[i*3+0]=pRGB[i].b;
            rgb[i*3+1]=pRGB[i].g;
            rgb[i*3+2]=pRGB[i].r;
        }
        BmpWrite3(outfile,rgb,n,n,0,0);
        fclose(outfile);
        free(rgb);
```

```
    }
    free(pRGB);

    return 0;
}

int main()
{
    main_test(1024);
    return 0;
}
```

附录B.2　cuda_1_0

```
__device__ inline
void ComputeColor(double x, double y, unsigned char *color_b,unsigned char
*color_g,unsigned char *color_r)
{
    double zx, zy;
    double tempx, tempy;
    int count;

    count=0;
    zx=0;
    zy=0;

    while((count<MAX_ITE) && ((zx*zx+zy*zy)<4.0))
    {
        tempx=zx*zx-zy*zy+x;
        tempy=2.0*zx*zy+y;
        zx=tempx;
        zy=tempy;
        count++;
    }

    color_b[0]=(count<=8 ? (count*16-1): (unsigned char)((count-8)/120*128
    +127));
    color_g[0]=(count<=32 ? (count*8-1): 255);
    color_r[0]=(count<=16 ? (count*16-1): (unsigned char)((count-16)/239*
    128+127));
}
```

```
__global__
void compute_color_gpu(unsigned char *color_b,unsigned char *color_g,unsigned
char *color_r,double x0,double dx,double y1,double dy,int n)
{
    unsigned int tid=threadIdx.x;
    unsigned int bid=blockIdx.x;
    double y,x;
    while(bid<n)
    {
        tid=threadIdx.x;
        y=y1-bid*dy;
        while(tid<n)
        {
            x=x0+tid*dx;
            ComputeColor(x, y, &color_b[bid*n+tid], &color_g[bid*n+tid],
            &color_r[bid*n+tid]);
            tid+=blockDim.x;
        }
        bid+=gridDim.x;
    }
}
```

```
    unsigned char *pR,*pG,*pB;
    pB= (unsigned char *)malloc(sizeof(char)*n*n);
    pG= (unsigned char *)malloc(sizeof(char)*n*n);
    pR= (unsigned char *)malloc(sizeof(char)*n*n);
    cudaHostRegister(pB,sizeof(char)*n*n,0);
    cudaHostRegister(pG,sizeof(char)*n*n,0);
    cudaHostRegister(pR,sizeof(char)*n*n,0);
    unsigned char *d_pR,*d_pG,*d_pB;
    cudaMalloc((void**) &d_pB, sizeof(char)*n*n);
    cudaMalloc((void**) &d_pG, sizeof(char)*n*n);
    cudaMalloc((void**) &d_pR, sizeof(char)*n*n);
        ⋮
    compute_color_gpu<<<threadnum,threadnum>>>(d_pB,d_pG,d_pR,x0,dx,y1,dy,
    n);
    cudaMemcpy(pB, d_pB, sizeof(char)*n*n, cudaMemcpyDeviceToHost);
    cudaMemcpy(pG, d_pG, sizeof(char)*n*n, cudaMemcpyDeviceToHost);
    cudaMemcpy(pR, d_pR, sizeof(char)*n*n, cudaMemcpyDeviceToHost);
        ⋮
```

附录 B.3　cuda_0_2

```cpp
__device__ inline
void ComputeColor(double x, double y, uchar4 *color)
{
    double zx, zy;
    double tempx, tempy;
    int count;
    count=0;
    zx=0;
    zy=0;
    while((count<MAX_ITE) && ((zx*zx+zy*zy)<4.0))
    {
        tempx=zx*zx-zy*zy+x;
        tempy=2.0*zx*zy+y;
        zx=tempx;
        zy=tempy;
        count++;
    }
    uchar4 rgb={0,0,0,0};
    rgb.x=(count<=8 ? (count*16-1): (unsigned char)((count-8)/120*128+
127));
    rgb.y=(count<=32 ? (count*8-1): 255);
    rgb.z=(count<=16 ? (count*16-1): (unsigned char)((count-16)/239*128+
127));
    *color=rgb;
}
__global__
void compute_color_gpu(uchar4 *pRGB,double x0,double dx,double y1,double dy,
int n)
{
    unsigned int tid=threadIdx.x;
    unsigned int bid=blockIdx.x;
    double y,x;
    while(bid<n)
    {
        tid=threadIdx.x;
        y=y1-bid*dy;
        while(tid<n)
        {
            x=x0+tid*dx;
```

```
            ComputeColor(x, y, &pRGB[bid * n+tid]);
            tid+=blockDim.x;
        }
        bid+=gridDim.x;
    }
}
```

```
    uchar4 *pRGB;
    pRGB=(uchar4 *) malloc (n * n * sizeof(uchar4));
    cudaHostRegister(pRGB,sizeof(uchar4) * n * n,0);
    uchar4 * d_pRGB;
    cudaMalloc((void **) &d_pRGB, sizeof(uchar4) * n * n);
    compute_color_gpu<<<blocknum,threadnum>>>(d_pRGB,x0,dx,y1,dy,n);
    cudaMemcpy(pRGB, d_pRGB, sizeof(uchar4) * n * n, cudaMemcpyDeviceToHost);
```

附录 B.4　cuda_zerocopy

```
#include "bmpimage.h"
#define MAX_ITE 255

typedef struct _LI_RGB
{
    unsigned char b,g,r;
}LI_RGB;
__device__ inline
void ComputeColor(double x, double y, LI_RGB *color)
{
    double zx, zy;
    double tempx, tempy;
    int count;

    count=0;
    zx=0;
    zy=0;

    while((count<MAX_ITE) && ((zx * zx+zy * zy)<4.0))
    {
        tempx=zx * zx-zy * zy+x;
        tempy=2.0 * zx * zy+y;
```

```
            zx=tempx;
            zy=tempy;
            count++;
    }

    (*color).b=(count<=8 ? (count * 16-1): (unsigned char)((count-8)/120 * 128
+127));
    (*color).g=(count<=32 ? (count * 8-1): 255);
    (*color).r=(count<=16 ? (count * 16-1): (unsigned char)((count-16)/239 *
128+127));
}

__global__
void compute_color_gpu(LI_RGB *pRGB,double x0,double dx,double y1,double dy,
int n)
{
    unsigned int tid=threadIdx.x;
    unsigned int bid=blockIdx.x;
    double y,x;
    while(bid<n)
    {
        tid=threadIdx.x;
        y=y1-bid * dy;
        while(tid<n)
        {
            x=x0+tid * dx;
            ComputeColor(x, y, &pRGB[bid * n+tid]);
            tid+=blockDim.x;
        }
        bid+=gridDim.x;
    }
}
```

```
    LI_RGB *pRGB;
    cudaHostAlloc((void **) &pRGB, sizeof(LI_RGB) * n * n,cudaHostAllocMapped);
    LI_RGB *d_pRGB;
    cudaHostGetDevicePointer((void **)&d_pRGB, (void *)pRGB, 0);
    compute_color_gpu<<<blocknum,threadnum>>>(d_pRGB,x0,dx,y1,dy,n);
        ⋮
```

附录 B.5　cuda_1_0_zerocopy

```
#include "bmpimage.h"
#define MAX_ITE 255
__device__ inline
void ComputeColor(double x, double y, unsigned char *color_b,unsigned char
*color_g,unsigned char *color_r)
{
    double zx, zy;
    double tempx, tempy;
    int count;

    count=0;
    zx=0;
    zy=0;

    while((count<MAX_ITE) && ((zx*zx+zy*zy)<4.0))
    {
        tempx=zx*zx-zy*zy+x;
        tempy=2.0*zx*zy+y;
        zx=tempx;
        zy=tempy;
        count++;
    }

    color_b[0]=(count<=8 ? (count*16-1): (unsigned char)((count-8)/120*128
    +127));
    color_g[0]=(count<=32 ? (count*8-1): 255);
    color_r[0]=(count<=16 ? (count*16-1): (unsigned char)((count-16)/239*
    128+127));
}

__global__
void compute_color_gpu(unsigned char *color_b,unsigned char *color_g,unsigned
char *color_r,double x0,double dx,double y1,double dy,int n)
{
    unsigned int tid=threadIdx.x;
    unsigned int bid=blockIdx.x;
    double y,x;
    while(bid<n)
    {
```

```
            tid=threadIdx.x;
            y=y1-bid*dy;
            while(tid<n)
            {
                x=x0+tid*dx;
                ComputeColor(x, y, &color_b[bid*n+tid], &color_g[bid*n+tid],
                &color_r[bid*n+tid]);

                tid+=blockDim.x;
            }
            bid+=gridDim.x;
    }
}
    unsigned char *pR,*pG,*pB;
    cudaHostAlloc((void **) &pB, sizeof(char) * n * n,cudaHostAllocMapped);
    cudaHostAlloc((void **) &pG, sizeof(char) * n * n,cudaHostAllocMapped);
    cudaHostAlloc((void **) &pR, sizeof(char) * n * n,cudaHostAllocMapped);
        ⋮
    unsigned char *d_pR,*d_pG,*d_pB;
    cudaHostGetDevicePointer((void **)&d_pB, (void *)pB, 0);
    cudaHostGetDevicePointer((void **)&d_pG, (void *)pG, 0);
    cudaHostGetDevicePointer((void **)&d_pR, (void *)pR, 0);
    compute_color_gpu<<<threadnum,threadnum>>>(d_pB,d_pG,d_pR,x0,dx,y1,dy,n);
        ⋮
    cudaFreeHost(pB);
    cudaFreeHost(pG);
    cudaFreeHost(pR);
```

附录 B.6　cuda_0_0_zerocopy

```
#include "bmpimage.h"
#define MAX_ITE 255

typedef struct _LI_RGB
{
    unsigned char b,g,r;
}LI_RGB;
__device__ inline
void ComputeColor(double x, double y, LI_RGB *color)
{
    double zx, zy;
```

```c
        double tempx, tempy;
        int count;

        count=0;
        zx=0;
        zy=0;

        while((count<MAX_ITE) && ((zx*zx+zy*zy)<4.0))
        {
            tempx=zx*zx-zy*zy+x;
            tempy=2.0*zx*zy+y;
            zx=tempx;
            zy=tempy;
            count++;
        }
        LI_RGB rgb;

        rgb.b=(count<=8 ? (count*16-1): (unsigned char)((count-8)/120*128+127));
        rgb.g=(count<=32 ? (count*8-1): 255);
        rgb.r=(count<=16 ? (count*16-1): (unsigned char)((count-16)/239*128+127));
        *color=rgb;
}

__global__
void compute_color_gpu(LI_RGB *pRGB,double x0,double dx,double y1,double dy,int n)
{
    unsigned int tid=threadIdx.x;
    unsigned int bid=blockIdx.x;
    double y,x;
    while(bid<n)
    {
        tid=threadIdx.x;
        y=y1-bid*dy;
        while(tid<n)
        {
            x=x0+tid*dx;
            ComputeColor(x, y, &pRGB[bid*n+tid]);
            tid+=blockDim.x;
        }
        bid+=gridDim.x;
```

```
    }
}
    LI_RGB *pRGB;
    cudaHostAlloc((void**) &pRGB, sizeof(LI_RGB) * n * n,cudaHostAllocMapped);
    ⋮
    LI_RGB *d_pRGB;
    cudaHostGetDevicePointer((void **)&d_pRGB, (void *)pRGB, 0);
    compute_color_gpu<<<blocknum,threadnum>>>(d_pRGB,x0,dx,y1,dy,n);
    ⋮
    cudaFreeHost(pRGB);
```

附录B.7 cuda_0_2_zerocopy

```
#include "bmpimage.h"
#define MAX_ITE 255

__device__ inline
void ComputeColor(double x, double y, uchar4 *color)
{
    double zx, zy;
    double tempx, tempy;
    int count;

    count=0;
    zx=0;
    zy=0;

    while((count<MAX_ITE) && ((zx * zx+zy * zy)<4.0))
    {
        tempx=zx * zx-zy * zy+x;
        tempy=2.0 * zx * zy+y;
        zx=tempx;
        zy=tempy;
        count++;
    }
    uchar4 rgb={0,0,0,0};

    rgb.x=(count<=8 ? (count * 16-1): (unsigned char)((count-8)/120 * 128+
127));
    rgb.y=(count<=32 ? (count * 8-1): 255);
```

```
    rgb.z=(count<=16 ? (count * 16-1): (unsigned char)((count-16)/239 * 128+
    127));
    *color=rgb;
}

__global__
void compute_color_gpu(uchar4 *pRGB,double x0,double dx,double y1,double dy,
int n)
{
    unsigned int tid=threadIdx.x;
    unsigned int bid=blockIdx.x;
    double y,x;
    while(bid<n)
    {
        tid=threadIdx.x;
        y=y1-bid * dy;
        while(tid<n)
        {
            x=x0+tid * dx;
            ComputeColor(x, y, &pRGB[bid * n+tid]);
            tid+=blockDim.x;
        }
        bid+=gridDim.x;
    }
}
```

```
    uchar4 *pRGB;
    cudaHostAlloc((void**) &pRGB, sizeof(uchar4) * n * n,cudaHostAllocMapped);
        ⋮
    uchar4 *d_pRGB;
    cudaHostGetDevicePointer((void **)&d_pRGB, (void *)pRGB, 0);
    compute_color_gpu<<<blocknum,threadnum>>>(d_pRGB,x0,dx,y1,dy,n);
        ⋮
    cudaFreeHost(pRGB);
```

附录 B.8 cuda_2

```
#include "bmpimage.h"
#define MAX_ITE 255
typedef struct _LI_RGB
{
```

```c
    unsigned char b,g,r;
}LI_RGB;
__device__ inline
void ComputeColor(double x, double y, LI_RGB *color)
{
    double zx, zy;
    double tempx, tempy;
    int count;
    count=0;
    zx=0;
    zy=0;
    while((count<MAX_ITE) && ((zx*zx+zy*zy)<4.0))
    {
        tempx=zx*zx-zy*zy+x;
        tempy=2.0*zx*zy+y;
        zx=tempx;
        zy=tempy;
        count++;
    }
    (*color).b=(count<=8 ? (count*16-1): (unsigned char)((count-8)/120*128
    +127));
    (*color).g=(count<=32 ? (count*8-1): 255);
    (*color).r=(count<=16 ? (count*16-1): (unsigned char)((count-16)/239*
    128+127));
}
__global__
void compute_color_gpu(LI_RGB *pRGB,double x0,double dx,double y1,double dy,
int n,int i)
{
    unsigned int tid=threadIdx.x;
    unsigned int bid=blockIdx.x+i;
    double y,x;
    //while(bid<n)
    {
        //tid=threadIdx.x;
        y=y1-bid*dy;
        while(tid<n)
        {
            x=x0+tid*dx;
            ComputeColor(x, y, &pRGB[bid*n+tid]);
            tid+=blockDim.x;
        }
```

```
        //bid+=gridDim.x;
    }
}
```

```
    LI_RGB *pRGB;
    pRGB= (LI_RGB *) malloc (n * n * sizeof(LI_RGB));
    cudaHostRegister(pRGB,sizeof(LI_RGB) * n * n,0);
    LI_RGB *d_pRGB;
    cudaMalloc((void**) &d_pRGB, sizeof(LI_RGB) * n * n);
    cudaStream_t stream0,stream1;
    cudaStreamCreate(&stream0);
    cudaStreamCreate(&stream1);
    ⋮
    for(i=0;i<n;i+=blocknum * 2)
    {
compute_color_gpu<<<blocknum,threadnum,0,stream0>>>(d_pRGB,x0,dx,y1,dy,n,i);
compute_color_gpu<<<blocknum,threadnum,0,stream1>>>(d_pRGB,x0,dx,y1,dy,n,i
+blocknum);
cudaMemcpyAsync(&pRGB[i * n], &d_pRGB[i * n], sizeof(LI_RGB) * n * blocknum,
cudaMemcpyDeviceToHost,stream0);
cudaMemcpyAsync(&pRGB[(i+blocknum) * n], &d_pRGB[(i+blocknum) * n], sizeof(LI_
RGB) * n * blocknum, cudaMemcpyDeviceToHost,stream1);
    }
    cudaStreamDestroy(stream0);
    cudaStreamDestroy(stream1);
    cudaFree(d_pRGB);
    ⋮
    cudaFreeHost(pRGB);
```

附录 B.9 cuda_1_2

```
#include<stdlib.h>
#include<stdio.h>
#include<cuda_runtime.h>
#include "bmpimage.h"
#define MAX_ITE 255
__global__
void compute_color_gpu(unsigned char *color_b,unsigned char *color_g,unsigned
char *color_r,double x0,double dx,double y1,double dy,int n,int j)
{
    unsigned int id=blockIdx.x * blockDim.x+threadIdx.x;
```

```
        double y,x;
        double zx, zy;
        double tempx, tempy;
        int count;
//      while(bid<n)
        y=y1-j*dy;
        if(id<n)
        {
            x=x0+id*dx;
            {
                count=0;
                zx=0;
                zy=0;
                while((count<MAX_ITE) && ((zx*zx+zy*zy)<4.0))
                {
                    tempx=zx*zx-zy*zy+x;
                    tempy=2.0*zx*zy+y;
                    zx=tempx;
                    zy=tempy;
                    count++;
                }
                color_b[j*n+id]=(count<=8 ? (count*16-1): (unsigned char)((count
                    -8)/120*128+127));
                color_g[j*n+id]=(count<=32 ? (count*8-1): 255);
                color_r[j*n+id]=(count<=16 ? (count*16-1): (unsigned char)
                    ((count-16)/239*128+127));
            }
        }
}
```

```
    unsigned char *pR,*pG,*pB;
    pB=(unsigned char*)malloc(sizeof(char)*n*n);
    pG=(unsigned char*)malloc(sizeof(char)*n*n);
    pR=(unsigned char*)malloc(sizeof(char)*n*n);
    cudaHostRegister(pB,sizeof(char)*n*n,0);
    cudaHostRegister(pG,sizeof(char)*n*n,0);
    cudaHostRegister(pR,sizeof(char)*n*n,0);
        ⋮
    unsigned char *d_pR,*d_pG,*d_pB;
    cudaMalloc((void**) &d_pB, sizeof(char)*n*n);
    cudaMalloc((void**) &d_pG, sizeof(char)*n*n);
    cudaMalloc((void**) &d_pR, sizeof(char)*n*n);
    int blocknum=(n+threadnum-1)/threadnum;
    for(j=0; j<n; j++)
```

```
    {
        compute_color_gpu<<<blocknum,threadnum>>>(d_pB,d_pG,d_pR,x0,dx,y1,
        dy,n,j);
    }
    cudaMemcpy(pB, d_pB, sizeof(char) * n * n, cudaMemcpyDeviceToHost);
    cudaMemcpy(pG, d_pG, sizeof(char) * n * n, cudaMemcpyDeviceToHost);
    cudaMemcpy(pR, d_pR, sizeof(char) * n * n, cudaMemcpyDeviceToHost);
    cudaFree(d_pB);
    cudaFree(d_pG);
    cudaFree(d_pR);
    ⋮
    cudaFreeHost(pB);
    cudaFreeHost(pG);
    cudaFreeHost(pR);
```

附录 C 几种图像处理完整源码

附录 C.1 BMP 图像读写头文件

```c
//bmpimage.h

#ifndef _BMPImage2_
#define _BMPImage2_

#include<stdio.h>
#include<stdlib.h>
#include<string.h>

#define CharLen        1 //sizeof(unsigned char)
#define LongLen        4 //sizeof(long)
#define IMGTYPE unsigned char


typedef struct
{
    int   BitCount;         //2 bits    8 or 24
    long Height;
    long Width;
    //long FileSize;
    unsigned char *red, *green, *blue, *gray;
}BMPIMAGE;

//为图像文件开辟内存空间
BMPIMAGE *ImageAlloc(int height, int width, int BitCount)
{
    BMPIMAGE *image;
    long LineBytes;

    if((image=(BMPIMAGE *) malloc (sizeof(BMPIMAGE)))==NULL)
```

```
{
    printf("Fail to allocate memory image.\n");
    return NULL;
}

LineBytes= (width * BitCount+31)/32 * 4;

image->BitCount=BitCount;
image->Height=height;
image->Width=width;

if(BitCount==8)
{
   if((image->gray= (unsigned char *) calloc (height * width, CharLen))==
   NULL)
    {
        printf("Fail to allocate image->gray.\n");
        return NULL;
    }
   //    image->FileSize=height * LineBytes+1078;
}
else if(BitCount==24)
{
   if((image->red= (unsigned char *) calloc (height * width, CharLen))==
   NULL)
    {
        printf("fail to allocate image->DataA.\n");
        return NULL;
    }

   if((image->green= (unsigned char *) calloc (height * width, CharLen))==
   NULL)
    {
        printf("Fail to allocate image->DataB.\n");
        return NULL;
    }

   if((image->blue= (unsigned char *) calloc (height * width, CharLen))==
   NULL)
    {
        printf("Fail to allocate image->DataC.\n");
        return NULL;
    }
```

```
        //      image->FileSize=height * LineBytes+54;
    }
    else
    {
        printf("Error parameter in ImageAlloc: %d.\n",BitCount);
        return NULL;
    }

    return image;
}
//释放图像文件占用的内存
void ImageDealloc(BMPIMAGE *image)
{
    free(image);
}

BMPIMAGE *BmpRead(char *filename)
{
    FILE *infile;      //打开的文件,filename 对应的文件的指针
    unsigned char *buffer, BmpHeader[54];   //buffer[width]每次取一行数据
    long i, j, width, height, BitCount, compression, LineBytes;
    unsigned short type;     //type="BM"
    BMPIMAGE *image=NULL;

    if(filename==NULL)
    {
        printf("Error in read: filename is NULL, assign it\n");
        return NULL;
    }
    infile=fopen (filename, "rb");
    if(infile==NULL)
    {
        printf("Unable to open file %s\n", filename);
        return NULL;
    }

    /* ---read BMP file header parameters --- */
    fread(BmpHeader, CharLen, 54, infile);
    type=(BmpHeader[1]<<8)+BmpHeader[0];//19712+66
    width= (BmpHeader[21]<<24) + (BmpHeader[20]<<16) + (BmpHeader[19]<<8) +
    BmpHeader[18];
    height= (BmpHeader[25]<<24) + (BmpHeader[24]<<16) + (BmpHeader[23]<<8) +
    BmpHeader[22];
```

```
BitCount=(BmpHeader[29]<<8)+BmpHeader[28];
compression=(BmpHeader[33]<<24)+(BmpHeader[32]<<16)+(BmpHeader[31]<<8)
+BmpHeader[32];
LineBytes=(width * BitCount+31)/32 * 4;
//padding=LineBytes-width * 3;

if(type!=((unsigned short)('M'<<8) | 'B'))//!!! 01001101 01000010=1978
=type
{
    printf("the file maybe is not a bmp image.\n");
    return NULL;
}
if(compression!=0)
{
    printf("Can not read a compressed file.\n");
    return NULL;
}
if((BitCount!=8)&&(BitCount!=24))
{
    printf("This program can only deal with the file with 256 colors or true
    color.\n");
    return NULL;
}

if(BitCount==8)
{
    if((image=ImageAlloc(height, width, BitCount))==NULL)
    {
        printf("Error when read call ImageAlloc.\n");
        return NULL;
    }
    buffer=new unsigned char [LineBytes];
    //到图像实际数据起始处
    //fseek(infile, 1078, SEEK_SET);
    fseek(infile, 1078, SEEK_SET);
    for(i=0; i<=height-1; i++)
    {
        fread(buffer, CharLen, LineBytes, infile);

        for(j=0; j<width; j++)
            //已经重新调整了数据的排放方式,从左上角开始,从左到右,从上到下
            image->gray[i * width+j]=buffer[j];
    }
```

```
    }//end of biBitCount==8

    if(BitCount==24)
    {
        if((image=ImageAlloc(height, width, BitCount))==NULL)
        {
            printf("Error when read call ImageAlloc.\n");
            return NULL;
        }

        buffer=new unsigned char [LineBytes]; //ordered by B、G、R

        //到图像实际数据起始处
        fseek(infile, 54, SEEK_SET);
        for(i=0; i<=height-1; i++)//i: 0 -->(biHeight-1)
        {
            fread (buffer, CharLen, LineBytes, infile);

            for(j=0; j<width; j++)      //以步进
            {
                image->blue[i*width+j]=buffer[3*j];      //Blue value
                image->green[i*width+j]=buffer[3*j+1];   //Green value
                image->red[i*width+j]=buffer[3*j+2];     //Red value
            }
        }
    }//end of biBitCount==24

    delete [] buffer;
    fclose(infile);

    return image;
}

void BmpWrite(char *filename, BMPIMAGE *image)
{
    FILE *outfile;
    long LineBytes, height, width, ImageBytes, FileBytes, OffBits; //扫描行长度
    int padding, BitCount;          //补零个数
    long i,j, tmp=0;
    unsigned char BmpHeader[54], *ColorMap;//,*red, *green, *blue, *gray; //
    ColorMap used in 256 colors bmp file

    outfile=fopen (filename, "wb");
```

```c
    if(outfile==NULL)
    {
        printf("Unable to create the file %s\n", filename);
        return;
    }
    
    height=image->Height;
    width=image->Width;
    
    BitCount=image->BitCount;
    LineBytes=(width*BitCount+31)/32*4;
    ImageBytes=LineBytes*height;
    OffBits=54+((BitCount==24) ? 0: ((1<<BitCount)*4));
    FileBytes=OffBits+ImageBytes;
    padding=LineBytes-((BitCount==8) ? width: (width*3));
    
    memset(BmpHeader, 0, 54);
    
    *(BmpHeader+0)='B';
    *(BmpHeader+1)='M';
    *(BmpHeader+2)=(unsigned char)(FileBytes);
    *(BmpHeader+3)=(unsigned char)(FileBytes>>8);//低位在前,高位在后
    *(BmpHeader+4)=(unsigned char)(FileBytes>>16);
    *(BmpHeader+5)=(unsigned char)(FileBytes>>24);
    *(BmpHeader+10)=(unsigned char)(OffBits);
    *(BmpHeader+11)=(unsigned char)(OffBits>>8);
    *(BmpHeader+12)=(unsigned char)(OffBits>>16);
    *(BmpHeader+13)=(unsigned char)(OffBits>>24);
    *(BmpHeader+14)=40;
    *(BmpHeader+18)=(unsigned char)(width);
    *(BmpHeader+19)=(unsigned char)(width>>8);
    *(BmpHeader+20)=(unsigned char)(width>>16);
    *(BmpHeader+21)=(unsigned char)(width>>24);
    *(BmpHeader+22)=(unsigned char)(height);
    *(BmpHeader+23)=(unsigned char)(height>>8);
    *(BmpHeader+24)=(unsigned char)(height>>16);
    *(BmpHeader+25)=(unsigned char)(height>>24);
    *(BmpHeader+26)=1;
    *(BmpHeader+28)=(unsigned char)(BitCount);
    *(BmpHeader+34)=(unsigned char)(ImageBytes);
    *(BmpHeader+35)=(unsigned char)(ImageBytes>>8);
    *(BmpHeader+36)=(unsigned char)(ImageBytes>>16);
    *(BmpHeader+37)=(unsigned char)(ImageBytes>>24);
```

```cpp
        fwrite(BmpHeader, CharLen, 54, outfile);

    if(BitCount==8)
    {
        ColorMap=new unsigned char [1024];
        for(i=0; i<256; i++)
        {
            for(j=0;j<3;j++)
                ColorMap[4*i+j]=(unsigned char)i;
            ColorMap[3+4*i]=0;
        }
        fwrite(ColorMap,CharLen,1024,outfile);
        tmp=0;
        for(i=0; i<=height-1; i++)
        {
            fwrite(&image->gray[i*width],CharLen,width,outfile);

            if(padding>0)
                fwrite(&tmp,CharLen,padding,outfile);
        }

        delete [] ColorMap;
        fclose(outfile);
    }

    if(BitCount==24)
    {
        tmp=0;
        for(i=0; i<=height-1; i++)
        {
            for(j=0; j<width; j++)
            {
                fwrite(&image->blue[i*width+j],CharLen,1,outfile);    //Blue
                fwrite(&image->green[i*width+j],CharLen,1,outfile);   //Green
                fwrite(&image->red[i*width+j],CharLen,1,outfile);     //Red
            }
            if(padding>0)
                fwrite(&tmp,CharLen,padding,outfile);
        }
        fclose(outfile);
    }
}
BMPIMAGE *BmpRead2(char *filename,long xs,long ys,long xt, long yt)
```

```c
{
    FILE *infile;        //打开的文件,filename 对应的文件的指针
    unsigned char *buffer, BmpHeader[54];    //buffer[width]每次取一行数据
    long i, j, width, height, BitCount, compression, LineBytes;
    unsigned short type;    //type="BM"
    BMPIMAGE *image=NULL;
    long w,h;            //小图像的宽和高

    if(filename==NULL)
    {
        printf("Error in read: filename is NULL, assign it\n");
        return NULL;
    }
    infile=fopen (filename, "rb");
    if(infile==NULL)
    {
        printf("Unable to open file %s\n", filename);
        return NULL;
    }

    /* ---read BMP file header parameters --- */
    fread(BmpHeader, CharLen, 54, infile);
    type= (BmpHeader[1]<<8)+BmpHeader[0];//19712+66
    width= (BmpHeader[21]<<24)+(BmpHeader[20]<<16)+(BmpHeader[19]<<8)+ BmpHeader[18];
    height= (BmpHeader[25]<<24)+(BmpHeader[24]<<16)+(BmpHeader[23]<<8)+ BmpHeader[22];
    BitCount= (BmpHeader[29]<<8)+BmpHeader[28];
    compression= (BmpHeader[33]<<24)+(BmpHeader[32]<<16)+(BmpHeader[31]<<8) +BmpHeader[32];
    LineBytes= (width * BitCount+31)/32 * 4;
    //padding=LineBytes-width * 3;

    if(type != ((unsigned short) ('M'<<8) | 'B'))//!!! 01001101 01000010=1978 =type
    {
        printf("The file maybe is not a bmp image.\n");
        return NULL;
    }
    if(compression!=0)
    {
        printf("Can not read a compressed file.\n");
        return NULL;
    }
```

```cpp
    if((BitCount!=8)&&(BitCount!=24))
    {
        printf("This program can only deal with the file with 256 colors or true
        color.\n");
        return NULL;
    }

    if(width<xt || height<yt || xs<0 || ys<0)
    {
        printf("Illegal coordinate(0<=xs,xt<=width, 0<=ys,yt<=height) .\n");
        return NULL;
    }
    w=xt-xs+1;
    h=yt-ys+1;

    if(BitCount==8)
    {
        //if((image=ImageAlloc(height, width, BitCount))==NULL)
        if((image=ImageAlloc(h, w, BitCount))==NULL)
        {
            printf("Error when read call ImageAlloc.\n");
            return NULL;
        }
        buffer=new unsigned char [LineBytes];
        //到图像实际数据起始处
        //fseek(infile, 1078, SEEK_SET);
        fseek(infile, 1078+LineBytes * ys+xs, SEEK_SET);
        //for(i=0; i<=height-1; i++)
        for(i=0; i<=h-1; i++)
        {
            fread (buffer, CharLen, LineBytes, infile);
            //for(j=0; j<width; j++)
            for(j=0; j<=w-1; j++)
                //已经重新调整了数据的排放方式,从左上角开始,从左到右,从上到下
                //image->gray[(height-1-i) * width+j]=buffer[j];
                image->gray[i * w+j]=buffer[j];
        }
    }//end of biBitCount==8

    if(BitCount==24)
    {
        //if((image=ImageAlloc(height, width, BitCount))==NULL)
        if((image=ImageAlloc(h, w, BitCount))==NULL)
```

```cpp
        {
            printf("Error when read call ImageAlloc.\n");
            return NULL;
        }

        buffer=new unsigned char [LineBytes]; //ordered by B,G,R

        //到图像实际数据起始处
        //fseek(infile, 54, SEEK_SET);
        fseek(infile, 54+LineBytes * ys+xs * 3, SEEK_SET);
        //for(i=0; i<=height-1; i++)//i: 0 -->(biHeight-1)
        for(i=0; i<=h-1; i++)
        {
            fread (buffer, CharLen, LineBytes, infile);

            //for(j=0; j<width; j++)      //以步进
            for(j=0; j<=w-1; j++)
            {
                //image->blue[(height-1-i) * width+j]=buffer[3 * j];
                    //Blue value
                image->blue[i * w+j]=buffer[3 * j];
                //image->green[(height-1-i) * width+j]=buffer[3 * j+1];
                    //Green value
                image->green[i * w+j]=buffer[3 * j+1];
                //image->red[(height-1-i) * width+j]=buffer[3 * j+2];
                    //Red value
                image->red[i * w+j]=buffer[3 * j+2];
            }
        }
    }//end of biBitCount==24

    image->Height=h;
    image->Width=w;

    delete [] buffer;
    fclose(infile);

    return image;
}
void SimpleTrans (double *fimage, unsigned char *image, long imagewidth, long imageheight)
```

```c
{
    long i;
    long imagesize;
    imagesize=imagewidth * imageheight;
    for(i=0;i<imagesize;i++)
    {
        if(fimage[i]<0.0)
            image[i]=0;
        else if(fimage[i]>255.0)
            image[i]=255;
        else
            image[i]=(unsigned char)fimage[i];
    }
}

unsigned char doubleToChar(double x)
{ return(x>255?255:(x<0 ? 0: (unsigned char)(x+0.5))); };

int BmpReadHeader(char *filename,long *imagewidth,long *imageheight)
{
    FILE *infile;       //打开的文件,filename 对应的文件的指针
    unsigned char BmpHeader[54]; //buffer[width]每次取一行数据
    long BitCount, compression;
    unsigned short type;    //type="BM"

    if(filename==NULL)
    {
        printf("Error in read: filename is NULL, assign it\n");
        return -1;
    }
    infile=fopen(filename, "rb");
    if(infile==NULL)
    {
        printf("Unable to open file %s\n", filename);
        return -1;
    }

    /* ---read BMP file header parameters --- */
    fread(BmpHeader, CharLen, 54, infile);
    type=(BmpHeader[1]<<8)+BmpHeader[0];//19712+66
    *imagewidth=(BmpHeader[21]<<24)+(BmpHeader[20]<<16)+(BmpHeader[19]<<8)
        +BmpHeader[18];
```

```c
    *imageheight=(BmpHeader[25]<<24)+(BmpHeader[24]<<16)+(BmpHeader[23]<<
    8)+BmpHeader[22];
    BitCount=(BmpHeader[29]<<8)+BmpHeader[28];
    compression=(BmpHeader[33]<<24)+(BmpHeader[32]<<16)+(BmpHeader[31]<<8)
    +BmpHeader[32];

    if(type != ((unsigned short)('M'<<8) | 'B'))//!!! 01001101 01000010=1978
    =type
    {
        printf("The file maybe is not a bmp image.\n");
        fclose(infile);
        return -1;
    }
    if(compression!=0)
    {
        printf("Can not read a compressed file.\n");
        fclose(infile);
        return -1;
    }
    if((BitCount!=8)&&(BitCount!=24))
    {
        printf("This program can only deal with the file with 256 colors or true
        color.\n");
        fclose(infile);
        return -1;
    }
    fclose(infile);
    return 0;
}
FILE * BmpWriteHeader(char * filename, long imagewidth, long imageheight, int
BitCount)
{
    FILE *outfile;
    long LineBytes, height, width, ImageBytes, FileBytes, OffBits; //扫描行长度
    int padding;              //补零个数
    long i,tmp;
    unsigned char BmpHeader[54];//ColorMap used in 256 colors bmp file

    outfile=fopen(filename, "wb+");
    if(outfile==NULL)
    {
        printf("BmpWriteHeader Error: Unable to create the file %s\n",
        filename);
```

```
    return NULL;
}

width=imagewidth;
height=imageheight;

LineBytes=(width*BitCount+31)/32*4;
ImageBytes=LineBytes*height;
OffBits=54+((BitCount==24) ? 0: ((1<<BitCount)*4));
FileBytes=OffBits+ImageBytes;
padding=LineBytes-((BitCount==8) ? width: (width*3));

memset(BmpHeader, 0, 54);

*(BmpHeader+0)='B';
*(BmpHeader+1)='M';
*(BmpHeader+2)=(IMGTYPE)(FileBytes);
*(BmpHeader+3)=(IMGTYPE)(FileBytes>>8);      //低位在前,高位在后
*(BmpHeader+4)=(IMGTYPE)(FileBytes>>16);
*(BmpHeader+5)=(IMGTYPE)(FileBytes>>24);
*(BmpHeader+10)=(IMGTYPE)(OffBits);
*(BmpHeader+11)=(IMGTYPE)(OffBits>>8);
*(BmpHeader+12)=(IMGTYPE)(OffBits>>16);
*(BmpHeader+13)=(IMGTYPE)(OffBits>>24);
*(BmpHeader+14)=40;
*(BmpHeader+18)=(IMGTYPE)(width);
*(BmpHeader+19)=(IMGTYPE)(width>>8);
*(BmpHeader+20)=(IMGTYPE)(width>>16);
*(BmpHeader+21)=(IMGTYPE)(width>>24);
*(BmpHeader+22)=(IMGTYPE)(height);
*(BmpHeader+23)=(IMGTYPE)(height>>8);
*(BmpHeader+24)=(IMGTYPE)(height>>16);
*(BmpHeader+25)=(IMGTYPE)(height>>24);
*(BmpHeader+26)=1;
*(BmpHeader+28)=(IMGTYPE)(BitCount);
*(BmpHeader+34)=(IMGTYPE)(ImageBytes);
*(BmpHeader+35)=(IMGTYPE)(ImageBytes>>8);
*(BmpHeader+36)=(IMGTYPE)(ImageBytes>>16);
*(BmpHeader+37)=(IMGTYPE)(ImageBytes>>24);

fwrite(BmpHeader, CharLen, 54, outfile);

if(BitCount==8)
```

```c
        {
            tmp=0;
            for(i=0;i<256;i++)
            {
                fwrite(&tmp, LongLen, 1, outfile);
                tmp+=0x00010101;
            }
        }

        return outfile;
}
void BmpWrite2(FILE *outfile,
               unsigned char *red,
               unsigned char *green,
               unsigned char *blue,
               long xLen, long yLen,long xBegin, long yBegin)
{
    long LineBytes, ImageWidth,ImageHeight, OffBits; //图像 outfile 的几个参数
    int BitCount;          //补零个数
    long i,j;
    long offset;

    //IMGTYPE *p;

    if(outfile==NULL)
    {
        printf("Error in BmpWrite2: The outfile is NULL!\n");
        return;
    }

    fseek(outfile, 10, SEEK_SET);
    fread(&OffBits, LongLen, 1, outfile);
    fseek(outfile, 18, SEEK_SET);
    fread(&ImageWidth, LongLen, 1, outfile);
    fread(&ImageHeight, LongLen, 1, outfile);
    fseek(outfile, 28, SEEK_SET);
    fread(&BitCount, LongLen, 1, outfile);

    if((BitCount!=8) && (BitCount!=24))
    {
        printf("BmpWrite2 Error: BitCount=%d!\n",BitCount);
        return;
    }
```

```c
        if((xBegin>=ImageWidth)||(yBegin>=ImageHeight))
        {
            printf("BmpWrite2 Error: The gaved begin location is out of range!\n");
            return;
        }
        if(((xBegin+xLen)>ImageWidth)||((yBegin+yLen)>ImageHeight))
        {
            printf("BmpWrite2 Error: The size of required data is out of range!\n");
            return;
        }
        //int padding=LineBytes-((BitCount==8) ? width: (width * 3));
        //go to begin location to write image data
        LineBytes=(ImageWidth * BitCount+31)/32 * 4;
        offset=OffBits+yBegin * LineBytes+xBegin * (BitCount/8);
        fseek(outfile, offset, SEEK_SET);

        for(i=0; i<yLen; i++)
        {
            //fwrite(p, 1, xLen * (BitCount/8), outfile);
            for(j=0; j<xLen; j++)
            {
                fwrite(&blue[i * xLen+j],CharLen,1,outfile);    //Blue
                fwrite(&green[i * xLen+j],CharLen,1,outfile);   //Green
                fwrite(&red[i * xLen+j],CharLen,1,outfile);     //Red
            //     p+=xLen * (BitCount/8);
            }
            offset+=LineBytes;
            fseek(outfile, offset, SEEK_SET);
        }
}
void BmpWrite3(FILE *outfile, IMGTYPE *image, long xLen, long yLen,long xBegin, long yBegin)
{
        long LineBytes, ImageWidth,ImageHeight, OffBits;    //图像 outfile 的几个参数
        int BitCount;           //补零个数
        long i;
        long offset;
        IMGTYPE *p;

        if(outfile==NULL)
        {
            printf("Error in BmpWrite2: The outfile is NULL!\n");
            return;
```

```
    }

    fseek(outfile, 10, SEEK_SET);
    fread(&OffBits, LongLen, 1, outfile);
    fseek(outfile, 18, SEEK_SET);
    fread(&ImageWidth, LongLen, 1, outfile);
    fread(&ImageHeight, LongLen, 1, outfile);
    fseek(outfile, 28, SEEK_SET);
    fread(&BitCount, LongLen, 1, outfile);

    if((BitCount!=8) && (BitCount !=24))
    {
        printf("BmpWrite3 Error: BitCount=%d!\n",BitCount);
        return;
    }
    if((xBegin>=ImageWidth)||(yBegin>=ImageHeight))
    {
        printf("BmpWrite3 Error: The gaved begin location is out of range!\n");
        return;
    }
    if(((xBegin+xLen)>ImageWidth)||((yBegin+yLen)>ImageHeight))
    {
        printf("BmpWrite3 Error: The size of required data is out of range!\n");
        return;
    }
    //go to begin location to write image data
    LineBytes= (ImageWidth * BitCount+31)/32 * 4;
    offset=OffBits+yBegin * LineBytes+xBegin * (BitCount/8);//适用于位和位中图像
    fseek(outfile, offset, SEEK_SET);
    p=image;

    for(i=0; i<yLen; i++)
    {
        fwrite(p, 1, xLen * (BitCount/8), outfile);

        p+=xLen * (BitCount/8);

        offset+=LineBytes;
        fseek(outfile, offset, SEEK_SET);
    }
}
void MergeRGB(IMGTYPE *red,IMGTYPE *green, IMGTYPE *blue, IMGTYPE *output, long length)
```

```
{
    long i;
    if(output==NULL)
    {
        printf("MergeRGB Error: Don't assign space for output!\n");
        return;
    }
    for(i=0; i<length; i++)
    {
        output[3*i]=blue[i];
        output[3*i+1]=green[i];
        output[3*i+2]=red[i];
    }
}
#endif
```

附录C.2　图像直方图串行代码

```
#include<stdlib.h>
#include<stdio.h>
#include "bmpimage.h"        //见附录C.1
int main(int argc, char *argv[])
{
    char inputfilename[50];
    long imageheight,imagewidth;
    sprintf(inputfilename,argv[1]);
    int header=BmpReadHeader(inputfilename,&imagewidth,&imageheight);
    long imagesize=imagewidth * imageheight;
    BMPIMAGE *Limage=BmpRead2(inputfilename,0,0,imagewidth-1,imageheight-1);
    unsigned char *img;
    img= (unsigned char *)malloc(sizeof(char)*imagesize);
    for(int i=0;i<imagesize;i++)
    {
        img[i]=Limage->gray[i];
    }
    int hist[256]={0};
#ifdef TIMETESTEVENT
    double time_0,time_1;
    time_0=dtime();
#endif
```

```
        for(int i=0;i<imagesize;i++)
        {
            hist[img[i]]++;
        }
#ifdef TIMETESTEVENT
        time_1=dtime();
        printf("time:\t%f\n ",(time_1-time_0) * 1000);
#endif
        FILE *outfile;
        outfile=fopen("hist_serial.txt","w");
        for(int i=0;i<256;i++)
        {
            fprintf(outfile,"%d\t%d\n",i,hist[i]);
        }
        fclose(outfile);
        free(img);
        return 0;
}
```

附录C.3 串行中值滤波代码

```
#include<stdio.h>
#include<stdlib.h>
#include "bmpimage.h"
#define TIMETESTEVENT
#ifdef TIMETESTEVENT
#include<sys/time.h>
double dtime()
{
    double tseconds=0.0;
    struct timeval mytime;
    gettimeofday(&mytime,(struct timezone *)0);
    tseconds= (double)(mytime.tv_sec+mytime.tv_usec * 1.0e-6);
    return tseconds;
}
#endif

int main(int argc, char *argv[])
{
    char inputfilename[50];
```

```c
    long imageheight,imagewidth;
    sprintf(inputfilename,argv[1]);
//inputfilename="pan_512x2.bmp";
    int header2=BmpReadHeader(inputfilename,&imagewidth,&imageheight);
    long imagesize=imagewidth * imageheight;
    BMPIMAGE *Himage=BmpRead2(inputfilename,0,0,imagewidth-1,imageheight-1);
    long i,j,k;
    unsigned char *img=(unsigned char *)malloc(sizeof(char) * imagesize);
    unsigned char *filter=(unsigned char *)malloc(sizeof(char) * imagesize);
    memset(filter, 0, sizeof(char) * imagesize);

    for(i=0;i<imagesize;i++)
    {
        img[i]=(Himage->gray[i]);
    }
    printf("imagewidth=%ld\timageheight=%ld\n",imagewidth,imageheight);

#ifdef TIMETESTEVENT
    double time_0,time_1;
    time_0=dtime();
#endif

    unsigned char p[9],temp;
    int l,m;
    float tmp_f;

    for(j=1;j<imageheight-1;j++)
    {
        for(k=1;k<imagewidth-1;k++)
        {
            p[0]=img[(j-1) * imagewidth+(k-1)];
            p[1]=img[(j-1) * imagewidth+(k)];
            p[2]=img[(j-1) * imagewidth+(k+1)];
            p[3]=img[(j) * imagewidth+(k-1)];
            p[4]=img[(j) * imagewidth+(k)];
            p[5]=img[(j) * imagewidth+(k+1)];
            p[6]=img[(j+1) * imagewidth+(k-1)];
            p[7]=img[(j+1) * imagewidth+(k)];
            p[8]=img[(j+1) * imagewidth+(k+1)];
            for(l=0;l<5;l++)
            {
                for(m=0;m<8-l;m++)
                {
                    if(p[m]>p[m+1])
```

```
                    {
                        temp=p[m];
                        p[m]=p[m+1];
                        p[m+1]=temp;
                    }
                }
            }
            filter[j*imagewidth+k]=p[4];
        }
    }

#ifdef TIMETESTEVENT
    time_1=dtime();
#endif
    FILE *outfile;
    outfile=BmpWriteHeader("out_gray.bmp",imagewidth,imageheight,8);
    if(outfile==NULL)
    {
        printf("ERROR: The outfile is NULL!\n");
        return 0;
    }
    BmpWrite3(outfile,filter,imagewidth,imageheight,0,0);
    fclose(outfile);

    free(img);
    free(filter);
#ifdef TIMETESTEVENT          //此处打印有点古怪,若放置在上一个ifdef,则输出滤波结果图
                              //像无法显示
    printf("time:\t%f\n",(time_1-time_0) * 1000);
#endif
    return 0;
}
```

附录 C.4 并行均值滤波相关代码

1. cuda_1d_1 版本代码

```
//cuda_1d_1
__global__ void mean_filter_1(unsigned char *img,unsigned char *filter,int imagewidth,int imageheight)
{
    __shared__ unsigned char srow0[514];
```

```
__shared__ unsigned char srow1[514];
__shared__ unsigned char srow2[514];
int bid=blockIdx.x;
int tid=threadIdx.x;
float tmp_f;
for(;bid<imageheight;bid+=gridDim.x)
{
    if((bid>0)&&(bid<(imageheight-1)))
    {
        for(tid=threadIdx.x;tid<imagewidth;tid+=blockDim.x)
        {
            __syncthreads();
            srow0[threadIdx.x+1]=img[(bid-1) * imagewidth+tid];
            srow1[threadIdx.x+1]=img[(bid) * imagewidth+tid];
            srow2[threadIdx.x+1]=img[(bid+1) * imagewidth+tid];
            if((tid==0)&&(threadIdx.x==0))
            {
                srow0[threadIdx.x]=0;
                srow1[threadIdx.x]=0;
                srow2[threadIdx.x]=0;
            }
            else if((tid>0)&&(threadIdx.x==0))
            {
                srow0[threadIdx.x]=img[(bid-1) * imagewidth+tid-1];
                srow1[threadIdx.x]=img[(bid) * imagewidth+tid-1];
                srow2[threadIdx.x]=img[(bid+1) * imagewidth+tid-1];
            }
            if(tid==(imagewidth-1))
            {
                srow0[threadIdx.x+2]=0;
                srow1[threadIdx.x+2]=0;
                srow2[threadIdx.x+2]=0;
            }
            else if((threadIdx.x+1)==blockDim.x)
            {
                srow0[threadIdx.x+2]=img[(bid-1) * imagewidth+tid+1];
                srow1[threadIdx.x+2]=img[(bid) * imagewidth+tid+1];
                srow2[threadIdx.x+2]=img[(bid+1) * imagewidth+tid+1];
            }
            __syncthreads();

            if((tid>0)&&(tid<(imagewidth-1)))
            {
```

```
                        tmp_f=0.0f;
                        tmp_f+=srow0[threadIdx.x];
                        tmp_f+=srow0[threadIdx.x+1];
                        tmp_f+=srow0[threadIdx.x+2];
                        tmp_f+=srow1[threadIdx.x];
                        tmp_f+=srow1[threadIdx.x+1];
                        tmp_f+=srow1[threadIdx.x+2];
                        tmp_f+=srow2[threadIdx.x];
                        tmp_f+=srow2[threadIdx.x+1];
                        tmp_f+=srow2[threadIdx.x+2];
                        tmp_f/=9;
                        filter[bid*imagewidth+tid]=(unsigned char)tmp_f;
                    }
                }
            }
        }
}
    median_filter_1<<<512,512>>>(d_img,d_filter,imagewidth,imageheight);
```

2. cuda_2d 版本代码

```
__global__ void mean_filter_2d(unsigned char *img,unsigned char *filter,int imagewidth,int imageheight)
{
    int yy=blockIdx.y*blockDim.y+threadIdx.y;
    int xx=blockIdx.x*blockDim.x+threadIdx.x;
    int xz=blockDim.x*gridDim.x;
    int yz=blockDim.y*gridDim.y;
    float tmp_f;
    for(;yy<imageheight;yy+=yz)
    {
        if((yy>0)&&(yy<(imageheight-1)))
        {
            for(xx=blockIdx.x*blockDim.x+threadIdx.x;xx<imagewidth;xx+=xz)
            {
                if((xx>0)&&(xx<(imagewidth-1)))
                {
                    tmp_f=0.0f;
                    tmp_f+=img[(yy-1)*imagewidth+(xx-1)];
                    tmp_f+=img[(yy-1)*imagewidth+(xx)];
                    tmp_f+=img[(yy-1)*imagewidth+(xx+1)];
                    tmp_f+=img[(yy)*imagewidth+(xx-1)];
```

```
                    tmp_f+=img[(yy) * imagewidth+ (xx)];
                    tmp_f+=img[(yy) * imagewidth+ (xx+1)];
                    tmp_f+=img[(yy+1) * imagewidth+ (xx-1)];
                    tmp_f+=img[(yy+1) * imagewidth+ (xx)];
                    tmp_f+=img[(yy+1) * imagewidth+ (xx+1)];
                    tmp_f/=9;
                    filter[yy * imagewidth+xx]=(unsigned char)tmp_f;
                }
            }
        }
    }
}
    dim3 threads1(32,32,1);
    dim3 blocks1(imagewidth/32,imageheight/32,1);
    median_filter_2d<<<blocks1,threads1>>>(d_img,d_filter,imagewidth,
imageheight);
```

3. cuda_2d_2 版本代码

```
__global__ void median_filter_2d_2(unsigned char *img,unsigned char *filter,
int imagewidth,int imageheight)
{
    __shared__ unsigned char shb[th_size+2][th_size+2];
    int yy=blockIdx.y * blockDim.y+threadIdx.y;
    int xx=blockIdx.x * blockDim.x+threadIdx.x;
    float tmp_f;

    __syncthreads();
    if(((yy-1)>=0)&&((xx-1)>=0))
    {
        shb[threadIdx.y][threadIdx.x]=img[(yy-1) * imagewidth+ (xx-1)];
    }
    if(threadIdx.y<2)
    {
        int tidy=threadIdx.y+th_size;
        if(((blockDim.y * blockIdx.y+tidy-1)<imageheight)&&((xx-1)>=0))
        {
            shb[tidy][threadIdx.x]=img[(blockDim.y * blockIdx.y+tidy-1) *
            imagewidth+ (xx-1)];
        }
    }
    else if((threadIdx.x<2)&&(threadIdx.y<6)&&(threadIdx.y>=4))
```

```
{
    int tidx=threadIdx.x+blockDim.x;
    int tidy=(threadIdx.y-4)+blockDim.x;
    if(((blockIdx.y*blockDim.y+tidy-1)<imageheight)&&((blockDim.x*
    blockIdx.x+tidx-1)<imagewidth))
    {
        shb[tidy][tidx]=img[(blockIdx.y*blockDim.y+tidy-1)*imagewidth
        +blockIdx.x*blockDim.x+tidx-1];
    }
}
if(threadIdx.x<2)//int yy=blockIdx.y*blockDim.y+threadIdx.y;替换就出错
{
    int tidx=threadIdx.x+th_size;
    int tidy=threadIdx.y;
    if(((yy-1)>=0)&&((blockDim.x*blockIdx.x+tidx-1)<imagewidth))
    {
        shb[tidy][tidx]=img[(yy-1)*imagewidth+blockDim.x*blockIdx.x+
        tidx-1];
    }
}
__syncthreads();
{
    if((yy>0)&&(yy<(imageheight-1)))
    {
        {
            if((xx>0)&&(xx<(imagewidth-1)))
            {
                tmp_f=0.0f;
                tmp_f+=shb[threadIdx.y][threadIdx.x];
                tmp_f+=shb[threadIdx.y][threadIdx.x+1];
                tmp_f+=shb[threadIdx.y][threadIdx.x+2];
                tmp_f+=shb[threadIdx.y+1][threadIdx.x];
                tmp_f+=shb[threadIdx.y+1][threadIdx.x+1];
                tmp_f+=shb[threadIdx.y+1][threadIdx.x+2];
                tmp_f+=shb[threadIdx.y+2][threadIdx.x];
                tmp_f+=shb[threadIdx.y+2][threadIdx.x+1];
                tmp_f+=shb[threadIdx.y+2][threadIdx.x+2];
                tmp_f/=9;
                filter[yy*imagewidth+xx]=(unsigned char)tmp_f;
            }
        }
    }
}
```

```
        }
}
    dim3 threads(th_size,th_size,1);
    dim3 blocks(imagewidth/th_size,imageheight/th_size,1);
    median_filter_2d_2<<<blocks,threads>>>(d_img,d_filter,imagewidth,
    imageheight);
```

4. cuda_2d_9 版本代码

```
__global__ void median_filter_2d_9(unsigned char *img,unsigned char *filter,
int imagewidth,int imageheight)
{
    __shared__ unsigned char shb[th_size+2][th_size+2];
    int yy=blockIdx.y*blockDim.y+threadIdx.y;
    int xx=blockIdx.x*blockDim.x+threadIdx.x;
    float tmp_f;
    int tidx,tidy;

    shb[threadIdx.y][threadIdx.x]=img[(yy)*(imagewidth+32)+(xx)];
    if(threadIdx.y<2)
    {
        tidy=threadIdx.y+blockDim.y;
        shb[tidy][threadIdx.x]=img[(blockIdx.y*blockDim.y+tidy)*
        (imagewidth+32)+xx];
    }
    else if(threadIdx.y<4)
    {
        tidy=threadIdx.x;
        tidx=(threadIdx.y-2)+blockDim.x;
        shb[tidy][tidx]=img[(blockIdx.y*blockDim.y+tidy)*(imagewidth+32)+
        blockIdx.x*blockDim.x+tidx];
    }
    else if((threadIdx.y<6)&&(threadIdx.x<2))
    {
        tidx=threadIdx.x+blockDim.x;
        tidy=(threadIdx.y-4)+blockDim.y;
        shb[tidy][tidx]=img[(blockIdx.y*blockDim.y+tidy)*(imagewidth+32)+
        blockIdx.x*blockDim.x+tidx];
    }
    __syncthreads();
//  filter[yy*imagewidth+xx]=shb[threadIdx.y+1][threadIdx.x+1];
    if((yy>0)&&(yy<(imageheight-1))&&(xx>0)&&(xx<(imagewidth-1)))
```

```
        {
            tmp_f=0.0f;
            tmp_f+=shb[threadIdx.y][threadIdx.x];
            tmp_f+=shb[threadIdx.y][threadIdx.x+1];
            tmp_f+=shb[threadIdx.y][threadIdx.x+2];
            tmp_f+=shb[threadIdx.y+1][threadIdx.x];
            tmp_f+=shb[threadIdx.y+1][threadIdx.x+1];
            tmp_f+=shb[threadIdx.y+1][threadIdx.x+2];
            tmp_f+=shb[threadIdx.y+2][threadIdx.x];
            tmp_f+=shb[threadIdx.y+2][threadIdx.x+1];
            tmp_f+=shb[threadIdx.y+2][threadIdx.x+2];
            tmp_f/=9;
            filter[yy*imagewidth+xx]=(unsigned char)tmp_f;
        }
}
    dim3 threads(th_size,th_size,1);
    dim3 blocks(imagewidth/th_size,imageheight/th_size,1);
    median_filter_2d_9<<<blocks,threads>>>(d_img,d_filter,imagewidth,
    imageheight);
```

附录 D nvprof 帮助菜单

```
[fangmq@cn18%yhstar test_in_book]$nvprof --help
Usage: nvprof [options] [CUDA - application] [application -
arguments]
Options:
  -o, --output-profile<file name>
          Output the result file which can be imported later
          or opened by the NVIDIA Visual Profiler.

          "%p" in the file name string is replaced with the
          process ID of the application being profiled.

          "%h" in the file name string is replaced with the
          hostname of the system.

          "%%" in the file name string is replaced with "%".

          Any other character following "%" is illegal.

          By default, this option disables the summary output.

          NOTE: If the application being profiled creates
          child processes, or if '--profile-all-processes' is
          used, the "%p" format is needed to get correct
          output files for each process.

  -i, --import-profile<file name>
          Import a result profile from a previous run.

  -s, --print-summary   Print a summary of the profiling result
on screen.

          NOTE: This is the default unless "--output-profile"
          or the print trace options are used.
```

```
--print-gpu-trace     Print individual kernel invocations (including CUDA
          memcpy's/memset's) and sort them in
          chronological order. In event/metric profiling mode,
          show events/metrics for each kernel invocation.

--print-api-trace     Print CUDA runtime/driver API trace.

--csv            Use comma-separated values in the output.

-u, --normalized-time-unit<s|ms|us|ns|col|auto>
          Specify the unit of time that will be used in the
          output.
          Allowed values:
          s-second, ms-millisecond, us-microsecond,
          ns-nanosecond
          col-a fixed unit for each column
          auto (default)-nvprof chooses the scale for
          each time value based on its length

-t, --timeout<seconds>   Set an execution timeout (in seconds) for the CUDA
          application.

          NOTE: Timeout starts counting from the moment the
          CUDA driver is initialized. If the application
          doesn't call any CUDA APIs, timeout won't be
          triggered.

    --demangling<on|off>
          Turn on/off C++name demangling of kernel names.
          Allowed values:
             on-turn on demangling (default)
             off-turn off demangling

    --events<event names>
          Specify the events to be profiled on certain
          device(s). Multiple event names separated by comma
          can be specified. Which device(s) are profiled is
          controlled by the "--devices" option. Otherwise
          events will be collected on all devices.
          For a list of available events, use
          "--query-events".
          Use "--devices" and "--kernels" to select a
          specific kernel invocation.
```

```
--metrics<metric names>
        Specify the metrics to be profiled on certain
        device(s). Multiple metric names separated by comma
        can be specified. Which device(s) are profiled is
        controlled by the "--devices" option. Otherwise
        metrics will be collected on all devices.
        For a list of available metrics, use
        "--query-metrics".
        Use "--devices" and "--kernels" to select a
        specific kernel invocation.

--analysis-metrics   Collect profiling data that can be imported to
        Visual Profiler's "analysis" mode.

        NOTE: Use "--output-profile" to specify an output
        file.

--devices<device ids>
        This option changes the scope of subsequent
        "--events", "--metrics", "--query-events" and
        "--query-metrics" options.
        Allowed values:
            all-change scope to all valid devices
            comma-separated device IDs-change scope to
            specified devices

--kernels<kernel path syntax>
        This option changes the scope of subsequent
        "--events", "--metrics" options
        The syntax is as following:
            <context id/name>:<stream id/name>:<kernel name>
            :<invocation>
        The context/stream IDs, names and kernel name can be
        regular expressions. Empty string matches any number
        or characters.
        If<context id/name>or<stream id/name>is a
        number, it's matched against both the context/stream
        id and name specified by the NVTX library. Otherwise
        it's matched against the context/stream name.
        The invocation count should be a positive number.
        Example: --kernels "1:foo:bar:2"
            profile any kernel whose name contains "bar"
            and was the 2nd instance on context 1 and on
```

```
                stream named "foo".

--query-events       List all the events available on the device(s).
        Device(s) queried can be controlled by the
        "--devices" option.

--query-metrics      List all the metrics available on the device(s).
        Device(s) queried can be controlled by the
        "--devices" option.

--concurrent-kernels<on|off>
        Turn on/off concurrent kernel execution.
        If concurrent kernel execution is off, all kernels
        running on one device will be serialized.
        Allowed values:
            on-turn on concurrent kernel execution
                (default)
            off-turn off concurrent kernel execution

--profile-from-start<on|off>
        Enable/disable profiling from the start of the
        application. If it's disabled, the application can
        use {cu,cuda}Profiler{Start,Stop} to turn on/off
        profiling.
        Allowed values:
            on-enable profiling from start (default)
            off-disable profiling from start

--aggregate-mode<on|off>
        This option turns on/off aggregate mode for events
        and metrics specified by subsequent "--events" and
        "--metrics" options. Those event/metric values will
        be collected for each domain instance, instead of
        the whole device.
        Allowed values:
            on-turn on aggregate mode(default)
            off-turn off aggregate mode

--system-profiling<on|off>
        Turn on/off power, clock, and thermal profiling.
        Allowed values:
            on-turn on system profiling
            off-turn off system profiling (default)
```

```
        --log-file<file name>
                Make nvprof send all its output to the specified
                file, or one of the standard channels. The file will
                be overwritten. If the file doesn't exist, a new
                one will be created.

                "%1" as the whole file name indicates standard
                output channel (stdout).

                "%2" as the whole file name indicates standard
                error channel (stderr).

                NOTE: This is the default.

                "%p" in the file name string is replaced with
                nvprof's process ID.

                "%h" in the file name string is replaced with
                the hostname of the system.

                "%%" in the file name is replaced with "%".

                Any other character following "%" is illegal.

        --quiet             Suppress all nvprof output.

        --profile-child-processes
                Profile the application and all child processes
                launched by it.

        --profile-all-processes
                Profile all processes launched by the same user who
                launched this nvprof instance.

                NOTE: Only one instance of nvprof can run with this
                option at the same time. Under this mode, there's
                no need to specify an application to run.
 -V     --version           Print version information of this tool.

 -h,    --help              Print this help information.
```

附录 E NVCC 帮助菜单

```
[fangmq@mn0%yhstar ~]$nvcc --help

Usage   : nvcc [options]<inputfile>

Options for specifying the compilation phase
============================================
More exactly, this option specifies up to which stage the input
files must be
compiled, according to the following compilation trajectories
for different
input file types:
    .c/.cc/.cpp/.cxx : preprocess, compile, link
    .o               : link
    .i/.ii           : compile, link
    .cu              : preprocess, cuda frontend, ptxassemble,
                       merge with host C code, compile, link
    .gpu             : cicc compile into cubin
    .ptx             : ptxassemble into cubin.

--cuda   (-cuda)
    Compile all .cu input files to .cu.cpp.ii output.

--cubin (-cubin)
    Compile all .cu/.ptx/.gpu input files to device-only .cubin
    files.
    This step discards the host code for each .cu input file.

--fatbin(-fatbin)
    Compile all .cu/.ptx/.gpu input files to ptx or device-only
    .cubin
    files (depending on the values specified for options '-arch'
    and/or
```

'-code') and place the result into the fat binary file specified with

option -o.
This step discards the host code for each .cu input file.

--ptx   (-ptx)
Compile all .cu/.gpu input files to device-only .ptx files. This step discards the host code for each of these input file.

--gpu   (-gpu)
Compile all .cu input files to device-only .gpu files. This step discards the host code for each .cu input file.

--preprocess   (-E)
Preprocess all .c/.cc/.cpp/.cxx/.cu input files.

--generate-dependencies   (-M)
Generate for the one .c/.cc/.cpp/.cxx/.cu input file(more than one input file is not allowed in this mode) a dependency file that can be included in a make file.

--compile   (-c)
Compile each .c/.cc/.cpp/.cxx/.cu input file into an object file.

--device-c   (-dc)
Compile each .c/.cc/.cpp/.cxx/.cu input file into an object file that contains relocatable device code. It is equivalent to
'--relocatable-device-code=true --compile'.

--device-w   (-dw)
Compile each .c/.cc/.cpp/.cxx/.cu input file into an object file that contains executable device code. It is equivalent to
'--relocatable-device-code=false --compile'.

--device-link   (-dlink)
Link object files with relocatable device code and .ptx/.cubin/.fatbin files into an object file with executable device code, which can be passed to the host linker.

--link (-link)
This option specifies the default behavior: compile and link all inputs.

--no-device-link   (-nodlink)

```
            Skip the device link step when linking object files.

--lib    (-lib)
            Compile all inputs into object files (if necessary) and add the results
            to the specified output library file.

--run    (-run)
            This option compiles and links all inputs into an executable, and
            executes it. Or, when the input is a single executable, it is executed
            without any compilation or linking. This step is intended for
            developers who do not want to be bothered with setting the necessary
            cuda dll search paths (these will be set temporarily by nvcc).

File and path specifications
============================

--x      (-x)
            Explicitly specify the language for the input files, rather than
            letting the compiler choose a default based on the file name suffix.
            Allowed values for this option:  'c','c++','cu'.

--output-file<file>                          (-o)
            Specify name and location of the output file. Only a single input file
            is allowed when this option is present in nvcc non-linking/archiving
            mode.

--pre-include<include-file>,…               (-include)
            Specify header files that must be preincluded during preprocessing.

--library<library>,…                         (-l)
            Specify libraries to be used in the linking stage without the library
            file extension. The libraries are searched for on the library search
            paths that have been specified using option '-L'.

--define-macro<macrodef>,…                   (-D)
            Specify macro definitions to define for use during preprocessing or
            compilation.

--undefine-macro<macrodef>,…                 (-U)
            Specify macro definitions to undefine for use during preprocessing or
            compilation.

--include-path<include-path>,…               (-I)
```

>   Specify include search paths.
>
> --system-include<include-path>,···   (-isystem)
>   Specify system include search paths.
>
> --library-path<library-path>,···   (-L)
>   Specify library search paths.
>
> --output-directory<directory>   (-odir)
>   Specify the directory of the output file. This option is intended for
>   letting the dependency generation step (option
>   '--generate-dependencies') generate a rule that defines the target
>   object file in the proper directory.
>
> --compiler-bindir<path>   (-ccbin)
>   Specify the directory in which the compiler executable(Microsoft
>   Visual Studio cl, or a gcc derivative) resides. By default, this
>   executable is expected in the current executable search path. For a
>   different compiler, or to specify these compilers with a different
>   executable name, specify the path to the compiler including the
>   executable name.
>
> --cudart(-cudart)
>   Specify the type of CUDA runtime library to be used: static CUDA
>   runtime library, shared/dynamic CUDA runtime library, or no CUDA
>   runtime library. By default, the static CUDA runtime library is used.
>   Allowed values for this option: 'none','shared','static'.
>   Default value: 'static'.
>
> --cl-version<cl-version-number>   --cl-version
>   <cl-version-number>
>   Specify the version of Microsoft Visual Studio installation. Note: this
>   option is to be used in conjunction with '--use-local-env', and is
>   ignored when '--use-local-env' is not specified.
>   Allowed values for this option: 2008,2010,2012.
>
> --use-local-env   --use-local-env
>
>   Specify whether the environment is already set up for the host compiler
>   .
>
> --libdevice-directory<directory>   (-ldir)

```
        Specify the directory that contains the libdevice library files when
        option '--dont-use-profile' is used. Libdevice library files are
        located in the 'nvvm/libdevice' directory in the CUDA toolkit.

Options for specifying behaviour of compiler/linker
===================================================

--profile                                       (-pg)
        Instrument generated code/executable for use by gprof(Linux only).

--debug (-g)
        Generate debug information for host code.

--device-debug                                  (-G)
        Generate debug information for device code.

--generate-line-info                            (-lineinfo)
        Generate line-number information for device code.

--optimize<level>                               (-O)
        Specify optimization level for host code.

--shared(-shared)
        Generate a shared library during linking. Note: when other linker
        options are required for controlling dll generation, use option
        -Xlinker.

--machine<bits>                                 (-m)
        Specify 32 vs 64 bit architecture.
        Allowed values for this option:  32,64.
        Default value:  64.

Options for passing specific phase options
==========================================
These allow for passing options directly to the intended compilation phase.
Using these, users have the ability to pass options to the lower level
compilation tools, without the need for nvcc to know about each and every such
option.

--compiler-options<options>,…                   (-Xcompiler)
        Specify options directly to the compiler/preprocessor.

--linker-options<options>,…                     (-Xlinker)
```

```
        Specify options directly to the host linker.

--archive-options<options>,…              (-Xarchive)
        Specify options directly to library manager.

--cudafe-options<options>,…               (-Xcudafe)
        Specify options directly to cudafe.

--ptxas-options<options>,…                (-Xptxas)
        Specify options directly to the ptx optimizing assembler.

--nvlink-options<options>,…               (-Xnvlink)
        Specify options directly to nvlink.

Miscellaneous options for guiding the compiler driver
=========================================================

--dont-use-profile                        (-noprof)
        Nvcc uses the nvcc.profiles file for compilation. When specifying this
        option, the profile file is not used.

--dryrun(-dryrun)
        Do not execute the compilation commands generated by nvcc. Instead,
        list them.

--verbose                                 (-v)
        List the compilation commands generated by this compiler driver, but do
        not suppress their execution.

--keep    (-keep)
        Keep all intermediate files that are generated during internal
        compilation steps.

--keep-dir                                (-keep-dir)
        Keep all intermediate files that are generated during internal
        compilation steps in this directory.

--save-temps                              (-save-temps)
        This option is an alias of '--keep'.

--clean-targets                           (-clean)
        This option reverses the behaviour of nvcc. When specified, none of the
        compilation phases will be executed. Instead, all of the non-temporary
        files that nvcc would otherwise create will be deleted.
```

```
--run-args<arguments>,…                         (-run-args)            Used in
combination with option -R, to specify command line arguments
        for the executable.

--input-drive-prefix<prefix>                    (-idp)
        On Windows platforms, all command line arguments that refer to file
        names must be converted to Windows native format before they are passed
        to pure Windows executables. This option specifies how the 'current'
        development environment represents absolute paths. Use '-idp /cygwin/'
        for CygWin build environments, and '-idp /' for Mingw.

--dependency-drive-prefix<prefix>               (-ddp)
        On Windows platforms, when generating dependency files (option -M), all
        file names must be converted to whatever the used instance of 'make'
        will recognize. Some instances of 'make' have trouble with the colon in
        absolute paths in native Windows format, which depends on the
        environment in which this 'make' instance has been compiled. Use '-ddp
        /cygwin/' for a CygWin make, and '-ddp /' for Mingw. Or leave these
        file names in native Windows format by specifying nothing.

--dependency-target-name<target>                (-MT)

        Specify the target name of the generated rule when generating a
        dependency file(option -M).

--drive-prefix<prefix>                          (-dp)
        Specifies<prefix>as both input-drive-prefix and
        dependency-drive-prefix.

--no-align-double                               --no-align-double

        Specifies that -malign-double should not be passed as a compiler
        argument on 32-bit platforms. WARNING: this makes the ABI incompatible
        with the cuda's kernel ABI for certain 64-bit types.

Options for steering GPU code generation
========================================

--gpu-architecture<gpu architecture name>       (-arch)

        Specify the name of the class of nVidia GPU architectures for which the
        cuda input files must be compiled.
        With the exception as described for the shorthand below, the
```

architecture specified with this option must be a virtual architecture (such as compute_10), and it will be the assumed architecture during the cicc compilation stage.

This option will cause no code to be generated (that is the role of nvcc option '--gpu-code', see below); rather, its purpose is to steer the cicc stage, influencing the architecture of the generated ptx intermediate.

For convenience in case of simple nvcc compilations the following shorthand is supported: if no value for option '--gpu-code' is specified, then the value of this option defaults to the value of '--gpu-architecture'. In this situation, as only exception to the description above, the value specified for '--gpu-architecture' may be a 'real' architecture(such as a sm_13), in which case nvcc uses the specified real architecture and its closest virtual architecture as effective architecture values. For example, 'nvcc -arch=sm_13' is equivalent to 'nvcc -arch=compute_13 -code=sm_13,compute_13'.

Allowed values for this option: 'compute_10','compute_11','compute_12', 'compute_13','compute_20','compute_30','compute_35','sm_10','sm_11', 'sm_12','sm_13','sm_20','sm_21','sm_30','sm_35'.

--gpu-code<gpu architecture name>,…   (-code)

Specify the names of nVidia gpus to generate code for.

nvcc will embed a compiled code image in the resulting executable for each specified 'code' architecture. This code image will be a true binary load image for each 'real' architecture(such as a sm_13), and ptx intermediate code for each virtual architecture(such as compute_10). During runtime, in case no better binary load image is found, and provided that the ptx architecture is compatible with the 'current' GPU, such embedded ptx code will be dynamically translated for this current GPU by the cuda runtime system.

Architectures specified for this option can be virtual as well as real, but each of these 'code' architectures must be compatible with the architecture specified with option '--gpu-architecture'.

For instance, 'arch'=compute_13 is not compatible with 'code'=sm_10, because the generated ptx code will assume the availability of compute_13 features that are not present on sm_10.

Allowed values for this option: 'compute_10','compute_11','compute_12', 'compute_13','compute_20','compute_30','compute_35','sm_10','sm_11', 'sm_12','sm_13','sm_20','sm_21','sm_30','sm_35'.

--generate-code   (-gencode)

This option provides a generalization of the '--gpu-architecture=<arch>
--gpu-code=code,…' option combination for specifying nvcc behavior
with respect to code generation. Where use of the previous options
generates different code for a fixed virtual architecture, option
'--generate-code' allows multiple cicc invocations, iterating over
different virtual architectures. In fact,
  '--gpu-architecture=<arch>--gpu-code=<code>,…'
is equivalent to
  '--generate-code arch=<arch>,code=<code>,…'.
'--generate-code' options may be repeated for different virtual
architectures.
Allowed keywords for this option: 'arch','code'.

--maxrregcount<N>   (-maxrregcount)
Specify the maximum amount of registers that GPU functions can use.
Until a function-specific limit, a higher value will generally
increase the performance of individual GPU threads that execute this
function. However, because thread registers are allocated from a global
register pool on each GPU, a higher value of this option will also
reduce the maximum thread block size, thereby reducing the amount of
thread parallelism. Hence, a good maxrregcount value is the result of a
trade-off.
If this option is not specified, then no maximum is assumed.
Value less than the minimum registers required by ABI will be bumped up
by the compiler to ABI minimum limit.

--ftz [true,false]   (-ftz)
When performing single-precision floating-point operations, flush
denormal values to zero or preserve denormal values. -use_fast_math
implies --ftz=true.
Default value: 0.

--prec-div [true,false]   (-prec-div)
For single-precision floating-point division and reciprocals, use IEEE
round-to-nearest mode or use a faster approximation. -use_fast_math
implies --prec-div=false.
Default value: 1.

--prec-sqrt [true,false]   (-prec-sqrt)
For single-precision floating-point square root, use IEEE
round-to-nearest mode or use a faster approximation. -use_fast_math
implies --prec-sqrt=false.

```
        Default value:  1.

--fmad [true,false]                           (-fmad)
        Enables (disables) the contraction of floating-point multiplies and
        adds/subtracts into floating-point multiply-add operations (FMAD, FFMA,
        or DFMA). This option is supported only when '--gpu-architecture' is
        set with compute_20, sm_20, or higher. For other architecture classes,
        the contraction is always enabled. -use_fast_math implies --fmad=true.
        Default value:  1.

--relocatable-device-code [true,false]        (-rdc)
        Enable(disable) the generation of relocatable device code. If
        disabled, executable device code is generated.
        Default value: 0.

Options for steering cuda compilation
=====================================

--target-cpu-architecture<cpu architecture name>  (-target-cpu-arch)
        Specify the name of the class of CPU architecture for which the input
        files must be compiled.
        Allowed values for this option:  'ARM','x86'.
        Default value:  'x86'.

--use_fast_math                               (-use_fast_math)
        Make use of fast math library. -use_fast_math implies -ftz=true
        -prec-div=false -prec-sqrt=false.

--entries entry,...                           (-e)
        In case of compilation of ptx or gpu files to cubin: specify the global
        entry functions for which code must be generated. By default, code will
        be generated for all entry functions.

Generic tool options
====================

--disable-warnings                            (-w)
        Inhibit all warning messages.

--source-in-ptx                               (-src-in-ptx)
        Interleave source in ptx.

--restrict                                    (-restrict)
```

>   Programmer assertion that all kernel pointer parameters are restrict
>   pointers.
>
> --Werror<kind>,…   (-Werror)
>   Make warnings of the specified kinds into errors. The following is the
>   list of warning kinds accepted by this option:
>
>   cross-execution-space-call
>   Be more strict about unsupported cross execution space calls.
>   The compiler will generate an error instead of a warning for a
>   call from a __host__ __device__ to a __host__ function.
>
>   Allowed values for this option: 'cross-execution-space-call'.
> --help   (-h)
>   Print this help information on this tool.
> --version   (-V)
>   Print version information on this tool.
> --options-file<file>,…   (-optf)
>   Include command line options from specified file.

附录 F

几种排序算法源代码

附录 F.1 bitonic_sort_block 函数

摘自《GPU 高性能运算之 CUDA》[2]一书,在 13.3.2 节被引用。

```
#define threadnum 1024
__global__ void bitonic_sort_gpu(DATATYPE *a,int *c, int n)
{
    const int tidx=threadIdx.x;
    const int t_n=blockDim.x;
    __shared__ DATATYPE arr[threadnum];
    __shared__ int crr[threadnum];
    arr[tidx]=a[tidx];
    crr[tidx]=c[tidx];
    __syncthreads();

    DATATYPE temp;
    int temp1;
    unsigned int k,j,ixj;
    for(k=2;k<=n;k*=2)
    {
        for(j=k/2;j>0;j/=2)
        {
            ixj=tidx^j;
            if(ixj>tidx)
            {
                if((tidx & k)==0 && (arr[tidx]>arr[ixj]))
                {
                    temp=arr[ixj];
                    arr[ixj]=arr[tidx];
                    arr[tidx]=temp;
                    temp1=crr[ixj];
                    crr[ixj]=crr[tidx];
```

```
                crr[tidx]=temp1;
            }
            else if(arr[tidx]<arr[ixj])
            {
                temp=arr[ixj];
                arr[ixj]=arr[tidx];
                arr[tidx]=temp;
                temp1=crr[ixj];
                crr[ixj]=crr[tidx];
                crr[tidx]=temp1;
            }
        }
        __syncthreads();
    }
    }
    a[tidx]=arr[tidx];
    c[tidx]=crr[tidx];
}
```

附录 F.2　GPU 快速排序完整代码

```
#define QSORT_BLOCKSIZE_SHIFT    9
#define QSORT_BLOCKSIZE          (1<<QSORT_BLOCKSIZE_SHIFT)
#define BITONICSORT_LEN          1024
#define QSORT_MAXDEPTH           16
typedef struct __align__(128) qsortAtomicData_t
{
    volatile unsigned int lt_offset;
    volatile unsigned int gt_offset;
    volatile unsigned int sorted_count;
    volatile unsigned int index;
} qsortAtomicData;
typedef struct qsortRingbuf_t
{
    volatile unsigned int head;
    volatile unsigned int tail;
    volatile unsigned int count;
    volatile unsigned int max;
    unsigned int stacksize;
    volatile void *stackbase;
```

```
} qsortRingbuf;
#define QSORT_STACK_ELEMS   1*1024*1024
static __device____forceinline__ unsigned int __btflo(unsigned int word)
{
    unsigned int ret;
    asm volatile("bfind.u32 %0, %1;": "=r"(ret): "r"(word));
    return ret;
}
__device____forceinline__ int qcompare(unsigned &val1, unsigned &val2)
{
    return (val1>val2) ? 1: (val1==val2) ? 0: -1;
}
static __device____forceinline__ unsigned int __qsflo(unsigned int word)
{
    unsigned int ret;
    asm volatile("bfind.u32 %0, %1;": "=r"(ret): "r"(word));
    return ret;
}
template<typename T>
static __device__ T *ringbufAlloc(qsortRingbuf *ringbuf)
{
    unsigned int loop=10000;
    while(((ringbuf->head-ringbuf->tail)>=ringbuf->stacksize) && (loop-->0));
    if(loop==0)
        return NULL;
    unsigned int index=atomicAdd((unsigned int *) &ringbuf->head, 1);
    T *ret= (T *)(ringbuf->stackbase)+(index & (ringbuf->stacksize-1));
    ret->index=index;
    return ret;
}
template<typename T>
static __device__ void ringbufFree(qsortRingbuf *ringbuf, T *data)
{
    unsigned int index=data->index;
    unsigned int count=atomicAdd((unsigned int *)&(ringbuf->count), 1)+1;
    unsigned int max=atomicMax((unsigned int *)&(ringbuf->max), index+1);
    if(max<(index+1)) max=index+1;
    if(max==count)
        atomicMax((unsigned int *)&(ringbuf->tail), count);
}
__global__ void bitonicsort_kernel(unsigned *indata, unsigned *outdata,
unsigned int offset, unsigned int len)
{
```

```
    __shared__ unsigned sortbuf[1024];
    unsigned int inside=(threadIdx.x<len);
    sortbuf[threadIdx.x]=inside ? indata[threadIdx.x+offset]: 0xffffffffu;
    __syncthreads();
    for(unsigned int k=2; k<=blockDim.x; k*=2)
    {
        for(unsigned int j=k>>1; j>0; j>>=1)
        {
            unsigned int swap_idx=threadIdx.x ^ j;
            unsigned my_elem=sortbuf[threadIdx.x];
            unsigned swap_elem=sortbuf[swap_idx];
            __syncthreads();
            unsigned int ascend=k * (swap_idx<threadIdx.x);
            unsigned int descend=k * (swap_idx>threadIdx.x);
            bool swap=false;
            if((threadIdx.x & k)==ascend)
            {
                if(my_elem>swap_elem)
                    swap=true;
            }
            if((threadIdx.x & k)==descend)
            {
                if(my_elem<swap_elem)
                    swap=true;
            }
            if(swap)
            {
                sortbuf[swap_idx]=my_elem;
            }
            __syncthreads();
        }
    }
    if(threadIdx.x<len)
        outdata[threadIdx.x+offset]=sortbuf[threadIdx.x];
}
__global__ void big_bitonicsort_kernel(unsigned *indata, unsigned *outdata,
unsigned *backbuf, unsigned int offset, unsigned int len)
{
    unsigned int len2=1<<(__btflo(len-1U)+1);
    if(threadIdx.x>=len2) return;
    for(unsigned int i=len; i<len2; i+=blockDim.x)
    {
        unsigned int index=i+threadIdx.x;
```

```
        if(index<len2)
        {
            if(index<len)
                indata[index+offset]=0xffffffffu;
            else
                backbuf[index+offset-len]=0xffffffffu;
        }
    }
    __syncthreads();
    for(unsigned int k=2; k<=len2; k*=2)
    {
        for(unsigned int j=k>>1; j>0; j>>=1)
        {
            for(unsigned int i=0; i<len2; i+=blockDim.x)
            {
                unsigned int index=threadIdx.x+i;
                unsigned int swap_idx=index ^ j;
                if(swap_idx>index)
                {
                    unsigned my_elem, swap_elem;
                    if(index<len)
                        my_elem=indata[index+offset];
                    else
                        my_elem=backbuf[index+offset-len];
                    if(swap_idx<len)
                        swap_elem=indata[swap_idx+offset];
                    else
                        swap_elem=backbuf[swap_idx+offset-len];
                    bool swap=false;
                    if((index & k)==0)
                    {
                        if(my_elem>swap_elem)
                            swap=true;
                    }
                    if((index & k)==k)
                    {
                        if(my_elem<swap_elem)
                            swap=true;
                    }
                    if(swap)
                    {
                        if(swap_idx<len)
                            indata[swap_idx+offset]=my_elem;
```

```
                    else
                        backbuf[swap_idx+offset-len]=my_elem;
                    if(index<len)
                        indata[index+offset]=swap_elem;
                    else
                        backbuf[index+offset-len]=swap_elem;
                }
            }
        }
        __syncthreads();
    }
}
if(outdata!=indata)
{
    for(unsigned int i=0; i<len; i+=blockDim.x)
    {
        unsigned int index=i+threadIdx.x;
        if(index<len)
            outdata[index+offset]=indata[index+offset];
    }
}
}
__global__ void qsort_warp(unsigned *indata,unsigned *outdata,
              unsigned int offset,unsigned int len,
              qsortAtomicData *atomicData,qsortRingbuf *atomicDataStack,
              unsigned int source_is_indata,unsigned int depth)
{
    unsigned int thread_id=threadIdx.x+(blockIdx.x*blockDim.x);
    unsigned int lane_id=threadIdx.x & (warpSize-1);
    if(thread_id>=len)
        return;
    unsigned pivot=indata[offset+len/2];
    unsigned data  =indata[offset+thread_id];
    unsigned int greater= (data>pivot);
    unsigned int gt_mask=__ballot(greater);
    if(gt_mask==0)
    {
        greater= (data>=pivot);
        gt_mask=__ballot(greater);
    }
    unsigned int lt_mask=__ballot(!greater);
    unsigned int gt_count=__popc(gt_mask);
    unsigned int lt_count=__popc(lt_mask);
```

```
    unsigned int lt_offset, gt_offset;
    if(lane_id==0)
    {
        if(lt_count>0)
lt_offset=atomicAdd((unsigned int *) &atomicData->lt_offset, lt_count);
        if(gt_count>0)
gt_offset=len-(atomicAdd((unsigned int *) &atomicData->gt_offset, gt_count)
+gt_count);
    }

    lt_offset=__shfl((int)lt_offset, 0);
    gt_offset=__shfl((int)gt_offset, 0);
    __syncthreads();
    unsigned lane_mask_lt;
    asm("mov.u32 %0, %%lanemask_lt;": "=r"(lane_mask_lt));
    unsigned int my_mask=greater ? gt_mask: lt_mask;
    unsigned int my_offset=__popc(my_mask & lane_mask_lt);
    my_offset+=greater ? gt_offset: lt_offset;
    outdata[offset+my_offset]=data;
    if(lane_id==0)
    {
        unsigned int mycount=lt_count+gt_count;
        if(atomicAdd((unsigned int *) &atomicData->sorted_count, mycount)+
        mycount==len)
        {
            unsigned int lt_len=atomicData->lt_offset;
            unsigned int gt_len=atomicData->gt_offset;
            cudaStream_t lstream, rstream;
            cudaStreamCreateWithFlags(&lstream, cudaStreamNonBlocking);
            cudaStreamCreateWithFlags(&rstream, cudaStreamNonBlocking);
            ringbufFree<qsortAtomicData>(atomicDataStack, atomicData);
            if(lt_len==0)
            {
                if(source_is_indata)
                    cudaMemcpyAsync(indata+offset, outdata+offset, gt_len *
                    sizeof(unsigned), cudaMemcpyDeviceToDevice, lstream);
                return;
            }
            if(lt_len>BITONICSORT_LEN)
            {
                if(depth>=QSORT_MAXDEPTH)
                {
```

```
                    big_bitonicsort_kernel<<<1, BITONICSORT_LEN, 0, lstream>>>
                    (outdata, source_is_indata ? indata: outdata, indata, offset,
                    lt_len);
                }
                else
                {
if((atomicData=ringbufAlloc<qsortAtomicData>(atomicDataStack))==NULL)
                    printf("Stack-allocation error. Failing left child
                    launch.\n");
                    else
                    {
atomicData->lt_offset=atomicData->gt_offset=atomicData->sorted_count=0;
unsigned int numblocks=(unsigned int)(lt_len+(QSORT_BLOCKSIZE-1))>>QSORT_
BLOCKSIZE_SHIFT;
qsort_warp<<<numblocks, QSORT_BLOCKSIZE, 0, lstream>>>(outdata, indata,
offset, lt_len, atomicData, atomicDataStack, !source_is_indata, depth+1);
                    }
                }
            }
            else if(lt_len>1)
            {
                unsigned int bitonic_len=1<<(__qsflo(lt_len-1U)+1);
                bitonicsort_kernel<<<1, bitonic_len, 0, lstream>>>(outdata,
                source_is_indata ? indata: outdata, offset, lt_len);
            }
            else if(source_is_indata && (lt_len==1))
                indata[offset]=outdata[offset];
            if(gt_len>BITONICSORT_LEN)
            {
                if(depth>=QSORT_MAXDEPTH)
                  big_bitonicsort_kernel<<<1, BITONICSORT_LEN, 0, rstream>>>
                  (outdata, source_is_indata ? indata: outdata, indata, offset+
                  lt_len, gt_len);
                else
                {
                        if((atomicData = ringbufAlloc < qsortAtomicData >
                        (atomicDataStack))==NULL)
                            printf("Stack allocation error! Failing right-side
                            launch.\n");
                        else
                        {
                            atomicData->lt_offset=atomicData->gt_offset=
                            atomicData->sorted_count=0;
```

```cpp
                            unsigned int numblocks=(unsigned int)(gt_len+(QSORT_
                                BLOCKSIZE-1))>>QSORT_BLOCKSIZE_SHIFT;
                            qsort_warp<<<numblocks, QSORT_BLOCKSIZE, 0, rstream>>>
                                (outdata, indata, offset+lt_len, gt_len, atomicData,
                                atomicDataStack, !source_is_indata, depth+1);
                    }
                }
            }
            else if(gt_len>1)
            {
                unsigned int bitonic_len=1<<(__qsflo(gt_len-1U)+1);
                bitonicsort_kernel<<<1, bitonic_len, 0, rstream>>>(outdata,
                    source_is_indata ? indata: outdata, offset+lt_len, gt_len);
            }
            else if(source_is_indata && (gt_len==1))
                indata[offset+lt_len]=outdata[offset+lt_len];
        }
    }
}
void run_quicksort_cdp(DATATYPE *d_a, DATATYPE *d_buff, unsigned int n)
{
    unsigned int stacksize=QSORT_STACK_ELEMS;
    qsortAtomicData *gpustack;
    cudaMalloc((void **)&gpustack, stacksize * sizeof(qsortAtomicData));
    cudaMemset(gpustack, 0, sizeof(qsortAtomicData));
    qsortRingbuf buf;
    qsortRingbuf * ringbuf;
    cudaMalloc((void **)&ringbuf, sizeof(qsortRingbuf));
    buf.head=1;
    buf.tail=0;
    buf.count=0;
    buf.max=0;
    buf.stacksize=stacksize;
    buf.stackbase=gpustack;
    cudaMemcpy(ringbuf, &buf, sizeof(buf), cudaMemcpyHostToDevice);
    unsigned int numblocks=(unsigned int)(n+(512-1))/512;
    qsort_warp<<<numblocks, 512>>>(d_a, d_buff, 0U, n, gpustack, ringbuf,
        true, 0);
    cudaFree(ringbuf);
    cudaFree(gpustack);
}
    ⋮
    DATATYPE *d_a,*d_buff;
```

```
    cudaMalloc((void**)&d_a, n*sizeof(DATATYPE));
    cudaMalloc((void**)&d_buff, n*sizeof(DATATYPE));
    cudaMemcpy(d_a, array_a, n*sizeof(DATATYPE), cudaMemcpyHostToDevice);
    run_quicksort_cdp(d_a,d_buff,n);
    cudaMemcpy(array_b, d_a, n*sizeof(DATATYPE), cudaMemcpyDeviceToHost);
    sort_err(array_b,n," quick sort");
    ⋮
```

附录F.3　GPU合并排序完整代码

```
#include<assert.h>
typedef unsigned int uint;
#define SHARED_SIZE_LIMIT 1024U
#define SAMPLE_STRIDE 128
static inline __host__ __device__ uint iDivUp(uint a, uint b)
{
    return ((a%b)==0) ? (a/b): (a/b+1);
}
static inline __host__ __device__ uint getSampleCount(uint dividend)
{
    return iDivUp(dividend, SAMPLE_STRIDE);
}
#define W (sizeof(uint) * 8)
static inline __device__ uint nextPowerOfTwo(uint x)
{
    return 1U<<(W-__clz(x-1));
}
template<uint sortDir>
static inline __device__ uint binarySearchExclusive(DATATYPE val, DATATYPE
*data, uint L, uint stride)
{
    if(L==0)
    {
        return 0;
    }
    uint pos=0;
    for(; stride>0; stride>>=1)
    {
        uint newPos=umin(pos+stride, L);
        if((sortDir && (data[newPos-1]<val)) || (!sortDir && (data[newPos-1]>
        val)))
```

```cpp
            {
                pos=newPos;
            }
        }
        return pos;
    }
    template<uint sortDir>
    static inline __device__ uint binarySearchInclusive(DATATYPE val, DATATYPE
    *data, uint L, uint stride)
    {
        if(L==0)
        {
            return 0;
        }
        uint pos=0;
        for(; stride>0; stride>>=1)
        {
            uint newPos=umin(pos+stride, L);
            if((sortDir && (data[newPos-1]<=val)) || (!sortDir && (data[newPos-1]
            >=val)))
            {
                pos=newPos;
            }
        }
        return pos;
    }
    template<uint sortDir>
    __global__ void mergeSortSharedKernel(DATATYPE *d_DstKey,DATATYPE *d_SrcKey,
    uint arrayLength)
    {
        __shared__ DATATYPE s_key[SHARED_SIZE_LIMIT];
        d_SrcKey+=blockIdx.x * SHARED_SIZE_LIMIT+threadIdx.x;
        d_DstKey+=blockIdx.x * SHARED_SIZE_LIMIT+threadIdx.x;
        s_key[threadIdx.x+0]=d_SrcKey[0];
        s_key[threadIdx.x+(SHARED_SIZE_LIMIT/2)]=d_SrcKey[(SHARED_SIZE_LIMIT/2)];
        for(uint stride=1; stride<arrayLength; stride<<=1)
        {
            uint     lPos=threadIdx.x & (stride-1);
            DATATYPE *baseKey=s_key+2 * (threadIdx.x-lPos);
            __syncthreads();
            DATATYPE keyA=baseKey[lPos+0];
            DATATYPE keyB=baseKey[lPos+stride];
            uint posA=binarySearchExclusive<sortDir>(keyA, baseKey+stride,
            stride, stride)+lPos;
```

```
        uint posB=binarySearchInclusive<sortDir>(keyB,baseKey+0,stride,
        stride)+lPos;
        __syncthreads();
        baseKey[posA]=keyA;
        baseKey[posB]=keyB;
    }
    __syncthreads();
    d_DstKey[0]=s_key[threadIdx.x+0];
    d_DstKey[(SHARED_SIZE_LIMIT/2)]=s_key[threadIdx.x+(SHARED_SIZE_LIMIT/2)];
}
static void mergeSortShared(DATATYPE *d_DstKey,DATATYPE *d_SrcKey,uint
batchSize,uint arrayLength,uint sortDir)
{
    if(arrayLength<2)
    {
        return;
    }
    assert(SHARED_SIZE_LIMIT%arrayLength==0);
    assert(((batchSize*arrayLength)%SHARED_SIZE_LIMIT)==0);
    uint  blockCount=batchSize*arrayLength/SHARED_SIZE_LIMIT;
    uint threadCount=SHARED_SIZE_LIMIT/2;
    if(sortDir)
    {
        mergeSortSharedKernel<1U><<<blockCount,threadCount>>>(d_DstKey,d_
        SrcKey,arrayLength);
    }
    else
    {
        mergeSortSharedKernel<0U><<<blockCount,threadCount>>>(d_DstKey,d_
        SrcKey,arrayLength);
    }
}
template<uint sortDir>
__global__ void generateSampleRanksKernel(uint *d_RanksA,uint *d_RanksB,
DATATYPE *d_SrcKey,uint stride,uint N,uint threadCount)
{
    uint pos=blockIdx.x*blockDim.x+threadIdx.x;
    if(pos>=threadCount)
    {
        return;
    }
    const uint    i=pos & ((stride/SAMPLE_STRIDE)-1);
```

```
        const uint segmentBase=(pos-i)*(2*SAMPLE_STRIDE);
        d_SrcKey+=segmentBase;
        d_RanksA+=segmentBase/SAMPLE_STRIDE;
        d_RanksB+=segmentBase/SAMPLE_STRIDE;
        const uint segmentElementsA=stride;
        const uint segmentElementsB=umin(stride,N-segmentBase-stride);
        const uint segmentSamplesA=getSampleCount(segmentElementsA);
        const uint segmentSamplesB=getSampleCount(segmentElementsB);
        if(i<segmentSamplesA)
        {
            d_RanksA[i]=i*SAMPLE_STRIDE;
            d_RanksB[i]=binarySearchExclusive<sortDir>(d_SrcKey[i*SAMPLE_
            STRIDE],d_SrcKey+stride,segmentElementsB,nextPowerOfTwo
            (segmentElementsB));
        }

        if(i<segmentSamplesB)
        {
            d_RanksB[(stride/SAMPLE_STRIDE)+i]=i*SAMPLE_STRIDE;
            d_RanksA[(stride/SAMPLE_STRIDE)+i]=binarySearchInclusive<sortDir>
            (d_SrcKey[stride+i*SAMPLE_STRIDE],d_SrcKey+0,segmentElementsA,
            nextPowerOfTwo(segmentElementsA));
        }
}
static void generateSampleRanks(uint *d_RanksA,uint *d_RanksB,DATATYPE *d_
SrcKey,uint stride,uint N,uint sortDir)
{
    uint lastSegmentElements=N%(2*stride);
    uint threadCount=(lastSegmentElements>stride) ? (N+2*stride-
    lastSegmentElements)/(2*SAMPLE_STRIDE): (N-lastSegmentElements)/(2*
    SAMPLE_STRIDE);

    if(sortDir)
    {
        generateSampleRanksKernel<1U><<<iDivUp(threadCount,256),256>>>(d_
        RanksA, d_RanksB, d_SrcKey, stride, N, threadCount);
    }
    else
    {
        generateSampleRanksKernel<0U><<<iDivUp(threadCount,256),256>>>(d_
        RanksA, d_RanksB, d_SrcKey, stride, N, threadCount);
    }
}
```

```cpp
__global__ void mergeRanksAndIndicesKernel(uint *d_Limits,uint *d_Ranks,uint
stride,uint N,uint threadCount)
{
    uint pos=blockIdx.x*blockDim.x+threadIdx.x;
    if(pos>=threadCount)
    {
        return;
    }
    const uint i=pos & ((stride/SAMPLE_STRIDE)-1);
    const uint segmentBase=(pos-i)*(2*SAMPLE_STRIDE);
    d_Ranks+=(pos-i)*2;
    d_Limits+=(pos-i)*2;
    const uint segmentElementsA=stride;
    const uint segmentElementsB=umin(stride,N-segmentBase-stride);
    const uint segmentSamplesA=getSampleCount(segmentElementsA);
    const uint segmentSamplesB=getSampleCount(segmentElementsB);
    if(i<segmentSamplesA)
    {
        uint dstPos = binarySearchExclusive < 1U > (d_Ranks[i], d_Ranks +
        segmentSamplesA, segmentSamplesB, nextPowerOfTwo(segmentSamplesB))+i;
        d_Limits[dstPos]=d_Ranks[i];
    }
    if(i<segmentSamplesB)
    {
        uint dstPos=binarySearchInclusive<1U>(d_Ranks[segmentSamplesA+i], d_
        Ranks, segmentSamplesA, nextPowerOfTwo(segmentSamplesA))+i;
        d_Limits[dstPos]=d_Ranks[segmentSamplesA+i];
    }
}
static void mergeRanksAndIndices (uint *d_LimitsA, uint *d_LimitsB, uint *d_
RanksA,uint *d_RanksB,uint stride,uint N)
{
    uint lastSegmentElements=N%(2*stride);
    uint threadCount= (lastSegmentElements>stride) ? (N+2*stride-lastSeg-
    mentElements)/(2*SAMPLE_STRIDE): (N-lastSegmentElements)/(2*SAMPLE_
    STRIDE);
    mergeRanksAndIndicesKernel<<<iDivUp(threadCount, 256), 256>>>(d_
    LimitsA,d_RanksA,stride,N,threadCount);
    mergeRanksAndIndicesKernel<<<iDivUp(threadCount, 256), 256>>>(d_
    LimitsB,d_RanksB,stride,N,threadCount);
}
template<uint sortDir>inline __device__ void merge(DATATYPE *dstKey,DATATYPE
*srcAKey,DATATYPE *srcBKey,uint lenA,uint nPowTwoLenA,uint lenB,uint
nPowTwoLenB)
```

```cpp
{
    uint keyA, keyB, dstPosA, dstPosB;
    if(threadIdx.x<lenA)
    {
        keyA=srcAKey[threadIdx.x];
        dstPosA = binarySearchExclusive < sortDir > (keyA, srcBKey, lenB,
            nPowTwoLenB)+threadIdx.x;
    }
    if(threadIdx.x<lenB)
    {
        keyB=srcBKey[threadIdx.x];
        dstPosB = binarySearchInclusive < sortDir > (keyB, srcAKey, lenA,
            nPowTwoLenA)+threadIdx.x;
    }
    __syncthreads();
    if(threadIdx.x<lenA)
    {
        dstKey[dstPosA]=keyA;
    }
    if(threadIdx.x<lenB)
    {
        dstKey[dstPosB]=keyB;
    }
}
template<uint sortDir>__global__ void mergeElementaryIntervalsKernel
(DATATYPE *d_DstKey,DATATYPE *d_SrcKey,uint *d_LimitsA,uint *d_LimitsB,
uint stride,uint N)
{
    __shared__ uint s_key[2*SAMPLE_STRIDE];
    __shared__ uint startSrcA, startSrcB, lenSrcA, lenSrcB, startDstA, startDstB;
    const uint intervalI=blockIdx.x & ((2*stride)/SAMPLE_STRIDE-1);
    const uint segmentBase=(blockIdx.x-intervalI) * SAMPLE_STRIDE;
    d_SrcKey+=segmentBase;
    d_DstKey+=segmentBase;
    if(threadIdx.x==0)
    {
        uint segmentElementsA=stride;
        uint segmentElementsB=umin(stride, N-segmentBase-stride);
        uint segmentSamplesA=getSampleCount(segmentElementsA);
        uint segmentSamplesB=getSampleCount(segmentElementsB);
        uint segmentSamples=segmentSamplesA+segmentSamplesB;
        startSrcA=d_LimitsA[blockIdx.x];
        startSrcB=d_LimitsB[blockIdx.x];
```

```cpp
            uint endSrcA=(intervalI+1<segmentSamples) ? d_LimitsA[blockIdx.x+1]:
            segmentElementsA;
            uint endSrcB=(intervalI+1<segmentSamples) ? d_LimitsB[blockIdx.x+1]:
            segmentElementsB;
            lenSrcA=endSrcA-startSrcA;
            lenSrcB=endSrcB-startSrcB;
            startDstA=startSrcA+startSrcB;
            startDstB=startDstA+lenSrcA;
        }
        __syncthreads();
        if(threadIdx.x<lenSrcA)
        {
            s_key[threadIdx.x+0]=d_SrcKey[0+startSrcA+threadIdx.x];
        }
        if(threadIdx.x<lenSrcB)
        {
            s_key[threadIdx.x+SAMPLE_STRIDE]=d_SrcKey[stride+startSrcB+
            threadIdx.x];
        }
        __syncthreads();
        merge<sortDir>(s_key,s_key+0,s_key+SAMPLE_STRIDE,lenSrcA,SAMPLE_
        STRIDE,lenSrcB,SAMPLE_STRIDE);
        __syncthreads();
        if(threadIdx.x<lenSrcA)
        {
            d_DstKey[startDstA+threadIdx.x]=s_key[threadIdx.x];
        }
        if(threadIdx.x<lenSrcB)
        {
            d_DstKey[startDstB+threadIdx.x]=s_key[lenSrcA+threadIdx.x];
        }
}
static void mergeElementaryIntervals(DATATYPE *d_DstKey,DATATYPE *d_SrcKey,
uint *d_LimitsA,uint *d_LimitsB,uint stride,uint N,uint sortDir)
{
    uint lastSegmentElements=N%(2*stride);
    uint mergePairs= (lastSegmentElements>stride) ? getSampleCount(N): (N-
    lastSegmentElements)/SAMPLE_STRIDE;
    if(sortDir)
    {
        mergeElementaryIntervalsKernel<1U><<<mergePairs,SAMPLE_STRIDE>>>
        (d_DstKey,d_SrcKey,d_LimitsA,d_LimitsB,stride,N);
    }
```

```cpp
        else
        {
            mergeElementaryIntervalsKernel<0U><<<mergePairs,SAMPLE_STRIDE>>>
                (d_DstKey,d_SrcKey,d_LimitsA,d_LimitsB,stride,N);
        }
}
void mergeSort(DATATYPE *d_DstKey,DATATYPE *d_BufKey,DATATYPE *d_SrcKey,uint N,uint sortDir)
{
    uint *d_RanksA,*d_RanksB,*d_LimitsA,*d_LimitsB;
    const uint max_sample_count=32768;
    cudaMalloc((void **)&d_RanksA, max_sample_count * sizeof(uint));
    cudaMalloc((void **)&d_RanksB, max_sample_count * sizeof(uint));
    cudaMalloc((void **)&d_LimitsA,max_sample_count * sizeof(uint));
    cudaMalloc((void **)&d_LimitsB,max_sample_count * sizeof(uint));
    uint stageCount=0;
    for(uint stride=SHARED_SIZE_LIMIT; stride<N; stride<<=1,stageCount++);
    DATATYPE *ikey,*okey;
    if(stageCount & 1)
    {
        ikey=d_BufKey;
        okey=d_DstKey;
    }
    else
    {
        ikey=d_DstKey;
        okey=d_BufKey;
    }
    assert(N<=(SAMPLE_STRIDE * max_sample_count));
    assert(N%SHARED_SIZE_LIMIT==0);
    mergeSortShared(ikey, d_SrcKey, N/SHARED_SIZE_LIMIT, SHARED_SIZE_LIMIT, sortDir);
    for(uint stride=SHARED_SIZE_LIMIT; stride<N; stride<<=1)
    {
        uint lastSegmentElements=N%(2 * stride);
        //Find sample ranks and prepare for limiters merge
        generateSampleRanks(d_RanksA, d_RanksB, ikey, stride, N, sortDir);
        //Merge ranks and indices
        mergeRanksAndIndices(d_LimitsA, d_LimitsB, d_RanksA, d_RanksB, stride, N);
        //Merge elementary intervals
        mergeElementaryIntervals(okey, ikey, d_LimitsA, d_LimitsB, stride, N, sortDir);
        if(lastSegmentElements<=stride)
```

```
            {
                cudaMemcpy(okey+(N-lastSegmentElements), ikey+(N-lastSegment-
                    Elements), lastSegmentElements * sizeof(uint), cudaMemcpyDevice-
                    ToDevice);
            }
            DATATYPE *t;
            t=ikey;
            ikey=okey;
            okey=t;
        }
    cudaFree(d_RanksA);
    cudaFree(d_RanksB);
    cudaFree(d_LimitsB);
    cudaFree(d_LimitsA);
}
        ⋮
    DATATYPE *d_a,*d_b,*d_buff;
    cudaMalloc((void **)&d_a, n * sizeof(DATATYPE));
    cudaMalloc((void **)&d_b, n * sizeof(DATATYPE));
    cudaMalloc((void **)&d_buff, n * sizeof(DATATYPE));
    cudaMemcpy(d_a, array_a, n * sizeof(DATATYPE), cudaMemcpyHostToDevice);
    mergeSort(d_b,d_buff,d_a,n,1);
    cudaMemcpy(array_b, d_b, n * sizeof(DATATYPE), cudaMemcpyDeviceToHost);
    sort_err(array_b,n,"merge sort");
        ⋮
```

参 考 文 献

[1] 陈国良. 并行计算——结构·算法·编程[M]. 北京：高等教育出版社, 2004.
[2] 张舒, 褚艳利, 赵开勇, 等. GPU 高性能运算之 CUDA[M]. 北京：中国水利水电出版社, 2009.
[3] Jason S, Edward K. GPU 高性能编程 CUDA 实战[M]. 聂雪军, 等译. 北京：机械工业出版社, 2011.
[4] David BK, Wen-mei WH. 大规模并行处理器编程实战[M]. 2 版. 赵开勇, 等译. 北京：清华大学出版社, 2013.
[5] Nicholas W. CUDA 专家手册[M]. 苏统华, 等译. 北京：机械工业出版社, 2014.
[6] Henry Wong, Misel-Myrto Papadopoulou, Maryam Sadooghi-Alvandi, etc. Demystifying GPU Microarchitecture through Microbenchmarking[C]. Performance Analysis of Systems & Software (ISPASS), 2010 IEEE International Symposium on. IEEE, 2010: 235-246.